Quantum Probability

This is a volume in
PROBABILITY AND MATHEMATICAL STATISTICS
Z. W. Birnbaum, editor

A list of titles in this series appears at the end of this volume.

Quantum Probability

Stanley P. Gudder
Department of Mathematics and Computer Science
University of Denver
Denver, Colorado

ACADEMIC PRESS, INC.
Harcourt Brace Jovanovich, Publishers
Boston San Diego New York
Berkeley London Sydney
Tokyo Toronto

Copyright © 1988 by Academic Press, Inc.
All rights reserved.
No part of this publication may be reproduced or
transmitted in any form or by any means, electronic
or mechanical, including photocopy, recording, or
any information storage and retrieval system, without
permission in writing from the publisher.

ACADEMIC PRESS, INC.
1250 Sixth Avenue, San Diego, CA 92101

United Kingdom Edition published by
ACADEMIC PRESS INC. (LONDON) LTD.
24–28 Oval Road, London NW1 7DX

Text preparation by Paula Spiegel Gudder with T_EX using *Textures*.
T_EX is a trademark of the American Mathematical Society.
Textures is produced by Kellerman & Smith.

Library of Congress Cataloging-in-Publication Data
Gudder, Stanley.
    Quantum probability / Stanley P. Gudder.
        p. cm. – (Probability and mathematical statistics)
    Bibliography: p.
    Includes index.
    ISBN 0-12-305340-4
    1. Probabilities. 2. Mathematical physics. 3. Quantum theory.
4. Statistics. 5. Stochastic processes. 6. Multivariate analysis.
I. Title. II. Series.
QC20.7.P7G83 1988
530.1'2 – dc19
                                                                88-21728
                                                                    CIP

Printed in the United States of America
88 89 90 91   9 8 7 6 5 4 3 2 1

TO OUR CHILDREN
DOLLY GAIL MICHAEL WALTER JOEY

AND TO CHARLEY'S MEMORY

Contents

	Preface	ix
Chapter 1	**Classical Probability Theory**	1
Section 1.1	Probability Spaces and Random Variables	1
Section 1.2	Independence and Conditional Probabilities	9
Section 1.3	Independent Random Variables	16
Section 1.4	Law of Large Numbers	21
Section 1.5	Central Limit Theorem	25
Section 1.6	Notes and References	34
Chapter 2	**Traditional Quantum Mechanics**	37
Section 2.1	Classical Mechanics	37
Section 2.2	Hilbert Space Quantum Mechanics	42
Section 2.3	Hilbert Space Probability	50
Section 2.4	Spin Systems	54
Section 2.5	Two-Hole Experiment	57
Section 2.6	Path-Integral Formalism	62
Section 2.7	Notes and References	68
Chapter 3	**Operational Statistics**	71
Section 3.1	Basic Definitions	72
Section 3.2	Operational Logics	75
Section 3.3	States	83
Section 3.4	Hidden Variables	89
Section 3.5	Products of Quasimanuals	93
Section 3.6	Simpson's Paradox	102
Section 3.7	Spin Manuals	106
Section 3.8	Notes and References	115

Chapter 4	**Amplitudes and Transition Amplitudes**	117
Section 4.1	Partial Hilbert Spaces	118
Section 4.2	Amplitude Spaces	122
Section 4.3	Frame Manuals	133
Section 4.4	Transition Amplitudes	137
Section 4.5	Transition Amplitude Spaces	141
Section 4.6	Isomorphisms and A-Forms	147
Section 4.7	Sums and Tensor Products	154
Section 4.8	Quantum Probability Spaces	160
Section 4.9	Notes and References	167
Chapter 5	**Generalized Probability Spaces**	169
Section 5.1	Embedding Generalized Measure Spaces	170
Section 5.2	Probability Models	177
Section 5.3	Measurable Functions and Compatibility	185
Section 5.4	Additivity of the Integral	189
Section 5.5	Bayes' Rule and Hidden Variables	199
Section 5.6	Notes and References	206
Chapter 6	**Probability Manifolds**	207
Section 6.1	Transfinite Induction	207
Section 6.2	Basic Results	215
Section 6.3	Deterministic Spin Model	224
Section 6.4	Phase Space Model	228
Section 6.5	Quantum Mechanics Derived	234
Section 6.6	Notes and References	242
Chapter 7	**Discrete Quantum Mechanics**	245
Section 7.1	Markov Amplitude Chains	246
Section 7.2	Random Phases and Quantum Processes	254
Section 7.3	Quantum Graphic Dynamics	263
Section 7.4	Discrete Spacetime Models	274
Section 7.5	One-Dimensional Lattices and Finite Graphs	283
Section 7.6	Discrete Feynman Amplitudes	292
Section 7.7	Notes and References	301
	Bibliography	303
	Index	311

Preface

It is sometimes stated that there is a kind of uncertainty principle in the foundations of physics. The greater the mathematical rigor, the less is the physical content and understanding [Geroch, 1985]. Although one can site evidence for this contention, we agree with R. Geroch that such an attitude causes more harm than good. The fact is that mathematical rigor not only guarantees the correctness of ones results, it provides insights for the direction to proceed to attain new results. Precise definitions and theorems eliminate the possibility of misunderstanding and provide a language in which physical concepts can be phrased. In order for physical intuition to be reliable, it must be guided by mathematical precision. In fact, it is probably true that physical intuition only comes after a mathematical model is developed. It is not our intention to downplay physical intuition and concepts. These are extremely important for constructing mathematical models describing physical systems. The point is that they can not replace mathematical rigor. Of course, the ultimate judge of a model is nature itself, and this judgement is manifested by experiment.

The time proven method for a rigorous foundational development is the axiomatic method. In constructing a physical theory, the axioms must be carefully and delicately chosen. A slight variation in the axioms at the foundation of a theory can result in huge changes at the frontier. Two different axiomatic structures may be relatively indistinguishable near their foundations but they must either diverge as one approaches the frontiers or one must entirely subsume the other.

There are many contending axiomatic bases for quantum mechanics. Historically, the first axiomatic framework to be developed was the traditional approach [Bohm, 1951; Born, 1937; Dirac, 1930; Eisberg and

Resnick, 1985; Heisenberg, 1930; Prugovecki, 1971; Schrödinger, 1928; von Neumann, 1955]. The traditional approach is based upon the structure of Hilbert space and its self-adjoint operators. It is the textbook framework and the one that is most used by the working physicist. There is the quantum logic approach [Beltrametti and Cassinelli, 1981; Birkhoff and von Neumann, 1936; Bodiou, 1964; Gudder, 1979; Jauch, 1968; Mackey, 1963; Piron, 1976; Varadarajan, 1968, 1970] which investigates the properties of experimental propositions. Although the word "logic" appears in its name, this framework is not intended as a modification of the logic of our thought processes. The algebraic approach [Emch, 1972; Haag and Kastler, 1964; Jordan, von Neumann and Wigner, 1934; Segal, 1974] studies the algebra of observables for a quantum system. Its methods are based on the structure of C^* and W^*-algebras. It is interesting to note the important role played by John von Neumann in all three of the above approaches. The convexity or operational approach [Davies, 1976; Davies and Lewis, 1970; Edwards, 1970; Ludwig, 1983, 1985; Mielnik, 1968, 1969] studies the convex set of states. It relies on the structure of base normed spaces and order unit spaces. The hidden variables approach [Belifante, 1973; Bohm and Bub, 1966a, 1966b; Einstein, Podolsky and Rosen, 1935; Groblicki, 1984; Gudder, 1970, 1979, 1985; Pitowski, 1982, 1983a, 1983b] goes back to the early days of quantum mechanics and has attained a recent revival. It is based on the existence of an underlying objective physical reality. Probably the latest attempt at an axiomatic framework for quantum mechanics is the operational statistics approach [Foulis and Randall, 1972, 1974, 1978, 1981, 1985; Randall and Foulis, 1973, 1981, 1983; Cook, 1985; Fischer and Rüttimann, 1978]. This framework develops the properties of physical operations and is designed to be a language of discourse for the empirical sciences.

Each of these approaches has its own school of investigators and its own vast literature. Each approach has its limited successes and advantages, yet none seem to clearly dominate the others. Is this telling us something? Should we once again venture into the foundations of quantum mechanics and attempt to discern its axiomatic content? Our main guide in such an attempt would be the following question. What is the essence of quantum mechanics? In other words, does quantum mechanics possess fundamental properties, concepts or relations which describe it uniquely? This guide has worked in the past. For example, classical mechanics is based upon Newton's laws, electrodynamics upon Maxwell's equations, special relativity upon Lorentz invariance. Each of the approaches in the previous paragraph has its primitive undefined elements and coresponding properties possessed by these elements. How do we know that these elements capture the essential features of quantum mechanics? Are there other possibilities that might be more successful?

There is yet another approach that we have not mentioned. This is a technique developed by R. Feynman and others, called the path integral formalism [Feynman, 1948, 1949; Feynman and Hibbs, 1965; Schulman, 1981; Schweber, 1961]. Unfortunately, except for certain special cases, this formalism is not mathematically rigorous, so it is not a true axiomatic model. Despite this, it has enjoyed some important successes. It has provided quantitative predictions in quantum electrodynamics that agree remarkably with experiment. It is still the most important computational tool for quantum systems with infinitely many degrees of freedom. The path integral formalism provides much shorter and cleaner methods than the more elegant and rigorous quantum field theory which is based upon the algebraic and traditional approaches. For these reasons, the path integral formalism must contain at least some of the essence of quantum mechanics.

According to Feynman, the basic features of quantum mechanics are the following. First, quantum mechanics must be described by a probabilistic theory. But unlike classical probability theory, quantum theory can possess outcomes that interfere with each other. Two outcomes are thought of as interfering if they can not be distinguished without disturbing the system. The main feature of quantum mechanics is the way that probabilities are computed. In classical probability theory, roughly speaking, the state of a system is determined by a probability function and one computes the probability of an event by summing the probabilities of the outcomes composing that event. However, in quantum probability theory, the state of a system is determined by a complex-valued function A on the outcome space (or sample space). We call A an amplitude function. If the outcomes of an event $E = \{x_1, x_2, \ldots\}$ interfere, the probability $P(E)$ of E is computed by

$$P(E) = \left|\sum A(x_i)\right|^2.$$

If the outcomes do not interfere, then

$$P(E) = \sum |A(x_i)|^2.$$

In the second case we essentially obtain classical probability theory. In the first case, $P(E)$ decomposes into the sum of two parts. One part is the classical counterpart and the other, which consists of the crossterms, provides a constructive or destructive interference which is similar to a wave phenomenon and is characteristic of quantum mechanics.

Thus, the basic axiom of the path integral formalism is that the state of a quantum system is determined by an amplitude function A (or more generally by a transition amplitude function) and that the probability of a set of interfering outcomes $\{x_1, x_2, \ldots\}$ is

$$\left|\sum A(x_i)\right|^2.$$

This point has been missed by the above axiomatic approaches and in this sense they have missed the essence of quantum mechanics. Of course, this essence can be regained by adjoining an amplitude function axiom to the other axioms of a system and we shall discuss such possibilities. But then the axiomatic system is stronger than necessary and might exclude important cases. It would be better to use the amplitude function and its properties as the only axioms of the system.

According to Feynman, nobody really understands quantum mechanics. By this, I think he means that nobody understands why nature has chosen to compute probabilities in this strange way. But nothing prevents us from accepting this fact and using it as our basic axiom. Paraphrasing Goethe, a creative person is someone that can change a problem into a postulate.

The reader is assumed to have a basic knowledge of measure theory and functional analysis. A knowledge of probability theory and quantum mechanics is not assumed and these subjects will be surveyed in Chapters 1 and 2. Chapter 3 presents operational statistics and this will provide us with a language upon which later discussions can be based. Chapter 4 develops the basic axiomatic model for our approach to quantum probability theory. The last three chapters (5, 6, 7) discuss applications of the model. These applications include generalized measure theory, probability manifolds, and discrete quantum mechanics.

1
Classical Probability Theory

As its name suggests, quantum probability theory has two main ingredients, probability theory and quantum mechanics. To understand how to combine these two ingredients we shall need a background knowledge of each. We begin with brief surveys of the orthodox approaches to these two subjects. The present chapter discusses classical probability theory and the next chapter discusses traditional quantum mechanics.

Classical probability theory is a vast subject in its own right. There are many monographs, textbooks, and journal articles devoted to its theoretical aspects and practical applications. Since it is not possible to survey the subject completely in one chapter, we shall concentrate mainly on those areas which will be of interest in our later work.

1.1 Probability Spaces and Random Variables

Although we assume that the reader is familiar with measure theory and the basics of functional analysis, we shall frequently review definitions and results from these fields as they are needed. If X is a set, a σ-*algebra* of subsets of X is a collection of subsets which contains the empty set $\emptyset$ and is closed under the formation of complements and countable unions. If X is a nonempty set and Σ is a σ-algebra of subsets of X, then the pair (X, Σ) is called a *measurable space*. In probability theory, X represents the set of individual outcomes for a probabilistic experiment and is called a *sample space*, while Σ represents the collection of outcome sets to which probabilities can be assigned. The sets in Σ are called *events*. A nonnegative function μ on Σ satisfying $\mu(X) = 1$ and $\mu(\cup A_n) = \Sigma \mu(A_n)$ whenever $A_n \in \Sigma$ and $A_n \cap A_m = \emptyset$ for $n \neq m$, is called a *probability measure*. We call $\mu(A)$ the *probability that the event A occurs*. If (X, Σ) is a measureable space and μ is a probability measure on Σ, then the triple (X, Σ, μ) is called a *probability space*.

Although a probability measure satisfies the countable additivity condition $\mu(\cup A_i) = \Sigma \mu(A_i)$ for mutually disjoint events A_i, what do we do if the A_i's are not disjoint? If there are only finitely many A_i's, then a useful result of elementary probability theory called the *inclusion-exclusion* formula can be applied. This formula, which can easily be proved by induction, states the following

$$\mu(A_1 \cup \cdots \cup A_n) = \sum_i \mu(A_i) - \sum_{i<j} \mu(A_i \cap A_j)$$
$$+ \sum_{i<j<k} \mu(A_i \cap A_j \cap A_k)$$
$$- \ldots (-1)^{n+1} \mu(A_1 \cap \cdots \cap A_n).$$

Example 1.1. The hats of n men are mixed and each man selects a hat at random. What is the probability that at least one man selects his own hat?

Solution. We first construct a mathematical model for this probabilistic experiment. Let us suppose that the men and their hats are numbered $1, 2, \ldots, n$. We represent an outcome of this experiment by a vector with n entries where the i-th entry is the number of the hat selected by the i-th man. The sample space X is the set whose elements are these $n!$ different vectors. We take the σ-algebra Σ of events to be the power set $P(X)$. Since hats are selected at random we take the probability of an outcome to be $1/n!$. It follows that the probability of any $A \in \Sigma$ is

$$\mu(A) = \frac{|A|}{n!}$$

where $|A|$ is the cardinality of A.

Now let A_i be the event that the i-th man selects his own hat, $i = 1, \ldots, n$. Thus A_i consists of all vectors in X whose i-th entry is i. Applying the inclusion-exclusion formula, the probability that at least one man selects his own hat becomes

$$\mu\left(\bigcup_{i=1}^{n} A_i\right) = \sum_{i} \mu(A_i) - \sum_{i<j} \mu(A_i \cap A_j)$$
$$+ \sum_{i<j<k} \mu(A_i \cap A_j \cap A_k)$$
$$- \ldots (-1)^{n+1} \mu(A_1 \cap \cdots \cap A_n).$$

The event $(A_{i_1} \cap \cdots \cap A_{i_j})$ that each of the j men $i_1, \ldots, i_j$ selects his own hat can occur in $(n-j)!$ ways since for the remaining $n-j$ men, the first can select any of $n-j$ hats, the second can select any of $n-(j+1)$ hats, and so forth. Hence,

$$\mu(A_{i_1} \cap \cdots \cap A_{i_j}) = \frac{(n-j)!}{n!}.$$

Since there are $\binom{n}{k} = n!/j!(n-j)!$ terms in the sum

$$\sum_{i_1 < \cdots < i_j} \mu(A_{i_1} \cap \cdots \cap A_{i_j})$$

this sum must equal $\binom{n}{k}(n-j)!/n! = 1/j!$. Hence,

$$\mu\left(\bigcup_{i=1}^{n} A_i\right) = 1 - \frac{1}{2!} + \frac{1}{3!} - \cdots (-1)^{n+1}\left(\frac{1}{n!}\right)$$

and this is the desired result. Notice that as $n \to \infty$ we have

$$\lim_{n \to \infty} \mu\left(\bigcup_{i=1}^{n} A_i\right) = \sum_{i=1}^{\infty} \frac{(-1)^{i+1}}{i!} = 1 - e^{-1} \approx 0.63245. \quad \square$$

Let (X, Σ, μ) be a probability space. A real valued function $f: X \to \mathbb{R}$ which is measurable with respect to Σ is called a *random variable*. If $\int |f| d\mu < \infty$, then the number $E(f) = \int f d\mu$ is called the *expectation* or *mean* of the random variable f. Given an $A \in \Sigma$ there is a particularly simple random variable corresponding to A. The *indicator function* ϕ_A is defined to be the random variable that is 1 on A and 0 on the complement A^c of A. Now let (X_1, Σ_1) be a measurable space and suppose $h: X \to X_1$

is measurable in the sense that $h^{-1}(A_1) \in \Sigma$ for every $A_1 \in \Sigma_1$. Define the probability measure ν_h on Σ_1 by setting

$$\nu_h(A_1) = \mu[h^{-1}(A_1)]$$

for all $A_1 \in \Sigma_1$. Then (X_1, Σ_1, ν_h) is another probability space. If f_1 is a random variable on (X_1, Σ_1, ν_h), then the composite function

$$f(\omega) = f_1[h(\omega)]$$

is a random variable on (X, Σ, μ). Our first theorem shows that f and f_1 have the same exectation.

Theorem 1.1. *If either side exists, then*

$$E(f) = \int f_1 \, d\nu_h. \tag{1.1}$$

Proof. If $f_1 = \phi_{A_1}$, $A_1 \in \Sigma_1$, then the right side of (1.1) is $\nu_h(A_1)$. We then have

$$f = \phi_{A_1} \circ h = \phi_{h^{-1}(A_1)}$$

so the left of (1.1) is $\mu[h^{-1}(A_1)]$. By the definition of ν_h, these two sides are equal so (1.1) holds for indicator functions. If follows that (1.1) holds for simple functions (finite linear combinations of indicator functions). Suppose next that $f_1 \geq 0$. Then there exists an increasing sequence of simple functions f_{1n} which converge to f_1. Moreover, the composites $f_n = f_{1n} \circ h$ are simple functions and they increase to f. Applying the monotone convergence theorem, we have

$$E(f) = \lim E(f_n) = \lim \int f_{1n} \, d\nu_h = \int f_1 \, d\nu_h.$$

Finally, the above argument can be applied separately to the positive and negative parts of f_1 to prove the general case. □

If $f: X \to \mathbb{R}$ is a random variable and $g: \mathbb{R} \to \mathbb{R}$ is a Borel function, then clearly the composite $g \circ f$ is a random variable. The same argument that was used in the proof of Theorem 1.1 gives the following.

PROBABILITY SPACES AND RANDOM VARIABLES   5

Corollary 1.2. *For a Borel function $g: \mathbf{R} \to \mathbf{R}$, if either side exists, then*

$$E(g \circ f) = \int g \circ f_1 \, d\nu_h. \tag{1.2}$$

Denote the σ-algebra of Borel subsets of $\mathbf{R}$ by $B(\mathbf{R})$. If f is a random variable on (X, Σ, μ) and $B \in B(\mathbf{R})$ then $f^{-1}(B) \in \Sigma$ is interpreted as the event that f has a value in B, and $\mu[f^{-1}(B)]$ gives the probability that f has a value in B. The measure μ_f on $(\mathbf{R}, B(\mathbf{R}))$ defined by $\mu_f(B) = \mu[f^{-1}(B)]$, $B \in B(\mathbf{R})$, is called the *distribution* of f. The function $F_f: \mathbf{R} \to \mathbf{R}$ defined by

$$F_f(\lambda) = \mu(\{x \in X : f(x) \le \lambda\}) = \mu_f((-\infty, \lambda]) \tag{1.3}$$

is called the *distribution function* of f. It is straightforward to show that F_f has the following properties:

(a) F_f is nondecreasing;
(b) F_f is continuous from the right;
(c) $\lim_{\lambda \to -\infty} F_f(\lambda) = 0$ and $\lim_{\lambda \to +\infty} F_f(\lambda) = 1$.

Conversely, for any function $F: \mathbf{R} \to \mathbf{R}$ satisfying (a), (b), and (c) there is a unique measure μ on $(\mathbf{R}, B(\mathbf{R}))$ satisfying the relation $\mu((-\infty, \lambda]) = F(\lambda)$ given in (1.3). To construct μ first define $\mathcal{B}_0$ to be the ring consisting of all finite unions of intervals $(a, b]$, $-\infty \le a < b < \infty$ and for each such interval let

$$\mu((a, b]) = F(b) - F(a).$$

Extend μ to $\mathcal{B}_0$ by defining

$$\mu(\cup (a_i, b_i]) = \sum \mu((a_i, b_i]), \quad (a_i, b_i] \cap (a_j, b_j] = \emptyset$$

$i \ne j$, $i, j = 1, 2, \ldots, n$. One can then verify that μ is well-defined and is countably additive on $\mathcal{B}_0$. By the Hahn extension theorem [Halmos, 1950], μ has a unique extension to a measure on $(\mathbf{R}, B(\mathbf{R}))$.

Now let f be a random variable on (X, Σ, μ). If, in Corollary 1.2, we let $h = f$ and let $f_1: \mathbf{R} \to \mathbf{R}$ be the identity function $f_1(\lambda) = \lambda$, then we obtain the following result.

Corollary 1.3. *For a Borel function $g: \mathbf{R} \to \mathbf{R}$, if either side exists, then*

$$E(g \circ f) = \int g \, d\mu_f. \tag{1.4}$$

In particular, if g is the identity function, then (1.4) becomes

$$E(f) = \int_{\mathbf{R}} \lambda \, d\mu_f(\lambda).$$

Furthermore, it is not hard to show that

$$E(f) = \int_{\mathbf{R}} \lambda \, d\mu_f(\lambda) = \int_{-\infty}^{\infty} \lambda \, dF_f(\lambda) \qquad (1.5)$$

where the right hand expression is a Riemann-Stieltjes integral. The above considerations show that the properties of a random variable f on (X, Σ, μ) can be studied using the concrete probability space $(\mathbf{R}, B(\mathbf{R}), \mu_f)$.

Example 1.2. A random variable f with values $0, 1, 2, \ldots$ is called a *Poisson random variable* with *parameter* λ if

$$\mu_f(i) = e^{-\lambda} \frac{\lambda^i}{i!}$$

$i = 0, 1, 2, \ldots$. The expectation of f becomes

$$E(f) = \sum_{i=1}^{\infty} i e^{-\lambda} \frac{\lambda^i}{i!} = \lambda e^{\lambda} \sum_{i=1}^{\infty} \frac{\lambda^{i-1}}{(i-1)!} = \lambda. \quad \square$$

Example 1.3. A random variable f is called a *normal random variable* if μ_f is absolutely continuous with respect to Lebesgue measure and there exist constants $m, \sigma \in \mathbf{R}$ such that

$$\frac{d\mu_f}{dx} = \frac{1}{(2\pi)^{1/2}\sigma} \exp\left[-\frac{(x-m)^2}{2\sigma^2}\right]$$

We call $d\mu_f/dx$ the *density* of f. The expectation of f becomes

$$\begin{aligned}E(f) &= \frac{1}{(2\pi)^{1/2}\sigma} \int_{-\infty}^{\infty} x \exp\left[-\frac{(x-m)^2}{2\sigma^2}\right] dx \\ &= \frac{1}{(2\pi)^{1/2}\sigma} \int_{-\infty}^{\infty} (x-m) \exp\left[-\frac{(x-m)^2}{2\sigma^2}\right] dx \\ &\quad + \frac{m}{(2\pi)^{1/2}\sigma} \int_{-\infty}^{\infty} \exp\left[-\frac{(x-m)^2}{2\sigma^2}\right] dx \\ &= 0 + m = m.\end{aligned}$$

PROBABILITY SPACES AND RANDOM VARIABLES

The *variance* $\sigma^2(f)$ is defined by $\sigma^2(f) = E[(f-E(f))^2]$. In this case, using the substitution $y = (x-m)/\sigma$ and an integration by parts, we obtain

$$\sigma^2(f) = \frac{1}{(2\pi)^{1/2}\sigma} \int_{-\infty}^{\infty} (x-m)^2 \exp\left[-\frac{(x-m)^2}{2\sigma^2}\right] dx$$

$$= \frac{\sigma^2}{(2\pi)^{1/2}} \int_{-\infty}^{\infty} y^2 \exp\left(-\frac{y^2}{2}\right) dy$$

$$= \frac{\sigma^2}{(2\pi)^{1/2}} \int_{-\infty}^{\infty} \exp\left(-\frac{y^2}{2}\right) dy = \sigma^2. \quad \Box$$

We now consider families of random variables. To begin, suppose f_1 and f_2 are two random variables on (X, Σ, μ). Define the function $h_{1,2}: X \to \mathbf{R}^2$ by $h_{1,2}(x) = (f_1(x), f_2(x))$. This function is Borel measurable so we can define a measure $\mu_{1,2}$ on $(\mathbf{R}^2, B(\mathbf{R}^2))$ as follows:

$$\mu_{1,2}(B) = \mu[h_{1,2}^{-1}(B)] = \mu[\{x: (f_1(x), f_2(x)) \in B\}] \quad (1.6)$$

where $B \in B(\mathbf{R}^2)$. We call $\mu_{1,2}$ the *joint distribution* of f_1 and f_2. In particular, if $B = A_1 \times A_2, A_1, A_2 \in B(\mathbf{R})$, then

$$\mu_{1,2}(A_1 \times A_2) = \mu[f_1^{-1}(A_1) \cap f_2^{-1}(A_2)].$$

In this case, $\mu_{1,2}(A_1 \times A_2)$ gives the probability that f_1 and f_2 simultaneously have values in the sets A_1 and A_2 respectively. The *joint distribution function* $F_{1,2}: \mathbf{R}^2 \to \mathbf{R}$ for f_1 and f_2 is defined as follows:

$$\begin{aligned} F_{1,2}(\lambda_1, \lambda_2) &= \mu_{1,2}((-\infty, \lambda_1] \times (-\infty, \lambda_2]) \\ &= \mu[f_1^{-1}((-\infty, \lambda_1]) \cap f_2^{-1}((-\infty, \lambda_2])]. \end{aligned} \quad (1.7)$$

The function $F_{1,2}$ satisfies two-dimensional analogs of equations (a), (b), and (c). Also, $\mu_{1,2}$ can be reconstructed from $F_{1,2}$ in essentially the same way as outlined previously for the one-dimensional case. If $\mu_{1,2}$ is the joint distribution of f_1 and f_2, then the *marginal distributions* $\mu_1 = \mu_{f_1}$, $\mu_2 = \mu_{f_2}$ are given by $\mu_1(A) = \mu_{1,2}(A \times \mathbf{R})$ and $\mu_2(A) = \mu_{1,2}(\mathbf{R} \times A)$.

Proceeding to the general situation, let $f_i: i \in I$ be a (not necessarily countable) family of random variables on (X, Σ, μ). For each n-tuple $(i_1, \ldots, i_n)$,

$$i_j \in I, \quad j = 1, \ldots, n,$$

let
$$h_{i_1,\ldots,i_n}: X \to \mathbb{R}^n$$
be the n-dimensional Borel function
$$h_{i_1,\ldots,i_n}(x) = (f_{i_1}(x),\ldots,f_{i_n}(x)), \quad n = 1,2,\ldots.$$
The *joint distribution* of $f_{i_1},\ldots,f_{i_n}$ is the measure on $(\mathbb{R}^n, B(\mathbb{R}^n))$ defined by
$$\mu_{i_1,\ldots,i_n}(B) = \mu[h_{i_1,\ldots,i_n}^{-1}(B)] \qquad (1.8)$$
where $B \in B(\mathbb{R}^n)$. The set of joint distributions for the family $\{f_i : i \in I\}$ satisfies two consistency conditions. Let π denote a permutation of the integers $1,\ldots,n$, and let $\tau_\pi: \mathbb{R}^n \to \mathbb{R}^n$ be the function
$$\tau_\pi(\lambda_1,\ldots,\lambda_n) = (\lambda_{\pi 1},\ldots,\lambda_{\pi n}).$$
Since
$$h_{i_{\pi 1},\ldots,i_{\pi n}}(x) = \tau_\pi[h_{i_1,\ldots,i_n}(x)]$$
the first consistency condition becomes
$$\mu_{i_{\pi 1},\ldots,i_{\pi n}}(B) = \mu_{i_1,\ldots,i_n}[\tau_\pi^{-1}(B)] \qquad (1.9)$$
for all $B \in B(\mathbb{R}^n)$. For an integer $m > n$ let $\sigma_{m,n}: \mathbb{R}^m \to \mathbb{R}^n$ be the function
$$\sigma_{m,n}(\lambda_1,\ldots,\lambda_m) = (\lambda_1,\ldots,\lambda_n).$$
Now it is clear that
$$h_{i_1,\ldots,i_n}^{-1}(B) = h_{i_1,\ldots,i_m}^{-1}[\sigma_{m,n}^{-1}(B)]$$
for every $B \in B(\mathbb{R}^n)$. Using (1.8), the second consistency condition becomes
$$\mu_{i_1,\ldots,i_n}(B) = \mu_{i_1,\ldots,i_m}[\sigma_{m,n}^{-1}(B)] \qquad (1.10)$$
for all $B \in B(\mathbb{R}^n)$.

We now come to an important existence problem. Let I be an arbitrary index set, and suppose that for each n-tuple $(i_1,\ldots,i_n)$ of elements in I, $n = 1,2,\ldots$, there exists a probability measure $\mu_{i_1,\ldots,i_n}$ on $(\mathbb{R}^n, B(\mathbb{R}^n))$, $n = 1,2,\ldots$. Does there exist a family $\{f_i : i \in I\}$ of random variables whose joint distributions are given by the measures $\mu_{i_1,\ldots,i_n}$? We have seen that (1.9) and (1.10) are necessary conditions for this to happen. An amazing theorem due to A. Kolmogorov shows that (1.9) and (1.10) are also sufficient. For a proof of this theorem, we refer the reader to the literature [Breiman, 1968; Doob, 1953; Kolmogorov, 1956; Lamperti, 1966; Loéve, 1963].

Theorem 1.4. *Let I be any set, and let $\mu_{i_1,\ldots,i_n}$ be a Borel probability measure on $\mathbb{R}^n$ for each n-tuple of elements in I, $n = 1, 2, \ldots$. If this set of measures $M = \{\mu_{i_1,\ldots,i_n}\}$ satisfies Conditions 1.9 and 1.10, then there exists a probability space (X, Σ, μ) and random variables $\{f_i : i \in I\}$ defined on it whose set of joint distributions equals M.*

1.2 Independence and Conditional Probabilities

Independence plays a central role in classical probability theory. It is one of the main characteristics which distinguishes probability theory from measure theory. We shall motivate the definition of independence by first considering conditional probabilities and conditional expectations.

Let (X, Σ, μ) be a probability space and let $B \in \Sigma$ satisfy $\mu(B) \neq 0$. For any $A \in \Sigma$, the *conditional probability* $\mu(A \mid B)$ of A given B is defined as

$$\mu(A \mid B) = \frac{\mu(A \cap B)}{\mu(B)}.$$

We may interpret $\mu(A \mid B)$ as the probability that the event A occurs given that the event B has occurred. Notice that the function $A \to \mu(A \mid B)$ is a probability measure on (X, Σ).

Now let $B_1, \ldots, B_n$ be mutually disjoint sets in Σ which satisfy

$$\cup B_i = X, \qquad \mu(B_i) \neq 0$$

$i = 1, \ldots, n$. We call $\{B_1, \ldots, B_n\}$ a *finite measurable partition* of X. If $A \in \Sigma$, we have

$$\mu(A) = \mu\left(\bigcup_{i=1}^n A \cap B_i\right) = \sum_{i=1}^n \mu(A \cap B_i) = \sum_{i=1}^n \mu(B_i)\mu(A \mid B_i). \quad (1.11)$$

Equation (1.11) states that the probability that A occurs is the sum of n terms where the i-th term is the probability that B_i occurs times the conditional probability that A occurs given B_i. Moreover, if $B \in \Sigma$ with $\mu(B) \neq 0$, we have

$$\begin{aligned}\mu(A \mid B) &= \frac{\mu(A \cap B)}{\mu(B)} \\ &= \frac{\mu(A)\mu(B \mid A)}{\mu(B)}.\end{aligned} \quad (1.12)$$

Combining (1.11) and (1.12) we obtain an important result of elementary probability theory called *Bayes' rule*.

Theorem 1.5. *If $\{B_1, \ldots, B_n\}$ is a finite measurable partition of X and $\mu(A) \neq 0$, then for $j = 1, \ldots, n$, we have*

$$\mu(B_j \mid A) = \frac{\mu(B_j)\mu(A \mid B_j)}{\sum_{i=1}^{n} \mu(B_i)\mu(A \mid B_i)}. \tag{1.13}$$

Again let $\{B_1, \ldots, B_n\}$ be a finite measurable partition of X and let Σ_0 be the σ-subalgebra of Σ generated by $B_1, \ldots, B_n$. For each $A \in \Sigma$ define the random variable $\mu(A \mid \Sigma_0)$ by

$$\mu(A \mid \Sigma_0)(x) = \mu(A \mid B_i)$$

where $x \in B_i$. We call $\mu(A \mid \Sigma_0)$ the *conditional probability* of A given Σ_0. Notice that $\mu(A \mid \Sigma_0)$ satisfies the following two properties.

(a) $\mu(A \mid \Sigma_0)$ is measurable with respect to Σ_0.
(b) $\int_B \mu(A \mid \Sigma_0)\, d\mu = \int_B \phi_A\, d\mu$ for all $B \in \Sigma_0$.

Property a is obvious. Property b clearly holds for $B = B_i$, and since every $B \in \Sigma_0$ is a union of B_i's, the result holds by additivity. Conversely, suppose f is a random variable on (X, Σ, μ) that satisfies

(c) f is measurable with respect to Σ_0.
(d) $\int_B f\, d\mu = \int_B \phi_A\, d\mu$ for all $B \in \Sigma_0$.

Applying (c), f is constant on B_i, and from (d) we have for $x \in B_i$

$$f(x) = [\mu(B_i)]^{-1} \int_{B_i} f\, d\mu = \frac{\mu(A \cap B_i)}{\mu(B_i)} = \mu(A \mid B_i).$$

Hence $f = \mu(A \mid \Sigma_0)$ so we see that (c) and (d) characterize $\mu(A \mid \Sigma_0)$.

Now let f be an arbitrary random variable on (X, Σ, μ) and again let $B \in \Sigma, \mu(B) \neq 0$. We define the *conditional expectation of f given B* as

$$E(f \mid B) = \int f\, d\mu(\cdot \mid B) = [\mu(B)]^{-1} \int_B f\, d\mu.$$

Let $B_1, \ldots, B_n$ and Σ_0 be defined as in the previous paragraph and define the random variable $E(f \mid \Sigma_0)(x) = E(f \mid B_i)$ where $x \in B_i$. We call $E(f \mid \Sigma_0)$ the *conditional expectation of f given Σ_0*. Just as before, $E(f \mid \Sigma_0)$ satisfies two properties.

(a') $E(f \mid \Sigma_0)$ is measurable with respect to Σ_0.
(b') $\int_B E(f \mid \Sigma_0)\, d\mu = \int_B f\, d\mu$ for all $B \in \Sigma_0$.

Moreover, as before, $E(f \mid \Sigma_0)$ is characterized by (a') and (b'). Notice that conditional probability is a special case of conditional expectation since $\mu(A \mid \Sigma_0) = E(\phi_A \mid \Sigma_0)$.

More generally, let Σ_0 be an arbitrary σ-subalgebra of Σ. The *conditional expectation* $E(f \mid \Sigma_0)$ of f given Σ_0 is a random variable satisfying (a') and (b'). It turns out that if $E(f)$ exists, then $E(f \mid \Sigma_0)$ exists and is unique almost everywhere. Indeed, if we define the measure

$$\nu(B) = \int_B f \, d\mu, \quad B \in \Sigma_0$$

then ν is absolutely continuous relative to μ restricted to Σ_0. By the Radon-Nikodym theorem [Halmos, 1950] there exists a unique (almost everywhere) function g which is measurable with respect to Σ_0 satisfying $\nu(B) = \int_B g \, d\mu$. Hence g satisfies (a') and (b'). The *conditional probability* is now defined as $\mu(A \mid \Sigma_0) = E(\phi_A \mid \Sigma_0)$.

Let A and B be events satisfying $\mu(B) \neq 0$, $\mu(A \mid B) = \mu(A)$. Notice that this is equivalent to $\mu(B) \neq 0$, $\mu(A \cap B) = \mu(A)\mu(B)$. Since the probability of A does not change when B has occurred we say that A and B are *independent*. More generally, we say that a finite family of events $A_1, \ldots, A_n$ is *independent* if every subfamily of them satisfies

$$\mu(A_{i_1} \cap \cdots \cap A_{i_k}) = \mu(A_{i_1})\mu(A_{i_2})\ldots\mu(A_{i_k}). \quad (1.14)$$

If $f_1, \ldots, f_n$ are random variables, they are said to be *independent* if the events $A_i = f_i^{-1}(B_i)$ satisfy (1.14) for every choice of sets $B_i \in B(\mathbf{R})$. We say that an infinite collection of events or random variables is *independent* if every finite subset is independent. Furthermore, a sequence of σ-subalgebras $\Sigma_1, \Sigma_2, \ldots$ is *independent* if (1.14) holds whenever $A_i \in \Sigma_i$.

Example 1.4. When a certain coin is flipped it has a probability p of landing heads and hence a probability $1-p$ of landing tails where $0 \leq p \leq 1$. We call such a coin a *p-coin* and if $p = \frac{1}{2}$ the coin is *fair*. Suppose a *p*-coin is flipped an infinite number of times. We now construct a mathematical model for this probabilistic experiment. We let X be the set of sequences of 0's and 1's where a 1(0) in the i-th term represents a head (tail) on the i-th flip. A *cylinder set* is a subset of X in which a finite number of terms are specified. We define Σ to be the σ-algebra generated by the cylinder sets. For a cylinder set A, we define $\mu_0(A) = p^i(1-p)^j$ where $i(j)$ is the number of specified 1's (0's). We then extend μ_0 to a probability measure μ

on Σ (which can be done by the Hahn extension theorem [Halmos, 1950]). Then (X, Σ, μ) becomes a probability space.

To show that (X, Σ, μ) gives a good model for the above experiment, let A be the event that the first i flips are heads and the next $n - i$ flips are tails, $i \leq n$. Then $\mu(A) = p^i(1-p)^{n-i}$. But this is exactly the answer we expect in this experiment. Indeed, let A_i be the event that a head appears on the i-th flip. Then $A_1, A_2, \ldots$ is a sequence of independent events in (X, Σ, μ) which is evident since physically the result of any flip does not depend on the results of other flips. Similarly, if we replace any of the A_i's by A_i^c we again get a sequence of independent events. Since

$$A = A_1 \cap \cdots \cap A_i \cap A_{i+1}^c \cap \cdots \cap A_n^c$$

by independence, the probability of A should be the product of the probabilities on the right hand side. But this product is again $p^i(1-p)^{n-i}$. Let g_i be the random variable which is $1(0)$ if the i-th flip results in a head (tail.) Then

$$g_i^{-1}(1) = A_i,$$
$$g_i^{-1}(0) = A_i^c$$

so $g_1, g_2, \ldots$ is a sequence of independent random variables. $\square$

Example 1.5. If a p-coin is flipped an infinite number of times, find the distribution for the number of heads in the first n flips.

Solution. Let f_n be the random variable that counts the number of heads in the first n flips. Since there are $\binom{n}{i}$ ways of obtaining i heads in n flips, $i \leq n$, it follows from Example 1.4 that

$$\mu_{f_n}(i) = \binom{n}{i} p^i (1-p)^{n-i} \qquad i = 0, 1, \ldots, n$$

and $\mu_{f_n}(i) = 0$, otherwise. A random variable with this distribution is called a *binomial* random variable with *parameters* (n, p). One would expect that $E(f_n) = np$. This is indeed the case since applying the binomial theorem gives

$$E(f_n) = \sum_{i=0}^{n} i \binom{n}{i} p^i (1-p)^{n-i}$$
$$= np \sum_{i=0}^{n-1} \binom{n-1}{i} p^i (1-p)^{n-1-i}$$
$$= np[p + (1-p)]^{n-1} = np. \quad \square$$

INDEPENDENCE AND CONDITIONAL PROBABILITIES

Example 1.6. If a p-coin is flipped an infinite number of times, find the distribution for the number of flips required until a total of r heads are accumulated.

Solution. Let f be the random variable that counts the number of flips required until r heads are accumulated. In order for the r-th head to occur at the i-th flip, there must be $r-1$ heads in the first $i-1$ flips and the i-th flip must be a head. Hence,

$$\mu_f(i) = \binom{i-1}{r-1} p^r (1-p)^{i-r}, \qquad i = r, r+1, \ldots$$

and $\mu_f(i) = 0$, otherwise. A random variable with this distribution is called a *negative binomial* random variable with *parameters* (r, p). In the particular case $r = 1$, f counts the number of flips required until the first head appears. We then have

$$\mu_f(i) = p(1-p)^{i-1}, \qquad i = 1, 2, \ldots.$$

Such a random variable is called a *geometric* random variable with *parameter* p.

If f is a geometric random variable with parameter $0 < p < 1$ and we let $q = 1 - p$, then

$$E(f) = \sum_{i=0}^{\infty} ipq^{i-1} = p \sum_{i=0}^{\infty} \frac{d}{dq} q^i = p \frac{d}{dq} \sum_{i=0}^{\infty} q^i$$

$$= p \frac{d}{dq} \frac{1}{1-q} = \frac{p}{(1-q)^2} = \frac{1}{p}. \qquad \square$$

The following is a useful theorem on independence.

Theorem 1.6. *Let $\Sigma_0, \Sigma_1, \Sigma_2, \ldots$ be independent σ-subalgebras. If S is the σ-algebra generated by any subfamily of $\Sigma_1, \Sigma_2, \ldots$, then Σ_0 and S are independent.*

Proof. There is no loss of generality in assuming that S is generated by all Σ_i, $i = 1, 2, \ldots$. Let $A \in \Sigma_0$; we must show that

$$\mu(A \cap B) = \mu(A)\mu(B) \tag{1.15}$$

for all $B \in S$. We may assume that $\mu(A) > 0$, as (1.15) is trivial otherwise. If

$$B = A_1 \cap \cdots \cap A_n, \quad A_i \in \Sigma_i \qquad (1.16)$$

then (1.15) holds by definition. The class F of finite unions of sets of the form (1.16) is an algebra and we now show that (1.15) holds for $B \in F$. Let

$$B = \bigcup_{i=1}^{k} B_i,$$

$$B_i = A_1^i \cap \cdots \cap A_{n(i)}^i$$

be an arbitrary set of F. By the inclusion-exclusion formula we have

$$\begin{aligned}\mu(A \cap B) &= \mu[\cup(A \cap B_i)] \\ &= \sum_i \mu(A \cap B_i) - \sum_{i<j} \mu(A \cap B_i \cap B_j) \\ &\quad + \sum_{i<j<k} \mu(A \cap B_i \cap B_j \cap B_k) - \cdots \end{aligned} \qquad (1.17)$$

where the series has k terms. Since (1.15) holds for the expressions after the summation symbols, we can factor $\mu(A)$ out of the right side of (1.17) and the remaining factor, by the inclusion-exclusion formula equals $\mu(\cup B_i)$. Hence (1.15) holds for all $B \in F$. We thus see that the measures μ and $\mu(\cdot \mid A)$ agree on F. By the uniqueness part of the Hahn extension theorem, they must agree on the σ-algebra S generated by F. □

Corollary 1.7. *Under the conditions of the theorem, if S_1 and S_2 are the σ-algebras generated by two disjoint subfamilies of the Σ_i, then S_1 and S_2 are independent.*

Proof. Each Σ_i which helps to generate S_1 is independent of S_2 by the theorem. Let S_2 play the role of Σ_0 and apply the theorem again. □

If R is a family of random variables on a probability space (X, Σ, μ) we denote the σ-algebra generated by R by $\Sigma(R)$; that is, $\Sigma(R)$ is the smallest σ-algebra with respect to which each $f \in R$ is measurable.

Corollary 1.8. *Let $f_1, \ldots, f_{n+m}$ be independent random variables and let $g: \mathbf{R}^n \to \mathbf{R}$ and $h: \mathbf{R}^m \to \mathbf{R}$ be Borel functions. Then $g(f_1, \ldots, f_n)$ and $h(f_{n+1}, \ldots, f_{n+m})$ are independent random variables.*

Proof. It follows from Corollary 1.7 that

$$\Sigma(\{f_1,\ldots,f_n\})$$

and

$$\Sigma(\{f_{n+1},\ldots,f_{n+m}\})$$

are independent σ-algebras. Since $g(f_1,\ldots,f_n)$ is measurable with respect to the first and

$$h(f_{n+1},\ldots,f_{n+m})$$

is measurable with respect to the second, these two random variables are independent. □

For a sequence of random variables $f_1, f_2, \ldots$, the σ-algebras

$$\Sigma_n = \Sigma(\{f_n, f_{n+1},\ldots\})$$

form a decreasing sequence and $\Sigma_\infty = \cap \Sigma_n$ is called the *tail σ-algebra* of the sequence. We now come to the important 0-1 law.

Corollary 1.9. (0-1 Law). *Let $f_1, f_2, \ldots$ be a sequence of independent random variables on (X, Σ, μ). If $A \in \Sigma_\infty$, then $\mu(A) = 0$ or 1.*

Proof. By Theorem 1.6, $\Sigma(f_n)$ is independent of

$$\Sigma(\{f_{n+1}, f_{n+2},\ldots\}) \supseteq \Sigma_\infty$$

so Σ_∞ is independent of $\Sigma(f_n)$ for every n. Again by Theorem 1.6, it follows that Σ_∞ is independent of $\Sigma(\{f_1, f_2,\ldots\})$. But this last σ-algebra contains Σ_∞ so Σ_∞ is independent of itself. Hence if $A \in \Sigma_\infty$, then $\mu(A) = \mu(A \cap A) = \mu(A)^2$, so that $\mu(A) = 0$ of 1. □

The next lemma, called the Borel-Cantelli lemma is another general result on independence which has important applications.

Lemma 1.10. (Borel-Cantelli). *Let $A_1, A_2, \ldots$ be events and let*

$$B = \limsup A_n = \bigcap_{k=1}^{\infty} \bigcup_{n=k}^{\infty} A_n.$$

(i) *If $\Sigma \mu(A_n) < \infty$, then $\mu(B) = 0$.*
(ii) *If $\Sigma \mu(A_n) = \infty$ and the A_n are independent, then $\mu(B) = 1$.*

Proof. (i) Since measures are subadditive, we have

$$\mu(B) \leq \mu\left(\bigcup_{n \geq k} A_n\right) \leq \sum_{n \geq k} \mu(A_n)$$

for each k. If $\Sigma \mu(A_n) < \infty$, the right side of the inequalities goes to 0 as $k \to \infty$. (ii) It is enough to show that $\mu(\bigcup_{n \geq k} A_n) = 1$ for every k. For each $K \geq k$ we have

$$1 - \mu\left(\bigcup_{n \geq k} A_n\right) \leq 1 - \mu\left(\bigcup_{n=k}^{K} A_n\right) = \mu\left(\bigcap_{n=k}^{K} A_n^c\right) = \prod_{n=k}^{K} [1 - \mu(A_n)].$$

If $\Sigma \mu(A_n) = \infty$, the above product diverges to 0. □

Notice that $\limsup A_n$ is the set of all x which belong to infinitely many of the A_n's. We hence interpret $\mu(B) = 0$ to mean that only finitely many A_n's occur (almost surely.) Part (ii) of the above lemma does not hold if we only require $\Sigma \mu(A_n) = \infty$. For example, suppose $A_n = A$, $n = 1, 2, \ldots$, and $\mu(A) \neq 0$ or 1.

1.3 Independent Random Variables

If f and g are independent random variables on (X, Σ, μ) then their joint distribution $\mu_{f,g}$ satisfies

$$\begin{aligned}
\mu_{f,g}(A \times B) &= \mu[f^{-1}(A) \cap f^{-1}(B)] \\
&= \mu[f^{-1}(A)]\mu[g^{-1}(B)] \\
&= \mu_f(A)\mu_g(B)
\end{aligned} \quad (1.18)$$

for all $A, B \in B(\mathbb{R})$. Hence $\mu_{f,g}$ is the product measure $\mu_f \times \mu_g$. It follows that their joint distribution function is the product of their individual distribution functions. That is,

$$F_{f,g}(\lambda_1, \lambda_2) = F_f(\lambda_1) F_g(\lambda_2) \quad (1.19)$$

for all $\lambda_1, \lambda_2 \in \mathbb{R}$. Conversely, if (1.18) or (1.19) holds then f and g are independent. The next result now follows easily.

Lemma 1.11. *Let f and g be independent random variables. Then $E(fg)$ exists if both $E(f)$ and $E(g)$ exist and in this case $E(fg) = E(f)E(g)$. Conversely, if $E(fg)$ exists and $f, g \neq 0$ a.e. then $E(f), E(g)$ exist.*

Proof. In Theorem 1.1, let $X_1 = \mathbb{R}^2$, $h(\omega) = (f(\omega), g(\omega))$, and
$$f_1(\lambda_1, \lambda_2) = \lambda_1 \lambda_2, \quad \lambda_1, \lambda_2 \in \mathbb{R}.$$

It then follows that
$$E(fg) = \int_{\mathbb{R}^2} \lambda_1 \lambda_2 \, d\mu_{f,g} = \int_{\mathbb{R}^2} \lambda_1 \lambda_2 \, d(\mu_f \times \mu_g).$$

Now apply Fubini's theorem and (1.6). □

Corollary 1.12. *If $f_1, \ldots, f_n$ are independent random variables with expectations, then $E(\Pi f_i) = \Pi E(f_i)$.*

Proof. By Corollary 1.8, f_1 and $\prod_{i=2}^{n} f_i$ are independent. Applying Lemma 1.11 gives
$$E\Big(\prod_{i=1}^{n} f_i\Big) = E(f_1) E\Big(\prod_{i=2}^{n} f_i\Big).$$

The result now follows by induction. □

Many of the results of probability theory concern sums of independent random variables. For this reason, it is useful to have an expression for the distribution function of the sum of two independent random variables. If F and G are distribution functions, the *convolution* of F and G is defined as
$$F * G(\lambda) = \int_{-\infty}^{\infty} F(\lambda - \alpha) \, dG(\alpha).$$

Lemma 1.13. *If f and g are independent random variables, then $F_{f+g} = F_f * F_g = F_g * F_f$.*

Proof. Define the Borel function $f_1 \colon \mathbb{R}^2 \to \mathbb{R}$ by $f_1(\lambda_1, \lambda_2) = 1$ if $\lambda_1 + \lambda_2 \leq \lambda$, $f_1(\lambda_1, \lambda_2) = 0$ otherwise. Applying Theorem 1.1 with $X_1 = \mathbb{R}^2$, $h(\omega) = (f(\omega), g(\omega))$, gives
$$E[f_1(f, g)] = \int_{\mathbb{R}^2} f_1 \, d\mu_{f,g}.$$

Since the left side is $F_{f+g}(\lambda)$, applying Fubini's theorem gives

$$F_{f+g}(\lambda) = \int f_1 \, d(\mu_f \times \mu_g) = \int\int f_1(\lambda_1, \lambda_2) \, d\mu_f(\lambda_1) \, d\mu_g(\lambda_2)$$

$$= \int\int f_1(\lambda_1, \lambda_2) \, d\mu_g(\lambda_1) \, d\mu_f(\lambda_2).$$

A change of variables now gives the result. □

Corollary 1.14. *Let f and g be independent random variables and suppose μ_f is absolutely continuous so that*

$$F_f(\lambda) = \int_{-\infty}^{\lambda} h(\alpha) \, d\alpha$$

for some density h. Then μ_{f+g} is absolutely continuous and its density is given by

$$\frac{d}{d\lambda}\mu(f+g \leq \lambda) = \int_{-\infty}^{\infty} h(\lambda - \alpha) \, dF_g(\alpha).$$

Proof. By Fubini's theorem we have

$$\int_{-\infty}^{\lambda}\int_{-\infty}^{\infty} h(t-s) \, dF_g(s) \, dt = \int_{-\infty}^{\infty}\int_{-\infty}^{\lambda} h(t-s) \, dt \, dF_g(s)$$

$$= \int_{-\infty}^{\infty} F_f(\lambda - s) \, dF_g(s) = \mu(f+g \leq \lambda). \quad □$$

Many theorems in probability theory involve a sequence of independent random variables whose distributions satisfy certain conditions. But how do we know that such a sequence exists? The following lemma is usually employed to answer such questions. It is a corollary of Kolmogorov's theorem (Theorem 1.4).

Lemma 1.15. *Let $\mu_1, \mu_2, \ldots$ be a sequence of probability measures on $(\mathbf{R}, B(\mathbf{R}))$. Then there exists a probability space (X, Σ, μ) with independent random variables $f_1, f_2, \ldots$ defined on it such that $\mu_{f_i} = \mu_i$, $i = 1, 2, \ldots$.*

Proof. Given an n-tuple of indices $(i_1, \ldots, i_n)$ define the probability measure $\mu_{i_1,\ldots,i_n} = \mu_{i_1} \times \cdots \times \mu_{i_n}$ on $(\mathbf{R}^n, B(\mathbf{R}^n))$. It is clear that these measures satisfy (1.9) and (1.10) so Theorem 1.4 can be applied. □

Of course, the above lemma can also be phrased in terms of distribution functions.

Although independent sequences of random variables are useful in many applications, there are important situations in which the independence assumption must be relaxed somewhat. Let $f_0, f_1, f_2, \ldots$ be a sequence of random variables and suppose that their ranges are all contained in a fixed countable set $E \subseteq \mathbf{R}$. It is customary to call E the "state space" of the "process" f and to refer to f_n as the state of the process f at time n. It is also customary to refer to the event $f_n^{-1}(j), j \in E$, by saying that the process is in state j at time n. We call $f = \{f_0, f_1, f_2, \ldots\}$ a *Markov chain* if

$$\mu(f_{n+1}^{-1}(j) \mid \Sigma(f_0, \ldots, f_n)) = \mu(f_{n+1}^{-1}(j) \mid \Sigma(f_n)) \qquad (1.20)$$

for all $j \in E$ and $n = 0, 1, 2, \ldots$.

A Markov chain, then, is a sequence of random variables such that for all n, f_{n+1} is conditionally independent of $f_0, \ldots, f_{n-1}$ given f_n. That is, the "next" state f_{n+1} of the process is independent of "past" states $f_0, \ldots, f_{n-1}$ provided that the "present" state f_n is known. We say that a Markov chain f is *stationary* if the conditional probability

$$T(i, j) = \mu(f_{n+1}^{-1}(j) \mid f_n^{-1}(i)), \qquad i, j \in E$$

is independent of n. We then call the probabilities $T(i, j)$ *transition probabilities* and we call the matrix T the *transition matrix* of the Markov chain f. Notice that the transition matrix T satisfies the following two properties:

(a) $T(i, j) \geq 0$ for all $i, j \in E$;
(b) for each $i \in E, \Sigma_j T(i, j) = 1$.

A matrix T which satisfies (a) and (b) is called a *stochastic matrix*. Thus, the transition matrix of a stationary Markov chain is a stochastic matrix. Conversely, for any stochastic matrix T, one can construct a stationary Markov chain whose transition matrix is T. The proof of the following result is straightforward.

Theorem 1.16. *Let f be a stationary Markov chain with transition matrix T and state space E. Then for any $i_0, \ldots, i_n \in E$ we have*

$$\mu(f_1^{-1}(i_1) \cap \cdots \cap f_n^{-1}(i_n) \mid f_0^{-1}(i_0)) = T(i_0, i_1) \ldots T(i_{n-1}, i_n).$$

Corollary 1.17. *If $p(i) = \mu(f_0^{-1}(i))$, $i \in E$, then*

$$\mu(f_0^{-1}(i_0) \cap \cdots \cap f_n^{-1}(i_n)) = p(i_0)T(i_0, i_1)\ldots T(i_{n-1}, i_n)$$

for any $i_0, \ldots, i_n \in E$.

This corollary shows that the joint distribution of $f_0, \ldots, f_n$ is completely specified for every n once the initial distribution p and the transition matrix T are known. The next corollary follows easily from Theorem 1.16.

Corollary 1.18. *For any $n = 0, 1, \ldots$,*

$$\mu(f_n^{-1}(j) \mid f_0^{-1}(i)) = T^n(i, j). \tag{1.21}$$

If $m < n$ are positive integers, since

$$T^n(i, j) = \sum_k T^m(i, k) T^{n-m}(k, j)$$

it follows from (1.21) that

$$\mu(f_n^{-1}(j) \mid f_0^{-1}(i)) = \sum_k \mu(f_m^{-1}(k) \mid f_0^{-1}(i)) \mu(f_n^{-1}(j) \mid f_m^{-1}(k)). \tag{1.22}$$

Equation 1.22 is called the *Chapman-Kolmogorov* equation. It says that starting at state i, in order for the process f to be in state j after n steps, it must be in some intermediate state k after the m-th step and then move from that state k into state j during the remaining $n - m$ steps.

We now give an important example of a stationary Markov chain. Let $g_1, g_2, \ldots$ be independent, identically distributed random variables with distribution $p(i) = p_i$, $i = 0, 1, 2, \ldots$. Define $f_0 = 0$, and $f_n = g_1 + \cdots + g_n$, $n \geq 1$. Since g_{n+1} is independent of $f_0, \ldots, f_n$ we have

$$\mu(f_{n+1}^{-1}(j) \mid f_0^{-1}(i_0) \cap \cdots \cap f_n^{-1}(i_n))$$
$$= \mu(\{x \colon f_n(x) + g_{n+1}(x) = j\} \mid f_0^{-1}(i_0) \cap \cdots \cap f_n^{-1}(i_n))$$
$$= \mu(g_{n+1}^{-1}(j - i_n) \mid f_0^{-1}(i_0) \cap \cdots \cap f_n^{-1}(i_n))$$
$$= p_{j-i_n} = \mu(f_{n+1}^{-1}(j) \mid f_n^{-1}(i_n)).$$

It follows that f is a stationary Markov chain whose transition probabilities are

$$T(i, j) = \mu(f_{n+1}^{-1}(j) \mid f_n^{-1}(i)) = p_{j-i}.$$

We thus see that successive sums of independent, identically distributed, discrete random variables form a stationary Markov chain.

LAW OF LARGE NUMBERS

Example 1.7. In Example 1.5, f_n counted the number of heads in the first n flips of a p-coin. Since $f_n = g_1 + \cdots + g_n$, where g_i is given in Example 1.4, we conclude that $\{f_0, f_1, \ldots\}$ is a stationary Markov chain with state space $\{0, 1, 2, \ldots\}$. The transition probabilities are

$$T(i,j) = \mu\bigl(f_{n+1}^{-1}(j) \mid f_n^{-1}(i)\bigr) = \begin{cases} p & \text{if } j = i+1 \\ 1-p & \text{if } j = 1 \\ 0 & \text{otherwise.} \end{cases}$$

It is easy to take powers of the transition matrix to find the m-step transition probabilities

$$\mu\bigl(f_{n+1}^{-1}(j) \mid f_n^{-1}(i)\bigr) = T^m(i,j) = \binom{m}{j-1} p^{j-i}(1-p)^{m-j+i}$$

where $j = i, \ldots, m+i$. □

Example 1.8. A p-coin is flipped an infinite number of times and f_r is the number of flips required until r heads appear. The distribution of f_r is given in Example 1.6. It is not hard to show that $\{f_0, f_1, \ldots\}$ is a stationary Markov chain with state space $\{0, 1, \ldots\}$. The transition probabilities are

$$T(i,j) = \mu\bigl(f_{n+1}^{-1}(j) \mid f_n^{-1}(i)\bigr) = \begin{cases} p(1-p)^{j-i-1} & \text{if } j \geq i+1 \\ 0 & \text{otherwise} \end{cases}$$

and the m-step transition probabilities are

$$\begin{aligned} T^m(i,j) &= \mu\bigl(f_{n+m}^{-1}(j) \mid f_n^{-1}(i)\bigr) \\ &= \mu\bigl(f_m^{-1}(j-i)\bigr) \\ &= \begin{cases} \binom{j-i-1}{m-1} p^m q^{j-i-m} & \text{if } j \geq i+m \\ 0 & \text{otherwise.} \end{cases} \end{aligned}$$ □

1.4 Law of Large Numbers

This section considers two types of limit theorems, the weak law of large numbers and the strong law of large numbers. The next section treats the central limit theorem. Although there are variations and refinements of these limit theorems, space permits only a representative example of each of the three types.

First we discuss the concept of convergence. A sequence of random variables $f_1, f_2, \ldots$ on (X, Σ, μ) *converges in probability* (also called *converges in measure*) to a random variable f if for each $\varepsilon > 0$

$$\lim_{n \to \infty} \mu[\{x \colon |f_n(x) - f(x)| > \varepsilon\}] = 0.$$

We say that f_n *converges almost surely* (abbreviated a.s., also called *almost everywhere*) to f if

$$\mu(\{x \colon \lim f_n(x) = f(x)\}) = 1.$$

Finally, if for some $p > 0$ $\lim E(|f_n - f|^p) = 0$ we have *convergence in mean* of order p. It is not hard to show that convergence a.s. or in the mean (of any order) implies convergence in probability but that no other implications hold among the three types.

If f is a random variable, the numbers $m_k = E(f^k)$, if they exist, are called the *moments* of f. In particular, the first moment m_1 is the mean of f, and the second moment of $f - m_1$ is called the *variance* of f. The variance of f is denoted $\sigma^2(f)$ and it is easy to show that $\sigma^2 = m_2 - m_1^2$. If f and g are independent random variables, a simple computation gives $\sigma^2(f + g) = \sigma^2(f) + \sigma^2(g)$. The following lemma, due to Chebyshev is frequently useful.

Lemma 1.19. (Chebyshev's inequality). *Let f be a random variable and let g be an increasing, nonnegative function on its range. If $g(\lambda) > 0$, then*

$$\mu(\{x \colon f(x) \geq \lambda\}) \leq \frac{E(g \circ f)}{g(\lambda)}.$$

Proof. Applying Corollary 1.3 we have

$$E(g \circ f) = \int_{-\infty}^{\infty} g \, d\mu_f.$$

Since g is nonnegative and increasing, we obtain

$$\int_{-\infty}^{\infty} g \, d\mu_f \geq \int_{\lambda}^{\infty} g \, d\mu_f \geq g(\lambda) \int_{\lambda}^{\infty} d\mu_f = g(\lambda) \mu(\{x \colon f(x) \geq \lambda\}). \qquad \square$$

By choosing different functions for g in Chebyshev's inequality we obtain some useful special cases. If we choose $g(\lambda) = \lambda^2$ and replace f by $|f - E(f)|$ we obtain

$$\mu(|f - E(f)| \geq \lambda) \leq \frac{\sigma^2(f)}{\lambda^2}, \qquad \lambda > 0. \tag{1.23}$$

Other cases of importance are

$$\mu(f \geq \lambda) \leq e^{-c\lambda} E(e^{cf}), \qquad c > 0. \tag{1.24}$$

$$\mu(|f| \geq \lambda) \leq \lambda^{-k} E(|f|^k), \qquad \lambda, k > 0. \tag{1.25}$$

Suppose we flip a fair coin a large number of times. As the number of flips n gets very large we would expect that the number of heads divided by n approaches $1/2$. How can we make this statement precise? First we make the reasonable assumption that the result of each flip is independent of the others. Next we represent "tails" by the number 0 and "heads" by the number 1. If μ_n is the probability associated with the n-th flip then μ_n is a probability measure on $(\mathbf{R}, B(\mathbf{R}))$ given by

$$\mu_n(\{0\}) = \tfrac{1}{2}$$
$$\mu_n(\{1\}) = \tfrac{1}{2}.$$

Now $\mu_1, \mu_2, \ldots$ is a sequence of probability measures on $(\mathbf{R}, B(\mathbf{R}))$ so by Kolmogorov's theorem (specifically Lemma 1.15) there exists a probability space (X, Σ, μ) and a sequence of independent random variables $f_1, f_2, \ldots$ on (X, Σ, μ) such that $\mu_{f_i} = \mu_i$. Thus $f_1, f_2, \ldots$ is a sequence of independent, identically distributed (abbreviated i.i.d.) random variables such that

$$\mu(f_i^{-1}(0)) = \tfrac{1}{2}$$
$$\mu(f_i^{-1}(1)) = \tfrac{1}{2}$$

$i = 1, 2, \ldots$. We may interpret f_i as giving the number of heads on the i-th flip. The number of heads in the first n flips divided by n is represented by the random variable

$$S_n = n^{-1}(f_1 + \cdots + f_n).$$

As previously stated, we would expect that $S_n \to 1/2$ in some sense. If $S_n \to 1/2$ in probability we have a weak law of large numbers and if $S_n \to 1/2$ a.s. we have a strong law of large numbers.

There are already mathematical indications that such laws should hold. For example, let $A \in \Sigma$ be the set for which S_n converges. Since the convergence of S_n does not depend on any finite number of f_i's it is almost obvious that A is in the tail field Σ_∞ of the sequence of independent random variables $f_1, f_2, \ldots$. By the 0-1 law (Corollary 1.9), $\mu(A) = 0$ or 1. If

$\mu(A) = 1$, then S_n converges a.s. . Thus if we can show that S_n converges on any set of nonzero measure, then S_n converges a.s. . Furthermore, suppose $S_n \to f$ a.s. . Then as above, f is measurable with respect to Σ_∞. Again by the $0-1$ law, f is a constant a.s. .

The following is a weak law of large numbers with a very simple proof.

Theorem 1.20. *Let $f_1, f_2, \ldots$ be a sequence of i.i.d. random variables and let*
$$S_n = n^{-1}[f_1 + \cdots + f_n].$$
If the f_i have finite second moments then $S_n \to E(f_1)$ in probability.

Proof. Letting $\sigma^2 = \sigma^2(f_i)$, by additivity, the mean and variance of S_n are $E(f_1)$ and σ^2/n respectively. Applying (1.23) with $f = S_n$ and $\lambda = \varepsilon$ gives
$$\mu\bigl(|S_n - E(f_1)| \geq \varepsilon\bigr) \leq \frac{\sigma^2}{n\varepsilon^2}. \quad \square$$

The above theorem can easily be generalized to the case in which the f_i's are not indentically distributed. Also, the existence of finite second moments can be relaxed to a weaker condition [Breiman, 1968; Lamperti, 1966; Loève, 1963]. We next prove a version of a strong law of large numbers.

Theorem 1.21. *Let $f_1, f_2, \ldots$ be a sequence of i.i.d. random variables with mean m_1 and let*
$$S_n = n^{-1}[f_1 + \cdots + f_n].$$
If the f_i's have finite fourth moment, then $S_n \to m_1$ a.s.

Proof. Schwarz's inequality shows that second moments exist. Letting $\sigma^2 = \sigma^2(f_i)$, a straightforward calculation using Corollary 1.12 gives
$$E\left(\left[\sum_{i=1}^n (f_i - m_1)\right]^4\right) = nE([f_1 - m_1]^4) + 6\binom{n}{2}\sigma^4 \leq cn^2.$$

Applying (1.25) yields
$$\mu\left(\left|\sum_{i=1}^n (f_i - m_1)\right| > \varepsilon n\right) \leq \frac{c}{\varepsilon^4 n^2} \qquad (1.26)$$

and the sum on n of the right side is finite. Let

$$A(n,\varepsilon) = \{x\colon |S_n(x) - m_1| > \varepsilon\}$$

and let

$$B(\varepsilon) = \limsup A(n,\varepsilon).$$

Using (1.26),

$$\Sigma\mu[A(n,\varepsilon)] < \infty$$

and hence by the Borel-Cantelli lemma (Lemma 1.10) $\mu[B(\varepsilon)] = 0$. If $S_n(x) \not\to m_1$ then there exists a positive integer k such that

$$|S_n(x) - m_1| > k^{-1}$$

for infinitely many n. Hence x is in infinitely many of the sets $A(n, k^{-1})$; that is $x \in B(k^{-1})$. Hence

$$B = \{x\colon S_n(x) \not\to m_1\} = \bigcup_{k=1}^{\infty} B(k^{-1})$$

so $\mu(B) = 0$. □

With quite a bit more work, Theorem 1.21 can be proved without requiring the existence of fourth moments. Also, there are generalizations which do not require the f_i's to be identically distributed [Breiman, 1968; Lamperti, 1966; Loève, 1963].

1.5 Central Limit Theorem

Before we treat the central limit theorem we need some preliminaries. Let $C(\mathbb{R})$ be the set of bounded, continuous, real-valued functions on $\mathbb{R}$. We say that a sequence of probability measures μ_n on $(\mathbb{R}, B(\mathbb{R}))$ converges weakly to a probability measure μ (denoted $\mu_n \Rightarrow \mu$) if

$$\lim_{n\to\infty} \int g\,d\mu_n = \int g\,d\mu \qquad (1.27)$$

for every $g \in C(\mathbb{R})$. Since probability measures on $(\mathbb{R}, B(\mathbb{R}))$ are regular [Halmos, 1950; Lamperti, 1966; Loève, 1963], it follows easily that weak limits are unique. We now characterize weak convergence on $\mathbb{R}$ in terms of convergence of distribution functions.

Theorem 1.22. *Let μ_n, μ be probability measures on $(\mathbf{R}, B(\mathbf{R}))$, and let F_n, F be their respective distribution functions. Then $\mu_n \Rightarrow \mu$ if and only if $\lim F_n(\lambda) = F(\lambda)$ for each λ at which F is continuous.*

Proof. Suppose $\mu_n \Rightarrow \mu$ and let $\varepsilon > 0$. Let $g_\varepsilon(t)$ be the continuous function which has value 1 for $t \leq \lambda, 0$ for $t \geq \lambda + \varepsilon$, and is linear between. We then obtain

$$\limsup_{n \to \infty} F_n(\lambda) = \limsup_{n \to \infty} \int_{-\infty}^{\lambda} g_\varepsilon \, d\mu_n \leq \lim_{n \to \infty} \int_{-\infty}^{\infty} g_\varepsilon \, d\mu_n$$

$$= \int_{-\infty}^{\infty} g_\varepsilon \, d\mu_n \leq F(\lambda + \varepsilon).$$

Letting $\varepsilon \to 0$, and using the right continuity of F yields

$$\limsup F_n(\lambda) \leq F(\lambda).$$

Let $h_\varepsilon(t)$ be 1 for $t \leq \lambda - \varepsilon, 0$ for $t \geq \lambda$, and linear between, and apply a similar argument to yield

$$\liminf F_n(\lambda) \geq F(\lambda - \varepsilon).$$

Letting $\varepsilon \to 0$, and using the assumed continuity of F at λ gives

$$\liminf F_n(\lambda) \geq F(\lambda).$$

Conversely, suppose $\lim F_n(\lambda) = F(\lambda)$ for each continuity point λ, and let $g \in C(\mathbf{R})$. Let $\varepsilon > 0$. Since F has at most a countable number of discontinuities, there exists a continuity point $-b \in \mathbf{R}$ such that $F(-b) \leq \varepsilon/4$. Then $F_n(-b) \leq \varepsilon/2$ for all n sufficiently large. Similarly, there exists a continuity point c such that $1 - F(c) \leq \varepsilon/4$ and hence $1 - F_n(c) \leq \varepsilon/2$ for all n sufficiently large. We can thus find an interval $[-a, a]$ such that

$$\mu_n([-a, a]^c), \quad \mu([-a, a]^c) \leq \varepsilon.$$

If $M = \sup |g(\lambda)|$, it follows that

$$\left| \int_{-\infty}^{\infty} g \, d\mu_n - \int_{-a}^{a} g \, d\mu_n \right| \leq M\varepsilon$$

and this also holds for μ. Hence to prove that $\mu_n \Rightarrow \mu$ it is sufficient to prove that

$$\lim \int_{a}^{a} g \, d\mu_n = \int_{a}^{a} g \, d\mu. \tag{1.28}$$

Let $\delta > 0$. Since g is uniformly continuous on $[-a, a]$ there exists a step function
$$s(\lambda) = \sum_{i=1}^{k-1} c_i \phi_{(a_i, a_{i+1}]}(\lambda)$$
such that $|g(\lambda) - s(\lambda)| < \delta$ for $\lambda \in [-a, a]$. Moreover, the points $a_1, \ldots, a_k$ can be chosen as continuity points of F, since there are only a countable number of discontinuities to be avoided. We then obtain

$$\left| \int_{-a}^{a} g \, d\mu_n - \int_{-a}^{a} g \, d\mu \right| \leq \int_{-a}^{a} |g - s| \, d\mu_n + \int_{-a}^{a} |g - s| \, d\mu$$
$$+ \left| \int_{-a}^{a} s \, d\mu_n - \int_{-a}^{a} s \, d\mu \right|$$
$$< 2\delta + \sum_{i=1}^{k-1} |c_i| \big[|F_n(a_{i+1}) - F(a_{i+1})|$$
$$+ |F_n(a_i) - F(a_i)| \big].$$

Since $a_1, \ldots, a_k$ are continuity points of F, each term is the summation goes to 0 as $n \to \infty$. This proves (1.28). □

We now consider the question of when a sequence of probability measures contains a weakly convergent subsequence. Let M be a family of probability measure on $(\mathbf{R}, B(\mathbf{R}))$. We call M *relatively compact* if every sequence of elements of M contains a weakly convergent subsequence (the subsequence need not converge to an element of M). We say that M is *tight* if for any $\varepsilon > 0$ there exists a compact set $K \subseteq \mathbf{R}$ such that $\mu(K) > 1 - \varepsilon$ for all $\mu \in M$. It can be shown that M is tight if and only if M is relatively compact [Billingsley, 1968]. However, for our purposes, we shall only need the following.

Theorem 1.23. *Let M be a family of probability measures on $(\mathbf{R}, B(\mathbf{R}))$. If M is tight, then it is relatively compact.*

Proof. Let $\mu_1, \mu_2, \ldots$ be a sequence in M. Since M is tight, for any $\varepsilon > 0$ we can find an $a_\varepsilon \in \mathbf{R}$ such that

$$\mu_i([-a_\varepsilon, a_\varepsilon]) > 1 - \varepsilon, \qquad i = 1, 2, \ldots. \tag{1.29}$$

Let F_i be the distribution function of μ_i, $i = 1, 2, \ldots$, and let $r_1, r_2, \ldots$ be an enumeration of the rationals. Since

$$0 \leq F_i(r_1) \leq 1, \qquad i = 1, 2, \ldots$$

we can choose a subsequence i_1 such that $F_{i_1}(r_1)$ converges. Then choose a subsubsequence i_2 such that $F_{i_2}(r_2)$ and $F_{i_2}(r_1)$ converge. Continue this process. Let i' be the diagonal sequence; that is, i' consists of the first element of i_1, the second element of i_2, and so on. Then the sequence of functions $F_{i'}$ converge for every rational number. Hence there exists a function F on the rationals such that

$$\lim_{i'\to\infty} F_{i'}(r) = F(r)$$

for every rational r. It follows that F is nondecreasing and $0 \leq F(r) \leq 1$ for every rational r. As an immediate consequence of (1.29) we have that $F_{i'}(\lambda) \to 0$ uniformly as $\lambda \to -\infty$ and $F_{i'}(\lambda) \to 1$ uniformly as $\lambda \to \infty$. Hence $F(r) \to 0$ as $r \to -\infty$ and $F(r) \to 1$ as $r \to \infty$, r rational. Now for any $\lambda \in \mathbb{R}$, define

$$F(\lambda) = \inf_{r \geq \lambda} F(r).$$

It is easy to check that F is a distribution function. Let μ be the corresponding probability measure on $(\mathbb{R}, B(\mathbb{R}))$. We shall show that $\mu_i \Rightarrow \mu$. By Theorem 1.22, it is sufficient to show that $F_{i'}(\lambda) \to F(\lambda)$ at continuity points of F. Let λ be a fixed continuity point of F. For any rational $r > \lambda$ we have

$$\lim_{i'\to\infty} \sup F_{i'}(\lambda) \leq \lim_{i'\to\infty} F_{i'}(r) = F(r).$$

For any rational $r_1 \leq \lambda$ we obtain

$$\lim_{i'\to\infty} \inf F_{i'}(\lambda) \geq \lim_{i'\to\infty} F_{i'}(r_1) = F(r_1).$$

Since F is continuous at λ, for any $\varepsilon > 0$ there exists rationals $r_1 \leq \lambda < r$ such that $F(r) \leq F(r_1) + \varepsilon$. Hence

$$\liminf F_{i'}(\lambda) \geq F(r_1) \geq F(r) - \varepsilon \geq \limsup F_{i'}(\lambda) - \varepsilon.$$

It follows that $F_{i'}(\lambda) \to F(\lambda)$ as required. $\square$

If f is a random variable, the complex-valued function $\psi_f : \mathbb{R} \to \mathbb{C}$ defined as $\psi_f(\lambda) = E(e^{i\lambda f})$ is called the *characteristic function* of f. By Corollary 1.3 we have

$$\psi_f(\lambda) = \int e^{i\lambda t}\, d\mu_f(t) = \int e^{i\lambda t}\, dF_f(t).$$

CENTRAL LIMIT THEOREM

Notice that $\psi_f(0) = 1$, and that ψ_f is continuous. Moreover, ψ_f has another important property which we now define. A function $\psi: \mathbb{R} \to \mathbb{C}$ is *positive definite* if for any finite sets

$$\{\lambda_1, \ldots, \lambda_n\} \subseteq \mathbb{R}, \quad \{c_1, \ldots, c_n\} \subseteq \mathbb{C}$$

we have

$$\sum_{i,j=1}^n c_i \bar{c}_j \psi(\lambda_i - \lambda_j) \geq 0 \tag{1.30}$$

where $^-$ is complex conjugation. A direct computation shows that ψ_f is positive definite. A well-known classical theorem due to Bochner [Billingsley, 1968; Dunford and Schwartz, 1958] states that the converse holds. That is, if ψ is a continuous, positive definite function with $\psi(0) = 1$, then there exists a unique probability measure μ on $(\mathbb{R}, B(\mathbb{R}))$ such that

$$\psi(\lambda) = \int \exp i\lambda t \, d\mu(t).$$

We say that ψ is the *characteristic function* (or *Fourier transform*) of μ.

As an example, suppose μ_σ is the *normal distribution* (with mean 0 and variance σ^2)

$$\mu_\sigma(A) = \frac{1}{\sqrt{2\pi}\,\sigma} \int_A e^{-u^2/2\sigma^2} du, \quad A \in B(\mathbb{R}). \tag{1.31}$$

Then the distribution function becomes

$$F_\sigma(t) = \frac{1}{\sqrt{2\pi}\,\sigma} \int_{-\infty}^t e^{-u^2/2\sigma^2} du \tag{1.32}$$

and a straightforward calculation gives the characteristic function

$$\psi_\sigma(\lambda) = e^{-\sigma^2 \lambda^2/2}. \tag{1.33}$$

Let $f_1, f_2, \ldots$ be independent random variables. We conclude from Corollary 1.8 and Lemma 1.13 that

$$F_{f_1 + \cdots + f_n} = F_{f_1} * F_{f_2} * \cdots * F_{f_n}. \tag{1.34}$$

Unfortunately, the convolutions in (1.34) are difficult to compute. One of the important properties of the characteristic function, however, is that

$$\psi_{f_1 + \cdots + f_n} = \psi_{f_1} \psi_{f_2} \cdots \psi_{f_n}. \tag{1.35}$$

Equation 1.35 follows from Corollary 1.12 and the fact that

$$e^{i\lambda f_1}, \ldots, e^{i\lambda f_2}$$

are independent.

We now prove the inversion formula.

Theorem 1.24. *Let ψ and F be the characteristic and distribution functions respectively of a probability measure on $\mathbb{R}$. If $a < b$ are two points of continuity for F, then*

$$F(b) - F(a) = \lim_{\sigma \to 0} \frac{i}{2\pi} \int_{-\infty}^{\infty} \psi(\lambda) e^{-\sigma^2 \lambda^2 / 2} \frac{e^{-ib\lambda} - e^{-ia\lambda}}{\lambda} \, d\lambda. \qquad (1.36)$$

Proof. There exist two independent random variables f, f_σ with distribution functions F and F_σ, respectively. Let

$$G_\sigma = F_{f+f_\sigma} = F * F_\sigma$$

and

$$\phi_\sigma = \phi_{f+f_\sigma} = e^{-\sigma^2 \lambda^2 / 2} \psi(\lambda).$$

Since $|\psi(\lambda)| \leq 1$, ϕ_σ is integrable and we have

$$(2\pi)^{-1} \int_{-\infty}^{\infty} e^{-i\lambda t} \phi_\sigma(\lambda) \, d\lambda = (2\pi)^{-1} \int_{-\infty}^{\infty} e^{-i\lambda t} e^{-\sigma^2 \lambda^2 / 2} \int_{-\infty}^{\infty} e^{i\lambda s} \, dF(s) d\lambda$$

$$= \int_{-\infty}^{\infty} (2\pi)^{-1} \int_{-\infty}^{\infty} e^{i\lambda(s-t)} e^{-\sigma^2 \lambda^2 / 2} \, d\lambda dF(s) \qquad (1.37)$$

where we can interchange the order of integration by Fubini's theorem. Now the inner integral on the right side of (1.37) is the characteristic function of a normal distribution with variance $1/\sigma^2$, evaluated at $(s-t)$ and multiplied by $1/(2\pi)^{\frac{1}{2}} \sigma$. Applying (1.33) we have

$$(2\pi)^{-1} \int_{-\infty}^{\infty} e^{-i\lambda t} \phi_\sigma(\lambda) \, d\lambda = \int_{-\infty}^{\infty} ((2\pi)^{\frac{1}{2}} \sigma)^{-1} e^{-(s-t)^2 / 2\sigma^2} \, dF(s). \qquad (1.38)$$

Applying Corollary 1.14, the right side of (1.38) is the density of G_σ. Integrating both sides of (1.38) from a to b (and again using Fubini's theorem) we obtain

$$G_\sigma(b) - G_\sigma(a) = \frac{i}{2\pi} \int_{-\infty}^{\infty} \psi(\lambda) e^{-\sigma^2 \lambda^2 / 2} \frac{e^{-ib\lambda} - e^{-ia\lambda}}{\lambda} \, d\lambda.$$

Hence it suffices to show that $\lim_{\sigma \to 0+} G_\sigma(t) = F(t)$ for each continuity point t of F. Let $\varepsilon > 0$. It follows from Chebyshev's inequality (1.23) that

CENTRAL LIMIT THEOREM   31

for σ sufficiently small, $\mu(|f_\sigma| \geq \varepsilon) \leq \varepsilon$. Hence for sufficiently small σ we have
$$G_\sigma(t) = \mu(f \leq t - f_\sigma) \leq \mu(f \leq t - f_\sigma, |f_\sigma| < \varepsilon)$$
$$+ \mu(|f_\sigma| \geq \varepsilon) \leq \mu(f \leq t - \varepsilon) + \varepsilon$$
$$= F(t + \varepsilon) + \varepsilon$$

moreover,
$$G_\sigma(t) = \mu(f \leq t - f_\sigma) \geq \mu(f \leq t - \varepsilon, |f_\sigma| \leq \varepsilon) \geq \mu(f \leq t - \varepsilon) - \varepsilon$$
$$= F(t - \varepsilon) - \varepsilon.$$

Since t is a continuity point of F, the result follows. □

Corollary 1.25. *Two probability measures on $(\mathbf{R}, B(\mathbf{R}))$ with the same characteristic function are equal.*

Corollary 1.26. *If ψ is the characteristic function of a probability measure μ on $(\mathbf{R}, B(\mathbf{R}))$ and if $\psi \in L^1(\mathbf{R}, d\lambda)$, then μ is absolutely continuous and its density is*
$$h(s) = (2\pi)^{-1} \int e^{-ist} \psi(t)\, dt.$$

Proof. Since $\psi \in L^1(\mathbf{R}, d\lambda)$, we can use the dominated convergence theorem to take the limit in (1.38) under the integral sign obtaining
$$F(b) - F(a) = \frac{i}{2\pi} \int \psi(\lambda) \frac{e^{-ib\lambda} - e^{-ia\lambda}}{\lambda}\, d\lambda.$$

If we integrate the continuous function h from a to b we obtain $\int_a^b h(s)ds = F(b) - F(a)$. Hence $h = F'$ everywhere. □

Lemma 1.27. *For any $a > 0$,*
$$\mu(|f| > 2a^{-1}) \leq a^{-1} \int_a^a [1 - \psi_f(\lambda)]\, d\lambda.$$

Proof. By Fubini's theorem we have
$$a^{-1} \int_a^a [1 - \psi_f(\lambda)]\, d\lambda = a^{-1} \int_a^a \int_{-\infty}^\infty (1 - e^{i\lambda t}) dF_f(t)\, d\lambda$$
$$= 2 \int_{-\infty}^\infty [1 - (at)^{-1} \sin at]\, dF_f(t).$$

Since $(at)^{-1}\sin at \leq 1$, we obtain

$$a^{-1}\int_a^a [1-\psi_f(\lambda)]\,d\lambda \geq 2\left[\int_{-\infty}^{-2/a} + \int_{2/a}^{\infty}\right][1-(at)^{-1}\sin at]\,dF_f(t)$$

$$\geq 2\left[\int_{-\infty}^{-2/a} + \int_{2/a}^{\infty}\right][1-|at|^{-1}]\,dF_f(t)$$

$$\geq F_f(-2a^{-1}) + [1 - F_f(2a^{-1})]. \quad \square$$

It is clear from the definition of weak convergence that if μ_n, μ are probability measures on $(\mathbf{R}, B(\mathbf{R}))$ with characteristic functions ψ_n, ψ then $\mu_n \Rightarrow \mu$ implies $\psi_n(\lambda) \to \psi(\lambda)$ for all $\lambda \in \mathbf{R}$. We now prove a converse.

Theorem 1.28. *Let μ_n be probability measures on $(\mathbf{R}, B(\mathbf{R}))$ and let ψ_n be their characteristic functions. Suppose that*

$$\lim_{n\to\infty} \psi_n(\lambda) = \psi(\lambda)$$

exists for all $\lambda \in \mathbf{R}$, and that ψ is continuous at 0. Then there exists a probability measure μ on $(\mathbf{R}, B(\mathbf{R}))$ which has ψ as its characteristic function and $\mu_n \Rightarrow \mu$.

Proof. We first show that the family $\{\mu_n\}$ is tight. Let $\varepsilon > 0$. Since ψ is continuous at 0 and $\psi(0) = 1$, there exists an $a > 0$ such that

$$0 \leq a^{-1}\int_a^a [1-\psi(\lambda)]\,d\lambda \leq \frac{\varepsilon}{2}$$

(the integral is nonnegative by Lemma 1.27). Applying the dominated convergence theorem yields

$$a^{-1}\int_a^a [1-\psi_n(\lambda)]\,d\lambda \leq \varepsilon$$

for n sufficiently large. Applying Lemma 1.27 gives

$$\mu_n\left([-2a^{-1}, 2a^{-1}]\right) > 1 - \varepsilon$$

for n sufficiently large.

We can now enlarge the interval $[-2a^{-1}, 2a^{-1}]$ to an interval $[-b, b]$ such that

$$\mu_n([-b, b]) > 1 - \varepsilon$$

CENTRAL LIMIT THEOREM   33

for $n = 1, 2, \ldots$. Hence $\{\mu_n\}$ is tight and by Theorem 1.23 μ_n has a subsequence which converges weakly to a probability measure μ. Since the characteristic functions of the subsequence converge to the characteristic function of μ and also converge to ψ we conclude that ψ is the characteristic function of μ. Now suppose $\mu_n \not\Rightarrow \mu$. Then there exists a $g \in C(\mathbb{R})$ such that $\int g \, d\mu_n \not\to \int g \, d\mu$. Since the sequence $\int g \, d\mu_n$ is bounded there exists a subsequence $\int g \, d\mu_{n'}$ such that

$$\int g \, d\mu_{n'} \to c \neq \int g \, d\mu.$$

Applying Theorem 1.23 there exists a subsubsequence which converges weakly to some measure ν. Since $\int g \, d\nu = c$, we conclude that $\nu \neq \mu$. But by the argument used above ψ is the characteristic function of ν. This contradicts Corollary 1.25. □

The central limit theorem gives a reason why the normal distribution μ_σ (1.31) plays a central role in probability and statistics. It is well-known empirically that in some sense the distribution of a sum of independent trials approaches the normal distribution as the number of trials increases. The central limit theorem provides a rigorous statement and proof for this intuition. We shall prove the simplest version of the central limit theorem.

Theorem 1.29. *Let $f_1, f_2, \ldots$ be i.i.d. random variables with mean 0 and variance $\sigma^2 > 0$. Let $S_n = f_1 + \cdots + f_n$ and let μ_n be the distribution of $n^{-\frac{1}{2}} S_n$. Then $\mu_n \Rightarrow \mu_\sigma$.*

Proof. Let ψ be the characteristic function of the f_i's. It is straightforward to show that the first two derivatives of ψ exist and

$$\psi'(0) = iE(f_1) = 0, \quad \psi''(0) = -E(f_1^2) = -\sigma^2.$$

By (1.35) the characteristic function ψ_n of $n^{-1/2} S_n$ is

$$\psi_n(\lambda) = E\big[\exp(i\lambda n^{-1/2} S_n)\big] = \big[\psi(n^{-1/2}\lambda)\big]^n.$$

The Taylor expansion of ψ at 0 yields

$$\psi(\lambda) = \psi(0) + \psi'(0)\lambda + 2^{-1}\psi''(0)\lambda^2 + o(\lambda^2) = 1 - 2^{-1}\sigma^2\lambda^2 + o(\lambda^2).$$

For fixed λ, as $n \to \infty$ we have

$$\psi_n(\lambda) = \big[\psi(n^{-1/2}\lambda)\big]^n$$
$$= \big[1 - (2n)^{-1}\sigma^2\lambda^2 + o(n^{-1})\big]^n \to e^{-\sigma^2\lambda^2/2} = \psi_\sigma(\lambda).$$

Applying Theorem 1.28 yields $\mu_n \Rightarrow \mu_\sigma$. □

By replacing S_n by $S_n - nm_1$ we can dispense with the mean 0 requirement in Theorem 1.29. Using similar, but more refined methods, one can prove the most general (in a certain sense) version of the central limit theorem due to Lindeberg [Billingsley, 1968; Lamperti, 1966; Loève, 1963].

1.6 Notes and References

In this chapter we have only scratched the surface of classical probability theory. Although this subject is at least 300 years old its modern axiomatic approach was not formulated until the 1930's in the pioneering work of A. Kolmogorov [1956]. Other texts of historical interest are [Fine, 1973; Jeffreys, 1948; Keynes, 1962]. For an elementary introduction we suggest [Ross, 1984] and for more advanced treatments [Billingsley, 1968; Breiman, 1968; Doob, 1953; Lamperti, 1966; Loève, 1963]. Markov chains are treated in [Chung, 1967].

The basic purpose of this axiomatic approach is to construct a mathematical model for describing some probabilistic or statistical experiment. There are critics who argue that this way of defining probabilities is not operational. That is, it is not the way that an experimentor or empirical investigator would proceed. Empirically, one would define the probability of an event in terms of its relative frequency. Such a definition usually goes as follows: suppose that an experiment, whose sample space is X, is repeatedly performed under exactly the same conditions. For each event A of the sample space X, we define $n(A)$ to be the number of times in the first n repetitions of the experiment that the event A occurs. Then $\mu(A)$, the probability of the event A, is defined by

$$\mu(A) = \lim_{n \to \infty} \frac{n(A)}{n}.$$

That is, $\mu(A)$ is defined as the limiting frequency of A [Fine, 1973].

Although the preceding definition describes experimental procedure it has a drawback. How do we know that $n(A)/n$ converges to some constant limiting value that will be the same for each possible sequence of repetitions of the experiment? For example, suppose that the experiment to be repeatedly performed consists of flipping a coin. How do we know that the proportion of heads obtained in the first n flips will converge to some value as n gets large? Also, even if it does converge to some value, how do we know that, if the experiment is repeatedly performed a second time, we shall again obtain the same limiting proportion of heads?

Proponents of the relative frequency definition of probability usually answer this objection by stating that the convergence of $n(A)/n$ is an axiom of the system. However, this is a very complex axiom which is not at all evident. The Kolmogorov approach assumes a set of simpler and more self-evident axioms about probability. From these, one then proves that such a constant limiting frequency does indeed exist. In fact, this is the content of laws of large numbers.

If classical probability is so useful and successful, why must it be altered in quantum mechanics? For one thing, it is too limited. It gives a model for a single experiment or a set of subexperiments of a *single* experiment. In order to gain information about a system we need to execute various experiments on the system. These experiments may interact or interfere with each other, but nevertheless it is important to model the totality of all possible experiments on the system. For example, flipping a coin a large number of times only gives us information about how biased the coin is to landing heads or tails. It does not tell us what percentage of the coin is silver. We then have to execute a second experiment by extracting a sample of material from the coin. This second experiment could easily interfere with the first by changing the bias of the coin. But couldn't we take such a small sample that the interference is negligible? For a large system like a coin, we could; but for a microscopic quantal system we may not.

If we consider the totality of experiments for a quantum system, the axioms of classical probability theory break down. For example, if A and B are events for two different experiments, then $A \cup B$ may not be an event for a third experiment. Even if $A \cup B$ is an event, we may not have

$$\mu(A \cup B) \leq \mu(A) + \mu(B)$$

and the latter subadditivity always holds in classical probability theory. We shall see examples of such behavior in later chapters.

A second problem is that probabilities are computed differently in quantum probability than in classical probability. As mentioned in the preface, in quantum mechanics probabilities are computed by summing amplitudes and squaring the resulting absolute value. For this reason, probability measures must be replaced by amplitude functions in an axiomatic development of quantum probability.

2
Traditional Quantum Mechanics

This chapter presents a brief survey of traditional quantum mechanics. The theory is based upon the structure of Hilbert space and its self-adjoint operators. Although quantum mechanics generalizes classical mechanics, in order to motivate and understand the former, it is helpful to have some knowledge of the latter. For this reason we begin with a presentation of classical mechanics. We also consider a technique called the path integral formalism. Although this formalism is closely related to traditional quantum mechanics, it places special emphasis upon the role of transition amplitudes. This concept will be the basis of our later development of quantum probability.

2.1 Classical Mechanics

Before we can attempt a formulation of traditional quantum mechanics it is important to understand some of the basic principles of classical mechanics. A mechanical system has N *degrees of freedom* if its configuration at time t can be completely described by N continuous real-valued functions of t called *coordinate functions* or *coordinates*, but not by $N-1$ continuous real-valued functions. For example, a particle moving on a circle or constrained to move in a straight line has one degree of freedom, and a particle moving on the surface of a sphere in three-dimensional space $\mathbb{R}^3$ has two degrees of freedom. If we have n interacting particles in $\mathbb{R}^3$, there are $3n$ degrees of

37

freedom and the configuration of the system is completely described by coordinate functions $x_1(t), y_1(t), z_1(t), \ldots, x_n(t), y_n(t), z_n(t)$. We shall always assume that the coordinate functions are twice differentiable with respect to t. We now present the fundamental principle of classical mechanics.

(C) The coordinate functions at time t_2 are determined by the coordinate functions and their first derivatives at time t_1 if $t_1 \leq t_2$.

Principle C is a generalization of Newton's second law. For example, if a particle is constrained to move in the x-direction, Newton's second law states that
$$m \frac{d^2 x}{dt^2} = F\left(t, x, \frac{dx}{dt}\right).$$

If F is sufficiently smooth, then there is a unique solution of the differential equation satisfying any initial conditions $x(t_0)$, $(dx/dt)(t_0)$.

We call the time derivatives of the coordinates *velocity functions* or *velocities*. If a mechanical system has N degrees of freedom with coordinates $q_1, \ldots, q_N$ and velocities $v_1, \ldots, v_N$, the *state* $s(t)$ of the system at time t is
$$s(t) = \bigl(q_1(t), \ldots, q_N(t), v_1(t), \ldots, v_N(t)\bigr).$$

A knowledge of the state at t_0 completely describes the system at t_0 and by (C) if we know $s(t_0)$ we may determine $s(t)$ for all $t \geq t_0$. Let $U_t(s)$ denote the state at time t when the state at time $t = 0$ is s. For each $t \geq 0$, U_t is a transformation from the set of states S to itself. It is clear that $U_0 = I$ and for $t_1, t_2 \geq 0$ we have
$$U_{t_1+t_2} = U_{t_1} U_{t_2}. \tag{2.1}$$

A collection of transformations $\{U_t : t \geq 0\}$ satisfying (2.1) is called a *one-parameter semigroup* of transformations. A mechanical system is *reversible* if U_t is bijective for all $t \geq 0$. For simplicity we shall only consider reversible systems. In this case U_t^{-1} exists and we define $U_{-t} = U_t^{-1}$. We then see that (2.1) holds for all $t_1, t_2 \in (-\infty, \infty)$ and we call $\{U_t : -\infty < t < \infty\}$ a *one-parameter group* of transformations. If $s_0 \in S$, an *orbit* in S is a set of the form $\{U_t(s_0) : -\infty < t < \infty\}$. The following lemma is easily checked.

Lemma 2.1. *If a mechanical system is reversible, then each $s \in S$ lies on a unique orbit.*

The one-parameter group of a mechanical system is also called the *dynamical group* of the system. The dynamical group together with an initial state $s(t_0)$ gives a complete description of a mechanical system at all times. However, since we usually only know the forces acting on the system (and the masses), we usually do not know the dynamical group U a priori but only the tangent vectors along its orbits. Hence, we must derive U from these. To be precise, we are given $2N$ functions A'_i, A_i, $i = 1, \ldots, N$, of $t, q_1, \ldots, q_N, v_1, \ldots, v_N$ such that

$$\frac{dq_i}{dt} = A'_i(t, q_1, \ldots, q_N, v_1, \ldots, v_N)$$
$$\frac{dv_i}{dt} = A_i(t, q_1, \ldots, q_N, v_1, \ldots, v_N). \qquad (2.2)$$

If the A'_i, A_i are differentiable with respect to $q_1, \ldots, v_N$, we say that U is *twice differentiable*. If U is twice differentiable, a standard theorem tells us that there is a unique curve through each point of $\mathbf{R}^{2N}$ satisfying (2.2). The vector field $(A'_1, \ldots, A'_N, A_1, \ldots, A_N)$ is called the *infinitesimal generator* of U. Hence, there is a bijection between twice differentiable dynamical groups and differentiable vector fields. Since we already know that $v_i = dq_i/dt$, (2.2) reduces to

$$\frac{dq_i}{dt} = v_i, \quad \frac{dv_i}{dt} = A_i(t, q_1, \ldots, q_N, v_1, \ldots, v_N). \qquad (2.3)$$

Now (2.3) is equivalent to the set of second order differential equations

$$\frac{d^2 q_i}{dt^2} = A_i\left(t, q_1, \ldots, q_N, \frac{dq_1}{dt}, \ldots, \frac{dq_N}{dt}\right), \quad i = 1, \ldots, N. \qquad (2.4)$$

We now make further assumptions about our physical system in order to restrict the nature of the functions A_i.

(a) The A_i's are functions of the q_j's alone.
(b) There are positive constants M_i and a function $V(q_1, \ldots, q_N)$ such that

$$A_i = \frac{-1}{M_i} \frac{\partial V}{\partial q_i}, \quad i = 1, \ldots, N.$$

Assumption a assumes that the forces are time and velocity independent, while Assumption b follows from requiring that there are no dissapative forces such as friction. We call M_i the *mass* associated with the i-th coordinate, $p_i = M_i v_i$ the *momentum* component conjugate to q_i, $M_i A_i$ the *force* in the i-th direction, and V the *potential energy* of the system.

Since the v_i and p_i determine each other uniquely we may regard the state of the system as described by the q_i's and p_i's instead of the q_i's and the v_i's. Making this change we then call S the phase space for the system. A *dynamical variable* is a real-valued function on S. These correspond to quantities that can be measured. An *integral* of the system is a dynamical variable that is constant on orbits of the dynamical group U.

Lemma 2.2. *A differentiable function ϕ on S is an integral if and only if*
$$\sum_{i=1}^{n} \frac{p_i}{M_i} \frac{\partial \phi}{\partial q_i} = \sum_{i=1}^{n} \frac{\partial V}{\partial q_i} \frac{\partial \phi}{\partial p_i}.$$

Proof. The function ϕ is an integral if and only if
$$\left(\frac{d}{dt}\right)\phi[U_t(s)] = 0, \quad -\infty < t < \infty.$$
This last condition holds if and only if
$$0 = \frac{d}{dt}\phi\bigl(q_1(t),\ldots,q_N(t),p_1(t),\ldots,p_N(t)\bigr)$$
$$= \sum \frac{\partial \phi}{\partial q_i} \frac{dq_i}{dt} + \sum \frac{\partial \phi}{\partial p_i} \frac{dp_i}{dt}$$
$$= \sum \frac{p_i}{M_i} \frac{\partial \phi}{\partial q_i} - \sum \frac{\partial V}{\partial q_i} \frac{\partial \phi}{\partial p_i}. \quad \square$$

It follows from Lemma 2.2 that a sufficient condition for a function H to be an integral is that
$$\frac{\partial H}{\partial q_i} = \frac{\partial V}{\partial q_i}, \quad \frac{\partial H}{\partial p_i} = \frac{p_i}{M_i}, \quad i = 1,\ldots,N. \tag{2.5}$$
It follows immediately that
$$H = \sum \frac{p_i^2}{2M_i} + V(q_1,\ldots,q_N). \tag{2.6}$$
The fundamental integral H is called the *energy* or *Hamiltonian* of the mechanical system. The fact that H is an integral gives us the law of conservation of energy. We call $\Sigma p_i^2/2M_i$ the *kinetic energy* of the system. We can also write (2.5) in the *Hamiltonian form*
$$\frac{dq_i}{dt} = \frac{\partial H}{\partial p_i}, \quad \frac{dp_i}{dt} = -\frac{\partial H}{\partial q_i}, \quad i = 1,\ldots,N. \tag{2.7}$$

CLASSICAL MECHANICS

Hamilton's equations (2.7) determine the dynamical group and hence, together with the initial conditions, they completely describe the system for all times.

We have just presented the *Hamiltonian formulation* for classical mechanics. There is another important method called the *Lagrangian formulation*. Suppose we have a mechanical system with N degrees of freedom satisfying Assumptions a and b. We call $C = \mathbb{R}^N$ *configuration space*. A point $q = (q_1, \ldots, q_N) \in C$ describes the configuration of the system and is called a *Lagrangian state*. The dynamics of the system is described by a *path* or *trajectory* $q(t) = (q_1(t), \ldots, q_N(t))$, $-\infty < t < \infty$, in C. In the Hamiltonian formulation, the state $s(t)$ is determined for time t once the initial condition $s(t_0)$ is given. As we shall see, in the Lagrangian formulation the Lagrangian state $q(t)$, $t_a \leq t \leq t_b$ is determined once the boundary conditions $q(t_a)$, $q(t_b)$ are given. We again define the velocity functions $v_i = dq_i/dt$, $i = 1, \ldots, N$, and we let $v(t) = (v_1(t), \ldots, v_N(t))$. The *Lagrangian* for the system is defined as

$$L(v, q) = \sum \left(\frac{M_i}{2}\right) v_i^2 - V(q). \qquad (2.8)$$

Thus, L is the kinetic energy minus the potential energy.

The *action* along a path $q(t)$ between times t_a, t_b where $t_a < t_b$ is defined as

$$S[q(t)] = \int_{t_a}^{t_b} L[v(t), q(t)] \, dt. \qquad (2.9)$$

Suppose the system is in the Lagrangian state q_a at time t_a and q_b at time t_b; that is, $q_a = q(t_a)$, $q_b = q(t_b)$. In the Lagrangian formulation, the path that the system traverses, called the *classical trajectory*, is determined by the *least action principle*. This principle states that among all possible paths from q_a to q_b, the classical trajectory $\hat{q}(t)$ is the one for which $S[q(t)]$ is an extremum. That is, the value of $S[\hat{q}(t)]$ is unchanged in the first order if the path $\hat{q}(t)$ is modified slightly.

The extremal path $\hat{q}(t)$ is now found by the procedures of the calculus of variations. Suppose the path is varied away from $\hat{q}(t)$ by an amount $\delta q(t)$. The condition that the endpoints of $\hat{q}$ are fixed requires

$$\delta q(t_a) = \delta q(t_b) = 0. \qquad (2.10)$$

The condition that $\hat{q}$ is an extremum of S gives

$$\delta S = S[\hat{q} + \delta q] - S[\hat{q}] = 0$$

to first order in δq. Using the definition of S in (2.9) we obtain

$$S[q + \delta q] = \int_{t_a}^{t_b} L(v + \delta v, q + \delta q)\, dt$$

$$= \int_{t_a}^{t_b} \left[L(v,q) + \sum \delta v_i \frac{\partial L}{\partial v_i} + \sum \delta v_i \frac{\partial L}{\partial q_i} \right] dt$$

$$= S[q] + \sum \int_{t_a}^{t_b} \left(\delta v_i \frac{\partial L}{\partial v_i} + \delta q_i \frac{\partial L}{\partial q_i} \right) dt.$$

Upon integrating by parts, the variation $\delta S = S[q+\delta q] - S[q]$ becomes

$$\delta S = \sum \left\{ \delta q_i \frac{\partial L}{\partial v_i} \bigg|_{t_a}^{t_b} - \int_{t_a}^{t_b} \delta q_i \left[\frac{d}{dt} \frac{\partial L}{\partial v_i} - \frac{\partial L}{\partial q_i} \right] dt \right\}.$$

Applying (2.10), the first term in the summation vanishes. Between the endpoints δq_i can take arbitrary values. Thus the extremum is that path along which the following condition is satisfied:

$$\frac{d}{dt} \frac{\partial L}{\partial v_i} - \frac{\partial L}{\partial q_i} = 0, \qquad i = 1, \ldots, N. \tag{2.11}$$

Equations 2.11 are the classical Lagrangian equations of motion. It is not hard to show that these equations are equivalent to the Hamiltonian form (2.7).

2.2 Hilbert Space Quantum Mechanics

Quantum mechanics was developed as a systematic axiomatic structure in the early 1930's by P. Dirac [1930] and J. von Neumann [1955] using earlier ideas of M. Born [1926]. Although their theories were similar and Dirac's approach was succint and elegant, we shall follow von Neumann since his methods are more mathematically rigorous.

As we have seen in Section 2.1, the three important components of classical mechanics are states, observables (dynamical variables), and dynamics. The states correspond to a theoretically complete description of the physical system, the observables correspond to physically measurable quantities such as position, momentum, and energy, and the dynamics gives a deterministic description of the evolution of states in time. The same three components are present in quantum mechanics, but they are described by mathematically different objects. In classical mechanics, they are described by points, functions and trajectories in a phase space, while in quantum mechanics they are described by entities in a complex Hilbert space. The resulting axiomatic structure is called traditional or Hilbert space quantum mechanics.

Von Neumann proposed the following axioms for Hilbert space quantum mechanics.

(A1) The states of a quantum system are unit vectors in a complex Hilbert space $\mathcal{H}$.

(A2) The observables are self-adjoint operators in $\mathcal{H}$.

(A3) The probability that an observable T has a value in a Borel set $A \subseteq \mathbb{R}$ when the system is in the state ψ is $\langle P^T(A)\psi, \psi \rangle$ where $P^T(\cdot)$ is the resolution of identity (spectral measure) for T.

(A4) If the state at time $t = 0$ is ψ, then the state at time t is $\psi_t = e^{-itH/\hbar}\psi$, where H is the energy observable and $\hbar$ is a constant (Planck's constant).

To be precise, von Neumann did not say that the states are unit vectors and the observables are self-adjoint operators, but that the states are described by unit vectors and the observables are described by self-adjoint operators. However, for a more concise presentation we shall use the former terminology. Also, it is only the *pure* states that are unit vectors; there are other types of states called mixed states which we shall discuss later. Notice that (A3) is unchanged if we replace ψ by $\alpha\psi$, where $\alpha \in \mathbb{C}$ with $|\alpha| = 1$. Hence, it is frequently said that pure states are described by unit rays.

Axioms A1, A2, and A4 are descriptive. They tell how the physical concepts can be described by mathematical constructs and how the evolution of the system can be described mathematically. Axiom A3 is extremely important since it gives the contact between the mathematical structure and the physical world. It is this axiom which gives the distribution of values for an observable. This distribution is precisely what is measured in the laboratory, thus enabling one to test the theory and to make predictions about the behavior of the physical system. Since (A3) is stochastic in nature, we obtain a "quantum probability theory". Not only are the above postulates elegant and general, they contain the previous ideas of the founders of quantum mechanics, Planck, Bohr, de Broglie, Born, Schrödinger, Heisenberg and others.

For generality, (A1) does not specify the particular Hilbert space in which the states are represented. Moreover, (A2) does not prescribe which self-adjoint operator represents a particular observable. These must be decided upon according to the specific situation being considered. For simplicity, let us consider a single particle of mass m in $\mathbb{R}^3 = \{(x,y,z): x,y,z \in \mathbb{R}\}$. The generalization to more degrees of freedom is straightforward. In this case, the Hilbert space is prescribed to be $\mathcal{H} = L^2(\mathbb{R}^3, dxdydz)$. The x-coordinate observable is represented by the multiplication operator $Q_x f(x,y,z) = xf(x,y,z)$ and the x-momentum observable is represented by the differential operator $P_x = -i\hbar(\partial/\partial x)$. The y,z-coordinate and momentum operators are defined analogously. Moreover, it is natural to assume that functions of these observables be represented by the same functions of the corresponding operators. (Of course, this may cause problems if noncommuting operators are mixed together and we shall say more about this later.) This is called the *correspondence principle*. The above operators are self-adjoint on suitable domains.

We now "diagonalize" the momentum operators P_x, P_y, P_z. This can be done with the Fourier transform

$$F: L^2(\mathbb{R}^3) \to L^2(\mathbb{R}^3)$$

defined by

$$(Ff)(x,y,z,) = (2\pi\hbar)^{-\frac{3}{2}} \int\int\int f(x',y',z')e^{-i(xx'+yy'+zz')/\hbar}dx'dy'dz'.$$

Using the notation

$$(Ff)(\mathbf{x}) = (2\pi\hbar)^{-\frac{3}{2}} \int f(\mathbf{x}')e^{-i\mathbf{x}\cdot\mathbf{x}'/\hbar}d\mathbf{x}'$$

the inverse F^* of the unitary operator F becomes

$$(F^*f)(\mathbf{x}') = (2\pi\hbar)^{-\frac{3}{2}} \int f(\mathbf{x})e^{i\mathbf{x}\cdot\mathbf{x}'/\hbar}d\mathbf{x}.$$

Hence,

$$(P_{x'}F^*f)(\mathbf{x}') = (2\pi\hbar)^{-\frac{3}{2}} \int xf(\mathbf{x})e^{i\mathbf{x}\cdot\mathbf{x}'/\hbar}d\mathbf{x}.$$

It follows that $FP_xF^* = Q_x$ which "diagonalizes" P_x. Similar expressions hold for P_y and P_z. Since $P_x = F^*Q_xF$ we say that the Fourier transform of position is momentum.

If the potential energy of the system is given by $V(x,y,z)$, then by the correspondence principle, the potential energy observable is represented by the operator of multiplication by the function $V(x,y,z)$. Moreover, since the classical kinetic energy is given by $(p_x^2 + p_y^2 + p_z^2)/2m$, it again follows from the correspondence principle that the kinetic energy observable K is represented by the operator

$$\frac{(P_x^2 + P_y^2 + P_z^2)}{2m} = -(2m)^{-1}\hbar^2 \nabla^2.$$

The energy observable is then given by the operator

$$H = K + V = -(2m)^{-1}\hbar^2 \nabla^2 + V(x,y,z). \tag{2.12}$$

Of course, the justification for these prescriptions is the resulting agreement with experiment.

We have now given a concrete model for Axioms A1 and A2. We next consider Axiom A4. Differentiating (A4) with respect to t gives

$$\frac{\partial \psi_t}{\partial t} = i\hbar^{-1} H e^{-itH/\hbar} \psi = -i\hbar^{-1} H \psi_t. \tag{2.13}$$

Equation 2.13 is the time-dependent Schrödinger equation which is the basic dynamical equation of nonrelativistic quantum mechanics. It describes the evolution of de Broglie's "pilot waves" which were later enterpreted by Born as "probability waves".

Now consider Axiom A3. The spectral measure for the x-coordinate operator is

$$f(x,y,z) \mapsto (\phi_A f)(x,y,z), \quad A \subset B(\mathbb{R})$$

where ϕ_A is the indicator function for A. Applying (A3), the probability that the x-coordinate of the particle is in A when the system is in the state ψ becomes

$$\langle \phi_A \psi, \psi \rangle = \int_A \int_{\mathbb{R}} \int_{\mathbb{R}} |\psi(x,y,z)|^2 dx dy dz.$$

Similar expressions hold for the y and z-coordinate operators. If $A \in B(\mathbb{R}^3)$, then the probability that the particle is in A when the system is in the state ψ is

$$\int\int\int_A |\psi(x,y,z)|^2 dx dy dz. \tag{2.14}$$

The fact that $|\psi|^2$ should be interpreted as a probability density was proposed by Born in the 1920's.

Suppose the energy observable H has the value E with certainty when the system is in the state ψ. We then have
$$1 = \langle P^H(\{E\})\psi, \psi\rangle = \|P^H(\{E\})\psi\|^2.$$
It follows that $P^H(\{E\})\psi = \psi$ and hence
$$H\psi = E\psi. \tag{2.15}$$
Equation 2.15 is Schrödinger's eigenvalue equation which shows that the allowed energy values are the eigenvalues of H. This has been substantiated to a high degree of accuracy for such systems as the hydrogen atom. This reasoning applies to other observables as well.

If T is an observable, it follows from (A3) that $A \mapsto \langle P^T(A)\psi, \psi\rangle$ is the distribution of T in the state ψ. Hence, by the spectral theorem [Dunford and Schwartz,1963], the expectation $E_\psi(T)$ of T in a state ψ in its domain becomes
$$E_\psi(T) = \int \lambda \langle P^T(d\lambda)\psi, \psi\rangle$$
$$= \left\langle \int \lambda P^T(d\lambda)\psi, \psi\right\rangle$$
$$= \langle T\psi, \psi\rangle.$$
The variance and standard deviation (or dispersion) of T in the state ψ are
$$\sigma_\psi^2(T) = E_\psi[(T - E_\psi(T))^2]$$
$$= E_\psi(T^2) - [E_\psi(T)]^2$$
and $\Delta_\psi(T) = [\sigma_\psi^2(T)]^{\frac{1}{2}}$, respectively. We interpret $\Delta_\psi(T)$ as the uncertainty or average error made in a measurement of T in the state ψ.

Suppose S and T are self-adjoint operators and suppose $\psi, S\psi, T\psi$ are in the domains of S and T. If
$$\langle S\psi, \psi\rangle = \langle T\psi, \psi\rangle = 0$$
we have
$$|\langle (ST - TS)\psi, \psi\rangle| = |\langle S\psi, T\psi\rangle - \langle T\psi, S\psi\rangle|$$
$$= 2|\operatorname{Im}\langle S\psi, T\psi\rangle| \leq 2\|S\psi\|\,\|T\psi\|$$
$$= 2\langle S^2\psi, \psi\rangle^{\frac{1}{2}} \langle T^2\psi, \psi\rangle^{\frac{1}{2}}$$
$$= 2\Delta_\psi(S)\Delta_\psi(T).$$

If $\langle S\psi, \psi\rangle, \langle T\psi, \psi\rangle \neq 0$, replace S by $S - \langle S\psi, \psi\rangle$ and T by $T - \langle T\psi, \psi\rangle$ to again obtain

$$\Delta_\psi(S)\Delta_\psi(T) \geq \frac{1}{2}|\langle (ST - TS)\psi, \psi\rangle|. \tag{2.16}$$

Equation 2.16 is a general form of the *Heisenberg uncertainty* relation. In particular, let $S = P_x$ and $T = Q_x$ be the x-momentum and x-coordinate observables, respectively. As is easily checked, the operators P_x, Q_x satisfy the Heisenberg commutation relation

$$P_x Q_x - Q_x P_x = -i\hbar I \tag{2.17}$$

on the intersection of their domains. Substituting (2.17) into (2.16) gives

$$\Delta_\psi(P_x)\Delta_\psi(Q_x) \geq \frac{1}{2}\hbar. \tag{2.18}$$

Equation 2.18 together with the analogous expressions in y and z give the usual Heisenberg uncertainty relations.

If T is an observable, an important problem is to find states ψ for which $\Delta_\psi(T) = 0$. When the system is in such a state, there will be no dispersion or error in the measured values of T. Then T will have the exact value $E_\psi(T)$. Indeed, it follows from Schwarz's inequality that for any state ψ in the domain of T^2 we have

$$\begin{aligned}[E_\psi(T)]^2 &= \left[\int \lambda \langle P^T(d\lambda)\psi, \psi\rangle\right]^2 \\ &\leq \int \lambda^2 \langle P^T(d\lambda)\psi, \psi\rangle \\ &= E_\psi(T^2)\end{aligned}$$

and equality holds if and only if $P^T(\{\lambda\}) = 1$, where $\lambda = E_\psi(T)$. Hence, $\Delta_\psi(T) = 0$ if and only if $T\psi = \lambda\psi$. We thus see that the eigenvalues of an observable T are precisely those values that T can attain exactly.

If $\psi, \phi \in \mathcal{H}$ are states, then $\langle \psi, \phi\rangle$ is called the (*instantaneous*) *transition amplitude* from ψ to ϕ and $|\langle \psi, \phi\rangle|^2$ is the (*instantaneous*) *transition probability* from ψ to ϕ. This does not mean that the system instantaneously transforms from one state to another. One interpretation of $|\langle \psi, \phi\rangle|^2$ is the following. A state ψ contains information about the properties of a system. In particular, it describes the properties which hold when the system is in a certain condition. Then $|\langle \psi, \phi\rangle|^2$ gives the probability that the properties which hold in the state ψ also hold in the state ϕ. In a sense, $|\langle \psi, \phi \rangle|^2$ is a measure of the information common to ψ and ϕ. In this way, an orthonormal basis ψ_i for $\mathcal{H}$ gives an information theoretic complete specification of every state. (For a more precise tratment of information-theoretic aspects of quantum theory, see [Holevo, 1982]). Indeed, different ψ_i's have no common information and for any state ϕ, Parseval's equality gives

$$\sum |\langle \phi, \psi_i \rangle|^2 = \|\phi\|^2 = 1.$$

Hence, the totality of the ψ_i's contains all the information contained in ϕ. The more general Parseval's equality

$$\langle \psi, \phi \rangle = \sum \langle \psi, \psi_i \rangle \langle \psi_i, \phi \rangle \tag{2.19}$$

has an interesting interpretation. It says that the transition amplitude from ψ to ϕ is the transition amplitude from ψ to ψ_i times the transition amplitude from ψ_i to ϕ summed over i. As mentioned earlier, the concept of transition amplitude will be basic to our later development.

Truly physical transitions are described by the *time transition amplitude*

$$A_t(\psi, \phi) = \langle e^{-itH/\hbar} \psi, \phi \rangle \tag{2.20}$$

which has the following interpretation. If the system is initially in the state ψ, then $A_t(\psi, \phi)$ is the probability amplitude that the system will be in the state ϕ after time t. The corresponding time transition probability is given by $|A_t(\psi, \phi)|^2$. Notice that the dynamics of the system is completely described by the numbers $A_t(\psi, \phi)$. Indeed, if ψ_i is an orthonormal basis for $\mathcal{H}$, then for any state ψ we have

$$e^{-itH/\hbar} \psi = \sum A_t(\psi, \psi_i) \psi_i.$$

We have seen that traditional quantum mechanics is described by certain constructs in a complex Hilbert space. Now it is well known that Hilbert spaces can be combined in various ways to form new Hilbert spaces. Do such combinations also have relevance to quantum mechanics? The most important ways of constructing new Hilbert spaces from old ones are by direct sums and tensor products. Both of these constructions have physical significance. The first corresponds to superselection rules and the second to combined (or composite) systems.

A superselection rule in physics represents a quantity or property that is conserved for all physical processes and no two state vectors with distinct eigenvalues associated with the rule in question is again a state vector. Examples of superselection rules are electric charge and baryon number. Suppose we have a physical system S whose states are unit vectors in a Hilber space $\mathcal{H}$, and suppose S obeys a superselection rule R. Let us assume, for simplicity, that R has two values, say ± 1. Now the states of S can be divided into two classes A_+, A_- where $A_+(A_-)$ consists of those states for which $R = 1(-1)$. Let $\mathcal{H}_+$, $\mathcal{H}_-$ be the closed subspaces generated by A_+, A_-, respectively. Since S conserves R, "transitions" between A_+ and A_- are forbidden. It follows that $\langle \phi_+, \phi_- \rangle = 0$ for all $\phi_+ \in A_+$, $\phi_- \in A_-$ and hence $\mathcal{H}_+ \perp \mathcal{H}_-$. We then conclude that the direct sum $\mathcal{H}_+ \oplus \mathcal{H}_-$ is the applicable Hilbert space and that a state of S is given by a unit vector which must be in either $\mathcal{H}_+$ or $\mathcal{H}_-$. Another way of saying this is that the states of S must be eigenvectors of the projections onto $\mathcal{H}_+$ and $\mathcal{H}_-$.

The above discussion can be extended to the case where R has more than two values and also where there are two or more superselection rules. The representation Hilbert space then becomes a direct sum $\mathcal{H} = \oplus \mathcal{H}_i$. Then states are unit vectors belonging to precisely one of the $\mathcal{H}_i$ or equivalently are eigenvectors of the projections P_i onto $\mathcal{H}_i$, $i = 1, 2, \ldots$. Moreover, if T is an observable of the system, then similar reasoning leads us to conclude that T commutes with the P_i, $i = 1, 2, \ldots$.

Suppose we have two physical systems S_1, S_2 with corresponding Hilbert spaces $\mathcal{H}_1, \mathcal{H}_2$. We can then consider these systems as being part of a single composite system $S_1 \times S_2$. It is then reasonable to assume that the Hilbert-space representing $S_1 \times S_2$ is the tensor product $\mathcal{H}_1 \otimes \mathcal{H}_2$. (It is not completely clear why this should be the case and we shall say more about this later.) For example, if S_1 and S_2 correspond to single particles in $\mathbb{R}^3$ so that $\mathcal{H}_1 = \mathcal{H}_2 = L^2(\mathbb{R}^3)$, then we assume that the composite system consisting of both particles is represented by the Hilbert space

$$\mathcal{H} = L^2(\mathbb{R}^6) = L^2(\mathbb{R}^3) \otimes L^2(\mathbb{R}^3).$$

(If the particles are identical we must take either the symmetric or antisymmetric subspaces of $\mathcal{H}$. This corresponds to the boson-fermion superselection rule.) The tensor product gives the physically reasonable product rule for transition amplitudes, namely

$$\langle \phi_1 \otimes \phi_2, \psi_1 \otimes \psi_2 \rangle = \langle \phi_1, \psi_1 \rangle \langle \phi_2, \psi_2 \rangle.$$

Of course, this can be extended to combining several physical systems S_i with corresponding Hilbert spaces $\mathcal{H}_i$, $i = 1, \ldots, n$. In this case the Hilbert space for the composite system becomes $\otimes \mathcal{H}_i$.

2.3 Hilbert Space Probability

Axioms A1 to A4 of Section 2.2 can be derived from a single more basic axiom if we begin immediately with a probabilistic structure on Hilbert space. This framework also shows that quantum mechanics is a generalization of classical probability theory. The basic axiom takes the events of a physical system as primitive axiomatic elements. The events correspond to physical phenomena which either occur or do not occur. (Events are also called propositions. In a later deeper development we shall distinguish between the two concepts.) For example, suppose the physical system consists of a single particle and let A be a region in $\mathbb{R}^3$. If a_A is a counter that activates if and only if the particle enters the region A, then a_A corresponds to an event. Notice that if we subdivide $\mathbb{R}^3$ into disjoint regions, then a position measurement can be considered to be a collection of events. In general, practically any measurement or experiment can be analyzed in terms of its associated collection of events. The basic axiom of Hilbert space probability is the following.

(A) The events of a quantum system can be represented by (self-adjoint) projections on a complex Hilbert space.

Using Axiom A we can derive the concepts of state and observable as well as their important properties. But first let us discuss the properties and relationships of the events themselves. The most important relationship for events is that of disjointness. The disjointness of two events means that if one occurs, then the other does not occur. How should we describe disjointness for events represented by projections? Consider our previous example where a_A is a counter for the region A. Physically, we would have a_A and a_B disjoint if and only if $A \cap B = \emptyset$. Since intersection has no significance for projections, we must rephrase this condition in terms that can be carried over to projections. This can be done by considering the indicator functions ϕ_A for regions A in $\mathbb{R}^3$. Then an equivalent definition is that a_A and a_B are disjoint if and only if $\phi_A \phi_B = 0$. This product does have meaning for projections. In fact, on $L^2(\mathbb{R}^3)$ the multiplication operators ϕ_A are projections and $\phi_A \phi_B$ is the product of the two projections.

Another important concept is the disjoint union. For two disjoint events, their union is the event that occurs if and ony if precisely one of the two events occurs. In our example, this would be the counter

$a_{A \cup B}$ where $A \cap B = \emptyset$. Again, the union has no significance for projections. However, in terms of indicator functions, this would correspond to $\phi_{A \cup B} = \phi_A + \phi_B$ which is the sum of the projections ϕ_A and ϕ_B in $L^2(\mathbf{R}^3)$. In a similar way, if A_i, $i = 1, 2, \ldots$, is a sequence of mutually disjoint sets, then $\phi_{\cup A_i} = \Sigma \phi_{A_i}$. Also, the certain event corresponds to $\phi_{\mathbf{R}^3} = 1$.

Let $L(\mathcal{H})$ denote the set of all projections on the complex Hilbert space $\mathcal{H}$, and suppose $L(\mathcal{H})$ represents the events of a quantum system. From our preceding discussion, it is natural to assume that $P_1, P_2 \in L(\mathcal{H})$ represent disjoint events if P_1, P_2 are mutually orthogonally; that is, $P_1 P_2 = 0$. Moreover, if $P_1 P_2 = 0$, then $P_1 + P_2 \in L(\mathcal{H})$ and $P_1 + P_2$ represents the disjoint union of the two events. Similarly, if P_i, $i = 1, 2, \ldots$, are mutually orthogonal, then ΣP_i, in the strong operator topology, represents the disjoint union of the corresponding events. Also, the certain event is represented by the identity operator I.

To shorten our terminology, we shall call elements of $L(\mathcal{H})$ *quantum events*. We now describe the set of states on $L(\mathcal{H})$. A state of a physical system gives a theoretically complete description of the system. Since quantum mechanics is a probabilistic theory, a complete description of a quantum system is given by a probability measure on its set of events. We can then define a *state* as a map $s: L(\mathcal{H}) \to [0, 1]$ satisfying:

(S1) $s(I) = 1$
(S2) $s(\Sigma P_i) = \Sigma s(P_i)$ if $P_i P_j = 0$, $i \neq j$.

We say that a state s is *atomic* or *pure* if there is a one-dimensional projection P such that $s(P) = 1$. Otherwise, we call s a *mixed* state.

The following is an example of a state on $L(\mathcal{H})$. Let T be a positive trace class operator (called a *density operator*) on $\mathcal{H}$ with $\text{tr}(T) = 1$. Then it is easy to show that the map $s: L(\mathcal{H}) \to [0, 1]$ given by $s(P) = \text{tr}(TP)$ is a state on $L(\mathcal{H})$. It turns out that if $\dim \mathcal{H} \geq 3$, then every state has this form. This characterization is due to A. Gleason [1957] (also, see [Gudder, 1979; Varadarajan, 1968]) and we refer the reader to the literature for a proof.

Theorem 2.3. *If $\dim \mathcal{H} \geq 3$ and s is a state on $L(\mathcal{H})$, then there exists a unique density operator T on $\mathcal{H}$ such that $s(P) = \text{tr}(TP)$ for all $P \in L(\mathcal{H})$.*

Now suppose s is an atomic state. Then $s(P_\psi) = 1$ where P_ψ is the projection onto the one-dimensional subspace spanned by some unit vector

$\psi \in \mathcal{H}$. If $\dim \mathcal{H} \geq 3$, it follows from Gleason's theorem that $s(P) = \text{tr}(TP)$ for every $P \in L(\mathcal{H})$ where T is a density operator. Then

$$\langle T\psi, \psi \rangle = \text{tr}(TP_\psi) = s(P_\psi) = 1.$$

It follows that $T = P_\psi$. Hence, for every $P \in L(\mathcal{H})$

$$s(P) = \text{tr}(P_\psi P) = \langle P\psi, \psi \rangle. \tag{2.21}$$

We thus have a correspondence between pure states and unit vectors in $\mathcal{H}$. The correspondence is injective except for a multiplicative constant of modulus 1. In this way Axiom A1 of Section 2.2 follows from Axiom A. Moreover, any mixed state is a (possibly infinite) convex combination of pure states. It follows that the pure states are extremal in the convex set of all states.

Let s, t be pure states corresponding to unit vectors ψ, ϕ in $\mathcal{H}$, respectively. Then P_ϕ can be considered to be an event, namely the event that occurs when the system is in the state ϕ. Hence, $s(P_\phi)$ is the probability that the system is in the state ϕ (or t) when it is known that the system is in the state ψ (or s). But $s(P) = |\langle \psi, \phi \rangle|^2$ is what we called the transition probability from the state ψ to the state ϕ in Section 2.2. We again have the fact that $\langle \psi, \phi \rangle$ is the transition amplitude from ψ to ϕ.

Let x be an observable for our physical system. Then corresponding to $A \in B(\mathbb{R})$ we have an event $x(A)$, namely, the event that occurs when a measurement of x results in a value in A. If Axiom A holds, we may think of $x(A)$ as an element of $L(\mathcal{H})$ for some Hilbert space $\mathcal{H}$. It is physically clear that the map $x \colon B(\mathbb{R}) \to L(\mathcal{H})$ must have the following properties:

(O1) $x(\mathbb{R}) = I$,
(O2) if $A \cap B = \emptyset$, then $x(A)x(B) = 0$,
(O3) if $A_i \in B(\mathbb{R})$ are mutually disjoint, then $x(\cup A_i) = \Sigma x(A_i)$, where convergence of the summation is in the strong operator topology.

Thus, $x \colon B(\mathbb{R}) \to L(\mathcal{H})$ is a spectral measure and it follows from the spectral theorem that there exists a unique self-adjoint operator T_x on $\mathcal{H}$ such that

$$T_x = \int \lambda x(d\lambda).$$

We thus have a bijection between observables and self-adjoint operators which gives Axiom A2.

To derive Axiom A3, let T be a self-adjoint operator corresponding to an observable and let $\psi \in \mathcal{H}$ be a unit vector corresponding to a state s. Then from the above, we conclude that the probability that T has a value in $A \in B(\mathbb{R})$ when the system is in the state ψ is $\langle P^T(A)\psi, \psi\rangle$.

To derive Axiom A4, suppose that the system is initially in the state ψ and let $\psi_t = U_t\psi$ be the state at time t. Physical considerations lead us to conclude that U_t is a one-parameter group of transformations. It is reasonable to assume that U_t can be extended to a linear map on $\mathcal{H}$ and since U_t preserves the unit sphere, it follows that U_t is a one-parameter group of unitary transformations. Adding the physically reasonable assumption that $t \mapsto U_t$ is weakly continuous, it follows from Stone's theorem [Dunford and Schwartz, 1963] that there exists a unique self-adjoint operator H such that

$$\psi_t = e^{-itH/\hbar}\psi.$$

We then define H to be the energy operator and Axiom A4 follows.

It is not hard to show that Hilbert space probability theory generalizes classical probability theory. In fact, let (X, Σ, μ) be a probability space. Define $\mathcal{H} = L^2(X, \Sigma, \mu)$ and define $h: \Sigma \to L(\mathcal{H})$ by $h(A) = \phi_A$ where ϕ_A is thought of as a multiplication operator. We can thus represent events in Σ by quantum events in $L(\mathcal{H})$. Moreover, it is easy to show that h preserves the σ-algebra event structure. We can also represent the measure μ by a unit vector state $\psi \in \mathcal{H}$. Indeed, if ψ is the constant function on X with value 1, then $\mu(A) = \langle h(A)\psi, \psi\rangle$ for all $A \in \Sigma$. Finally, if f is a random variable on X, then $A \mapsto h[f^{-1}(A)]$ is a spectral measure on $\mathcal{H}$ so random variables correspond to self-adjoint operators.

It follows from the Heisenberg uncertainty relation (2.16) that observables which commute can be simultaneously measured with any degree of accuracy. For this reason, we call observables or events that commute *compatible*. Now it is easy to show that a collection of quantum events in $L(\mathcal{H})$ mutually commute if and only if they are contained in a Boolean σ-algebra in $L(\mathcal{H})$ [Gudder, 1979]. The imbedding $h: \Sigma \to L(\mathcal{H})$ of the previous paragraph shows that the events and random variables of classical probability theory correspond to compatible observables and quantum events. Thus, classical probability theory treats only single experiments and can not deal with incompatibility. However, Hilbert space probability describes all the possible experiments that can be performed on a physical system whether they are incompatible or not.

We close this section with a brief discussion of conditional probabilities. Let T be a density operator which represents the state of a physical system. Of course, the state is pure if and only if T is a one-dimensional projection. Suppose it is known that a quantum event $P_0 \in L(\mathcal{H})$ has occurred. Now it is easy to show that $T_0 = P_0 T P_0 / \text{tr}(TP_0)$ is again a density operator, if $\text{tr}(TP_0) \neq 0$. According to von Neumann, T_0 is the *conditional state given that P_0 has occurred*. For any $P \in L(\mathcal{H})$, the conditional probability of P given P_0 is computed from *von Neumann's formula*

$$\text{tr}(T_0 P) = \frac{\text{tr}(TP_0 P P_0)}{\text{tr}(TP_0)}. \tag{2.22}$$

It is not hard to show that (2.22) reduces to the usual definition of conditional probability in the classical probability theory case.

2.4 Spin Systems

We have seen a typical example of Hilbert space quantum mechanics in which the Hilbert space was $L^2(\mathbb{R}^3)$. In this case the states were unit vectors in $L^2(\mathbb{R}^3)$ and the observables were self-adjoint operators in $L^2(\mathbb{R}^3)$. There are also interesting quantum systems which are described by finite dimensional Hilbert spaces. We shall consider some in this section and others in later chapters.

A physical system can have an angular momentum independent of its motion and this is called intrinsic angular momentum or spin. Spin is a nonclassical degree of freedom and is characterized by a number j which can take only the values $0, \frac{1}{2}, 1, \frac{3}{2}, \ldots$. The system is then called a *spin-j* system. For instance, pions are spin-0 systems; electrons, protons, neutrons, and muons are spin-$\frac{1}{2}$ systems; photons are spin-1 systems; and gravitons are spin-2 systems. For a spin-j system, the component of spin along any axis in the three-dimensional configuration space can take only the values $-j, -j+1, \ldots, j-1, j$, a total of $2j+1$ values.

If we disregard the translational degrees of freedom of a spin-j system and only describe its spin, the associated Hilbert space becomes $\mathcal{H} = \mathbb{C}^{2j+1}$ with the standard inner product. We then have three observables J_1, J_2, J_3 which give the components of spin along the three coordinate axes x_1, x_2, x_3. In the representation in which J_3 is diagonal, these

SPIN SYSTEMS

observables are given by self-adjoint matrices with entries

$$[J_1]_{m,m'} = \frac{1}{2}[j(j+1) - m(m-1)]^{\frac{1}{2}}\delta_{m,m'+1}$$
$$+ \frac{1}{2}[j(j+1) - m(m+1)]^{\frac{1}{2}}\delta_{m,m'-1}$$
$$[J_2]_{m,m'} = \frac{i}{2}[j(j+1) - m(m-1)]^{\frac{1}{2}}\delta_{m,m'+1}$$
$$- \frac{i}{2}[j(j+1) - m(m+1)]^{\frac{1}{2}}\delta_{m,m'-1}$$
$$[J_3]_{m,m'} = m\delta_{m,m'}$$

where the row and column indices m, m' run from $-j$ to j. (Actually, the entries are multiplied by $\hbar$ but we take $\hbar = 1$ for simplicity.) The above operators satisfy the commutation relations

$$J_1 J_2 - J_2 J_1 = iJ_3$$

and the ones obtained by cyclic permutations of indices. The total spin operator J is obtained from $J^2 = J_1^2 + J_2^2 + J_3^2$ and equals the constant operator $[j(j+1)]^{1/2}I$. Even though the operators J_1, J_2, J_3 do not commute there is no uncertainty relation for them since they are bounded operators. In fact, $\Delta_\psi(J_1)\Delta_\psi(J_2)$ can be made arbitrarily small by properly choosing ψ.

Let us consider the simplest nontrivial case, namely the spin-$\frac{1}{2}$ system. Then the component of spin along any axis can take only the values $\frac{1}{2}, -\frac{1}{2}$ and the associated Hilbert space is $\mathcal{H} = \mathbb{C}^2$. We can write the three spin operators as $J_i = \frac{1}{2}\sigma_i$, $i = 1, 2, 3$, where the σ_i's are the *Pauli matrices*

$$\sigma_1 = \begin{bmatrix} 0 & 1 \\ 1 & 0 \end{bmatrix}, \quad \sigma_2 = \begin{bmatrix} 0 & -i \\ i & 0 \end{bmatrix}, \quad \sigma_3 = \begin{bmatrix} 1 & 0 \\ 0 & -1 \end{bmatrix}.$$

In this case, any observable (self-adjoint matrix) T on $\mathcal{H}$ has the form

$$T = a_1\sigma_1 + a_2\sigma_2 + a_3\sigma_3 + a_4 I \tag{2.23}$$

where $a_i \in \mathbb{R}$, $i = 1, \ldots, 4$. Notice that the J_i's each have eigenvalues $\pm\frac{1}{2}$. The corresponding eigenvectors for J_3 are $\binom{1}{0}, \binom{0}{1}$. The corresponding eigenvectors for J_1 are $1/\sqrt{2}\binom{1}{1}, 1/\sqrt{2}\binom{1}{-1}$ and for J_2 they are $1/\sqrt{2}\binom{1}{i}, 1/\sqrt{2}\binom{1}{-i}$.

Once an observable is written in the form (2.23), we can give its spectral decomposition in terms of the a_i's. The proof of the following lemma is straightforward.

Lemma 2.4. *An observable in the form (2.23) has spectral decomposition*

$$T = \lambda^+ P^+ + \lambda^- P^-$$

with

$$\lambda^{\pm} = a_4 \pm (a_1^2 + a_2^2 + a_3^2)^{\frac{1}{2}}$$
$$P^{\pm} = \tfrac{1}{2}\left[I \pm (a_1^2 + a_2^2 + a_3^2)^{-\frac{1}{2}}(a_1\sigma_1 + a_2\sigma_2 + a_3\sigma_3)\right].$$

Using Lemma 2.4, we can obtain an interesting correspondence between states and points of the unit ball in $\mathbb{R}^3$. We shall identify states with their corresponding density operators.

Theorem 2.5. *D is a state if and only if*

$$D = \tfrac{1}{2}(r_1\sigma_1 + r_2\sigma_2 + r_3\sigma_3 + I), \qquad r_i \in \mathbb{R}, \, i = 1, 2, 3 \tag{2.24}$$

where $r_1^2 + r_2^2 + r_3^2 \leq 1$. The state D is pure if and only if $r_1^2 + r_2^2 + r_3^2 = 1$.

Proof. Let $D = a_1\sigma_1 + a_2\sigma_2 + a_3\sigma_3 + a_4 I$ be an arbitrary self-adjoint matrix. Then its eigenvalues $\lambda^{\pm}$ are given in Lemma 2.4. Now D is a density operator if and only if $\lambda^{\pm} \geq 0$ and $\lambda^+ + \lambda^- = 1$. The second condition holds if and only if $a_4 = \tfrac{1}{2}$. The first condition is then equivalent to $a_1^2 + a_2^2 + a_3^2 \leq \tfrac{1}{4}$. Letting $a_i = r_i/2$, $i = 1, 2, 3$, then gives the first result. For the second result, D is pure if and only if D is a one-dimensional projection. This is equivalent to $\lambda^+ = 1$. The second result now follows. □

It follows from Theorem 2.5 that there is a bijection between points of the unit sphere in $\mathbb{R}^3$ and pure states. The north pole $(0, 0, 1)$ gives the eigenvector ψ^+ of J_3 corresponding to eigenvalue $\tfrac{1}{2}$ and is called the *spin up state* for J_3 while the south pole $(0, 0, -1)$ gives the eigenvector ψ^- of J_3 corresponding to eigenvalue $-\tfrac{1}{2}$ and is called the *spin down state* for J_3. It is easy to check that diametrically opposed points on the unit sphere correspond to mutually orthogonal pure states. Moreover, there is a bijection between mixed states and points of the interior of the unit ball. It is clear that mixed states are convex combinations of pure states and that these convex combinations are not unique. (To be completely precise, we are only considering states which are given by density operators. There are other pathological states which are not so given. This does not contradict Gleason's theorem since $\dim \mathcal{H} = 2$.)

Now suppose the system is in a pure state with density operator D given by (2.24). Then the probability that the spin has value $\frac{1}{2}$ in the x_3-direction becomes:

$$\mu_D\left(\frac{1}{2}\right) = \text{tr}(DP_{\psi+}) = \langle D\psi^+, \psi^+\rangle = \frac{(r_3+1)}{2}.$$

Similarly, $\mu_D(-\frac{1}{2}) = (1-r_3)/2$. If we identify D with the corresponding point (r_1, r_2, r_3) on the unit sphere, and write $r_3 = \cos\theta$ where θ is the polar angle, we obtain

$$\mu_D\left(\frac{1}{2}\right) = \frac{\cos\theta + 1}{2} = \cos^2\frac{\theta}{2} \qquad (2.25)$$

and in a similar way

$$\mu_D\left(-\frac{1}{2}\right) = \frac{1-\cos\theta}{2} = \sin^2\frac{\theta}{2}. \qquad (2.26)$$

We can interpret (2.25), (2.26) geometrically as follows. Suppose the system is in a spin-$\frac{1}{2}$ state in a direction that has an angle θ with the x_3-axis. Then the probability that the spin is $\frac{1}{2}$ in the x_3-direction is $\cos^2\theta/2$ and the probability that the spin is $-\frac{1}{2}$ in the x_3-direction is $\sin^2\theta/2$. We can also compute the expected spin, $E_D(J_3)$ in the x_3-direction. This becomes

$$\begin{aligned} E_D(J_3) = \text{tr}(DJ_3) &= \frac{1}{2}\cos^2\frac{\theta}{2} - \frac{1}{2}\sin^2\frac{\theta}{2} \\ &= \frac{1}{2}\cos\theta. \end{aligned} \qquad (2.27)$$

2.5 Two-Hole Experiment

Richard Feynman has emphasized that the method of computing probabilities involving subatomic particles is different than that of classical probability theory [Feynman and Hibbs, 1965]. As the size of objects increase, the quantum effects begin to cancel each other so that the quantum probabilities approach their classical counterparts. Thus, the classical theory is perfectly adequate for describing statistical systems of everyday life but not the behavior of very small particles. The meaning of probability is not changed in quantum mechanics. When we say that the probability of a

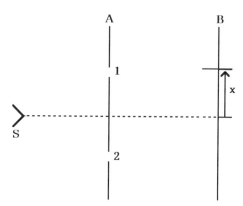

Figure 2.1. Two-Hole Experiment.

certain outcome of an experiment is p, we mean the usual thing; if the experiment is repeated many times, we expect that the fraction of those which give this outcome is roughly p. What is changed, and changed radically, is the method of calculating these probabilities.

To illustrate this, we consider a simple idealized experiment called the two-hole experiment. This experiment has not been actually carried out, but similar more complex experiments (which are harder to describe) have been, and there is no doubt that this experiment would give the results which we now discuss. Suppose we have a source of electrons S. The electrons all have the same energy but they leave S in all directions and many of them impinge on a planar screen A. The screen A has two holes, 1 and 2, through which the electrons may pass. Behind the screen A at a plane B we have an electron detector which can be placed at various distances x from the center of the screen (Figure 2.1). The detector records each passage of a single electron traveling from S through a hole in A to the point x.

After performing this experiment many times with many different values of x, we obtain a probability distribution (or more accurately a probability density) $P(x)$ that the electron passes from S to x as a function of x. If we analyze the problem of finding P classically, we would proceed as follows. Since an electron must pass through either hole 1 or hole 2, $P(x) = P_1(x) + P_2(x)$ where P_1 is the chance of arrival coming through 1 and P_2 is the chance of arrival coming through 2. Now we can easily find P_1 and P_2 experimentally. We simply close hole 2 and measure the chance of arrival at x with only hole 1 open. This gives the chance P_1 of arrival at x for electrons coming through 1. The result is given in Figure 2.2a. Similarly, we can find P_2 given in Figure 2.2b. The sum of these are shown in Figure 2.2c.

The actual experimental result is quite different. (This is fortunate, since otherwise we would not have the fascinating world of quantum mechanics.) The actual P is shown in Figure 2.3. We are then forced to conclude that $P \neq P_1 + P_2$.

We have now arrived at an apparent paradox. Since an electron must go through one hole or the other, we can classify all the arrivals at x into two disjoint classes A_1 and A_2, those arriving via hole 1 and those arriving via hole 2, respectively. The frequency of arrival at x would surely be the sum of the frequency P_1 of particles in A_1 and P_2 of those in A_2. We might try various methods to extricate ourselves from this difficulty. We might say that the electron does not go through either hole 1 or hole 2. Perhaps the electron spreads out somehow and passes partly through both holes. Perhaps the electron acts like a wave to produce the interference pattern in Figure 2.3 which is typical of waves. Or maybe the electron travels along a complex path going through hole 1, then back through hole 2 and finally through hole 1 again until it reaches x. All of these artifices, and other similar ones that have been thought of, can be shown to be experimentally false. The electron is a particle; it has a definite and extremely small extent. It definitely goes through either hole 1 or hole 2; one can "watch" it, by scattering light waves from it, to see which hole it goes through. It follows the simplest possible trajectory.

The way out of the paradox is to admit that we have not computed the probability correctly. We have computed it according to our everyday experience, but this is not the way nature does it. A new concept has arisen, namely that of interfering alternatives (or outcomes, or events). Two alternatives are interfering if they can be distinguished only by disturbing the system. Such a concept is not present in classical reasoning. The two alternatives A_1 and A_2 of the previous paragraph are interfering. If we try to distinguish between the two alternatives, we must disturb the system. In fact, we disturb the system to such an extent that the probability distribution changes radically. If we "watch" an electron by placing a light source behind screen A, then the light will scatter either near hole 1 or hole 2 so we can tell through which hole the electron went. But the electron will recoil from a light ray (photon) and will impinge upon the detection screen at a different point then it would otherwise. In fact, the resulting

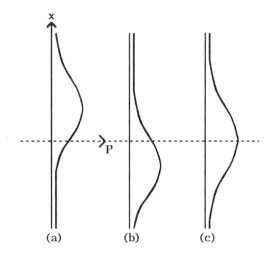

Figure 2.2. Classical Analysis. (a) P_1; (b) P_2; (c) $P_1 + P_2$.

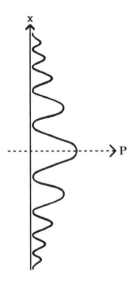

Figure 2.3. Actual P.

distribution will resemble the classical distribution in Figure 2.2c. If we do not "watch" the electron, the distribution will be that of Figure 2.3. But doesn't this effect depend upon the intensity and wave length of the light? Yes, of course it does. As the intensity decreases so does the number of photons, and more and more electrons are missed. As the wave length increases, the resolution decreases and it becomes harder to distinguish between the two holes. In both cases, the distribution changes slowly from Figure 2.2c to Figure 2.3. In summary, if we have mutually exclusive, noninterfering alternatives, their probabilities are added in the usual way. If we have mutually exclusive interfering alternatives, the probability of their union must be computed differently.

How *do* we compute the probabilities when we have interfering alternatives? The answer lies in the wave-like quality of the interference pattern in Figure 2.3. If waves were to impinge upon screen A, then the resulting intensities of the waves on screen B would resemble this pattern. Now the easiest way to represent a wave at a fixed time is by a complex function $\phi(x)e^{ikx}$ where $\phi(x)$ is the wave amplitude and e^{ikx} is the phase. The intensity of the wave is given by $|\phi(x)|^2$. By analogy, we postulate that $P(x)$ is the absolute square of a certain complex quantity $\psi(x)$ which we call the *probability amplitude* of arrival at x. Moreover, $\psi(x)$ is the sum of two contributions: ψ_1, the amplitude of arrival via hole 1, plus ψ_2, the amplitude of arrival via hole 2. It now follows that

$$P_1(x) = |\psi_1(x)|^2, \quad P_2(x) = |\psi_2(x)|^2$$

and

$$P(x) = |\psi_1(x) + \psi_2(x)|^2 = P_1(x) + P_2(x) + 2\operatorname{Re}\psi_1(x)\bar{\psi}_2(x). \quad (2.28)$$

The first two terms on the right hand side of (2.28) gives the classical result, the last term is the quantum interference term. The interference term can add to or subtract from the classical term and the sum will give an interference pattern.

Let A and B be two mutually exclusive alternatives. Then A and B can be interfering or noninterfering alternatives. If we can distinguish between them without disturbing the system, they are noninterfering and otherwise they are interfering. We frequently have both kinds of alternatives present. For example, in the two-hole experiment, suppose we want the probability that the electron arrives within 1cm of the center of the detection screen. We interpret this as the probability that if counters are arranged all over the screen, then the counter that activates is within 1cm of the center. The various alternatives are that the electron arrives at some counter via some hole. We know that the holes represent interfering alternatives. That the counters represent noninterfering alternatives can be seen as follows. If a counter is activated by an electron it will click, or flash a light, or something. If A and B are two counters, then we can tell which counter clicks or does not click without disturbing the sytem. This is because the click occurs after the experiment is over so it can not effect the electron before it arrives at the screen. Thus, to compute the desired probability, we first add $\psi_1(x) + \psi_2(x)$ for a fixed x, square the absolute value, and then sum (or integrate) this over all x within 1cm of the center. For other examples of interfering and noninterfering alternatives, as well as detailed mathematical analyzes, we refer the reader to [Feynman and Hibbs, 1965].

2.6 Path-Integral Formalism

Traditional quantum mechanics has been successful both in the conceptual understanding of quantum phenomena and in the numerical prediction of experimental results. It has been substantiated by many experiments in the laboratory and the universe at large. However, it leaves unanswered questions and has limited applicability for certain problems in modern physics such as the description of elementary particles and interactions. This may be because its axiomatic basis is at the wrong starting point. Although the concept of probability amplitude (or transition amplitude), which we called the essence of quantum mechanics, can be derived from the axioms of traditional quantum mechanics, this concept is not at the axiomatic foundation. An attempt has been made to start with the transition amplitude as the basic axiomatic element. This attempt has been called the path-integral formalism and it was initiated by R. Feynman [Feynman, 1948, 1949; Feynman and Hibbs, 1965; Schulman, 1981; Schweber, 1961]. Although this formalism has proved to be a useful tool, it has certain limitations due to its mathematically nonrigorous nature. Nevertheless, its usefulness shows that it is correct in some intuitive sense, and it may provide a basis for a mathematically sound theory.

Suppose a quantum particle (say an electron) begins at time t_a at a point x_a and then moves, possibly under the influence of some force. We

PATH-INTEGRAL FORMALISM 63

want to compute the probability $P(a,b)$ that the particle will arrive at a point x_b at time t_b. For simplicity, we suppose that the motion is one-dimensional. Now the particle can move along many alternative paths but we can not distinguish one from another without disturbing the system. If we "watch" to see which path the particle takes, its motion will be radically altered. The situation is like the two-hole experiment except there are infinitely many holes and an infinite number of screens. If the particle were classical, there would only be one possible path, the classical trajectory $\hat{x}(t)$ considered in Section 2.1. But in quantum mechanics, all possible paths must be considered. Each path $x(t)$ contributes an amplitude $\phi[x(t)]$ to the total amplitude $K(a,b)$ and $P(a,b) = |K(a,b)|^2$. In symbolic notation

$$K(a,b) = \sum \{\phi[x(t)]: \text{ all } x(t) \text{ with } x(t_a) = x_a, x(t_b) = x_b\}. \quad (2.29)$$

The basic axiom of the path-integral formalism is

(PI) $\phi[x(t)] = Ae^{(i/\hbar)S[x(t)]}$

where

$$S[x(t)] = \int_{t_a}^{t_b} L[v(t), x(t)]dt \quad (2.30)$$

is the action for the path $x(t)$ and A is a normalization constant. Recall that L is the Lagrangian of the system and that S is the action discussed in Section 2.1 with regard to the Lagrangian formulation of classical mechanics.

Before we consider the consequences of Axiom PI, let us compare this to the classical situation. From (PI) we see that each path contributes equally in modulus although their phases vary. Hence, it is not clear that some particular path becomes most important. However, in the classical approximation, S becomes very large compared with the small number $\hbar = 1.05 \times 10^{-27}$ erg-sec. Now if an arbitrary path $x(t)$ is changed by a small amount $\delta x(t)$ on the classical scale, although δS is also small on the classical scale, it is not small when measured in the tiny unit $\hbar$. These small changes in path will, in general, make enormous changes in phase which will oscillate rapidly. Hence, if the neighboring paths of $x(t)$ have a different action, their contributions to $K(a,b)$ in (2.29) will cancel each other, giving no net contribution. But for the special path $\hat{x}$, for which S is an extremum, a small change in path produces, in the first order at least, no change in S. Thus, paths in the vicinity of $\hat{x}$ do not give cancelling contributions. In this way, $\hat{x}$ is singled out and the classical laws of motion arise from the quantum laws. However, at the atomic level, when S becomes comparable to $\hbar$, no particular path is singled out. In this case all paths (more precisely, continuous paths) must be considered in computing $K(a,b)$ in (2.29).

One encounters intractable mathematical difficulties in giving a rigorous formulation for (2.29). The main problem is to find a suitable measure on the space of paths so that (2.29) can be formulated in terms of an integral over this space. There is an extensive literature on this [Schulman, 1981] and certain special cases have been solved. However, a general solution has not been attained. We shall follow the simple heuristic argument in [Feynman and Hibbs, 1965].

Let $t_a = t_0 < t_1 < \cdots < t_n = t_b$ be a partition of $[t_a, t_b]$ where

$$t_{i+1} - t_i = \varepsilon = \frac{(t_b - t_a)}{n}, \qquad i = 0, \ldots, n-1.$$

Let P_n be the set of polygonal paths of the form $x(t)$ where

$$x(t_a) = x_a, \quad x(t_b) = x_b, \quad x(t_i) = x_i, \qquad i = 1, \ldots, n-1$$

and for $t_i < t < t_{i+1}$, $(t, x(t))$ is the straight line segment from (t_i, x_i) to (t_{i+1}, x_{i+1}). We take the normalization constant to be

$$A_n = \left(\frac{m}{2\pi i \hbar \varepsilon}\right)^{n/2}$$

where m is the mass of the particle and define

$$K_n(a,b) = \sum \{\phi[x(t)] : x(t) \in P_n\}.$$

If we write this as an integral over the values of x_i, $i = 1, \ldots, n-1$, we obtain

$$K_n(a,b) = A_n \int \cdots \int e^{(i/\hbar)S[x(t)]} \, dx_1, \ldots, dx_{n-1}. \qquad (2.31)$$

We then define $K(a,b)$ by

$$K(a,b) = \lim_{n \to \infty} K_n(a,b).$$

There is no guarantee that this limit exists (and in many cases, it does not). When the limit does exist, it does not depend crucially on the polygonal nature of the paths. We shall use the following intuitive notation

$$K(a,b) = \int_a^b e^{(i/\hbar)S[x(t)]} D[x(t)] \qquad (2.32)$$

PATH-INTEGRAL FORMALISM

and shall call (2.32) a *path integral.*

Although (2.32) does not have a rigorous mathematical sense in general, it does have important physical meaning. Moreover, it can be used to discover properties that $K(a, b)$ should have if a rigorous formulation is possible. For example, let $t_c = t_i$ be an intermediate time for a fixed $i \in \{1, \ldots, n-1\}$. Writing the action for the time intervals $[t_a, t_c]$, $[t_c, t_b]$ as $S_a^c[x(t)]$, $S_c^b[x(t)]$, respectively, we obtain from (2.30) that

$$S_a^b[x(t)] = S_a^c[x(t)] + S_c^b[x(t)].$$

Hence, (2.31) gives

$$K_n(a,b) = \int K_{i-1}(a,c) K_{n-i+1}(c,b) \, dx_c$$

taking the limit as $n \to \infty$ we obtain the important formula

$$K(a,b) = \int K(a,c) K(c,b) \, dx_c. \tag{2.33}$$

We now evaluate the path integral for a simple specific case. For a free particle, the Lagrangian becomes $L = mv^2/2$. For $x(t) \in P_n$ we have

$$S[x(t)] = \int_{t_a}^{t_b} L[v(t), x(t)] \, dt$$

$$= \frac{m}{2\varepsilon} \sum_{i=1}^{n} (x_1 - x_{i-1})^2.$$

Hence,

$$K_n(a,b) = \Lambda_n \int \cdots \int \exp\left[\frac{im}{2\hbar\varepsilon} \sum_{i=1}^{n} (x_i - x_{i-1})^2\right] dx_1, \ldots, dx_{n-1}.$$

This gives a product of Gaussian integrals which can be evaluated individually to obtain

$$K_n(a,b) = \left(\frac{m}{2\pi i \hbar n \varepsilon}\right)^{\frac{1}{2}} \exp\left[\frac{im}{2\hbar n \varepsilon} (x_b - x_a)^2\right].$$

Since $n\varepsilon = t_b - t_a$, the K_n's are equal for all n. It follows that

$$K(a,b) = \left[\frac{m}{2\pi i \hbar (t_b - t_a)}\right]^{\frac{1}{2}} \exp\left[\frac{im(x_b - x_a)^2}{2\hbar(t_b - t_a)}\right]. \tag{2.34}$$

One can use similar methods to find $K(a,b)$ when L has a potential energy which is quadratic in the position.

We now show that Schrödinger's equation can be "derived" from the path-integral formalism. Of course, this must be taken with a grain of salt since the path-integral formalism itself is lacking in mathematical rigor. At the very least, it shows that the path-integral formalism is similar to traditional quantum mechanics and consistent with it.

If (x_0, t_0) is a fixed spacetime point, we define the *wave function* $\psi(x,t')$ to be

$$\psi(x,t') = K((x_0,t_0),(x,t')). \tag{2.35}$$

We can view $\psi(x,t')$ as the amplitude (or amplitude density) that a particle is at the point x at time t' when the initial condition is fixed. Applying (2.33) we obtain for any $t_0 < t < t'$

$$\begin{aligned}\psi(x,t') &= \int K((x_0,t_0),(x',t))K((x',t),(x,t'))\,dx' \\ &= \int K((x',t),(x,t'))\psi(x',t)\,dx'.\end{aligned} \tag{2.36}$$

Now suppose $t' = t + \varepsilon$ where ε is small. We then have the approximation

$$S_{(x',t)}^{(x,t')}[x(t)] = \int_t^{t'} L[v(t), x(t)]\,dt \approx \varepsilon L\left(\frac{x-x'}{\varepsilon}, \frac{x+x'}{2}\right)$$

and the approximation gets better as $\varepsilon \to 0$. We then obtain

$$\psi(x, t+\varepsilon) \approx A \int_{-\infty}^{\infty} \exp\left[\frac{i\varepsilon}{\hbar}L\left(\frac{x-x'}{\varepsilon}, \frac{x+x'}{2}\right)\right]\psi(x',t)\,dx'. \tag{2.37}$$

We now apply this to the case of a particle moving under a potential $V(x)$ in one dimension. We then have $L = mv^2/2 - V(x)$ so (2.37) becomes

$$\psi(x,t+\varepsilon) \approx A \int_{-\infty}^{\infty} \exp\left[\frac{im}{2\hbar\varepsilon}(x-x')^2 - \frac{i\varepsilon}{\hbar}V\left(\frac{x+x'}{2}\right)\right]\psi(x',t)\,dx'. \tag{2.38}$$

If we make the substitution $x' = x + \eta$, then (2.38) becomes

$$\psi(x,t+\varepsilon) \approx A \int_{-\infty}^{\infty} \exp\left(\frac{im\eta^2}{2\hbar\varepsilon}\right)\exp\left[\frac{-i\varepsilon}{\hbar}V\left(x+\frac{\eta}{2}\right)\right]\psi(x+\eta,t)\,d\eta. \tag{2.39}$$

PATH-INTEGRAL FORMALISM   67

If η is large compared to $(\hbar\varepsilon/m)^{\frac{1}{2}}$, then the first exponential oscillates very rapidly as η varies. When this factor oscillates rapidly, the integral over η gives a very small value since the other factors behave smoothly. Only if η is near zero, where the first exponential changes more slowly, do we get important contributions. In fact, the phase of the first exponential changes by the order of one radian when η is of the order $(\hbar\varepsilon/m)^{\frac{1}{2}}$, so most of the integral is contributed by values of η of this order.

We now expand ψ in a power series keeping only terms of order ε. This implies keeping second order terms in η. After the expansion, (2.39) has the form

$$\psi(x,t) + \varepsilon\frac{\partial\psi}{\partial t}$$
$$= A\int_{-\infty}^{\infty}\exp\left(\frac{im\eta^2}{2\hbar\varepsilon}\right)\left[1 - \frac{i\varepsilon}{\hbar}V(x)\right]\left[\psi(x,t) + \eta\frac{\partial\psi}{\partial x} + \frac{\eta^2}{2}\frac{\partial^2\psi}{\partial x^2}\right]d\eta.$$
(2.40)

If the leading terms of both sides of (2.40) are to be equal, we must have

$$1 = A\int_{-\infty}^{\infty}\exp\left(\frac{im\eta^2}{2\hbar\varepsilon}\right)d\eta = A\left(\frac{2\pi i\hbar\varepsilon}{m}\right)^{\frac{1}{2}}$$

and hence $A = (m/2\pi i\hbar\varepsilon)^{\frac{1}{2}}$. This agrees with our previous choice of A_n. In fact, this method can be used to determine the correct normalization used there. Using the integrals

$$\int_{-\infty}^{\infty}\eta\exp\left(\frac{im\eta^2}{2\hbar\varepsilon}\right)d\eta = 0, \quad A\int_{-\infty}^{\infty}\eta^2\exp\left(\frac{im\eta^2}{2\hbar\varepsilon}\right)d\eta = \frac{i\hbar\varepsilon}{m}$$

we obtain from (2.40)

$$\psi + \varepsilon\frac{\partial\psi}{\partial t} = \psi - \frac{i\varepsilon}{\hbar}V\psi - \frac{\hbar\varepsilon}{2im}\frac{\partial^2\psi}{\partial x^2}.$$

This equation holds to order ε if ψ satisfies the differential equation

$$i\hbar\frac{\partial\psi}{\partial t} = -\frac{\hbar^2}{2m}\frac{\partial^2\psi}{\partial x^2} + V\psi.$$

This is, of course, Schrödinger's equation for a one-dimensional particle. Similar methods work for a particle moving in three dimensions and for more complicated potentials.

2.7 Notes and References

Section 2.1 discussed the Hamiltonian and Lagranian formulations of classical mechanics. It was later shown that the Hamiltonian formulation was needed to develop Hilbert space quantum mechanics while the Lagrangian formulation was needed to develop the path-integral formalism. The Lagrangian formulation is also useful in the modern theories of quantum electrodynamics and chromodynamics. More detailed accounts of classical mechanics can be found in [Goldstein, 1957; Mackey, 1963].

Some recent investigations on the two-hole experiment are reported in [Wheeler and Zurek, 1983]. Our presentation in Section 2.5 is greatly simplified and does not consider the intricacies and subtleties of this important experiment. The experiment has recently been performed with photons and is reported in [Mittlestaedt, Prieur and Schrieder, 1988].

As we have seen, the main goal of classical probability theory is the construction of a mathematical model for a single probabilistic experiment or subexperiments of a single experiment. Hilbert space probability theory is much more ambitious. It seeks a mathematical model for the class of all possible experiments that can be performed on a physical system. But why can't this be done in classical probability theory? Why can't we construct a classical model for each experiment and then "paste" all the models together? The problem is we don't know how to do the pasting since we don't know how the various experiments interact or interfere. In Hilbert space probability theory, the pasting is automatically done by the Hilbert space structure. Although Hilbert space probability gives a model for a physical system, the physical system and hence the model might need to be enlarged. For example, we have used $\mathcal{H} = L^2(\mathbb{R}^3)$ to describe a particle with three degrees of freedom. But now that we know that a particle can have spin, we would like to include it in our description. This can be done by treating the particle as a composite system in which one component describes the spatial variables and the other component describes the spin variables. In fact, for a spin-j particle the corresponding Hilbert space in $L^2(\mathbb{R}^3) \otimes \mathbb{C}^{2j+1}$. References on Hilbert space quantum mechanics include [Bohm, 1951; Eisberg and Resnick, 1985; Prugovecki, 1981].

We have not discussed such topics as conditional expectations, joint distributions, limit theorems, and stochastic processes in Hilbert space probability theory. These topics have been developed and we refer the reader to [Accardi, 1981; Beltrametti and Cassinelli, 1981; Cycon and Hellwig, 1977; Davies, 1976; Gudder, 1979]. An interesting book on "stochastic quantum mechanics" is [Prugovecki, 1984], and further references on this subject may be found there. Some references on the path-integral formalism are [Feynman, 1948, 1949; Feynman and Hibbs, 1965; Schulman, 1981; Schweber, 1961].

The main problem with the axiomatics of traditional quantum mechanics is the question: where does the Hilbert space come from? For example, why are states represented by unit vectors? There have been various attempts to answer such questions [Mackey, 1963; Piron, 1976; Varadarajan, 1968]. One of the earliest is due to Dirac [1930]. According to Dirac, if ψ_1 and ψ_2 are two (pure) states and if $\alpha_1, \alpha_2 \in \mathbb{C}$, then $\alpha_1\psi_1 + \alpha_2\psi_2$ should correspond to another pure state in a certain sense. Dirac calls this the *superposition principle*. Although some fairly convincing arguments have been made for this principle, the case is far from closed. Nevertheless, many researchers feel that the superposition principle has an important place in quantum mechanics. If we accept this principle (in the absence of superselection rules) we come closer to Hilbert space.

3
Operational Statistics

In axiomatic quantum mechanics one attempts to establish a rigorous foundation for quantum theory by postulating physically motivated axioms. The axioms must be mathematically consistent and results derived from them must describe a large class of physical phenomena, have predictive power, and agree with experiment. In the past, various approaches have been taken towards axiomatizing quantum mechanics. Traditional Hilbert space quantum mechanics is one such approach. However, there are more basic approaches which do not rely on the unmotivated Hilbert space structure. The most basic approaches are the operational statistics approach and the quantum logic approach. The present chapter is devoted to these. Two other important approaches are the algebraic approach and the convexity approach. We shall not stress these here and refer the reader to the literature for a detailed discussion (see preface for references).

Operational statistics was first formulated by D. Foulis and C. Randall [Foulis and Randall, 1972, 1974, 1978, 1981, 1985; Randall and Foulis, 1973, 1981, 1983]. They have refined and developed their methods over a period of twenty years and have been joined by their students and others [Cook, 1985; Fischer and Rüttimann, 1978; Foulis, Piron and Randall, 1983; Gudder, 1982, 1986b; Wright, 1978]. They stress that they are not attempting to develop or advocate any particular theory, rather they are formulating a precise "language" in which such theories can be expressed, compared, evaluated, and related to laboratory experiments. In principle, any serious activity which is based upon the scientific method can be expressed in the language of operational statistics. The reason for this is that such activities are operationally (that is, experimentally) based and this language takes operations (experiments) as its primitive concept.

3.1 Basic Definitions

By a physical operation we shall mean a set of instructions for executing a well-defined, physically realizable, reproducible procedure. These instructions include the specification of a set of symbols corresponding to the possible outcomes of the procedure and each execution of the procedure must result in precisely one symbol. The symbols used to represent outcomes must be so constituted that the operation (or operations) capable of yielding these outcomes can be determined from the symbols themselves. In this way, a physical operation can be represented by its own outcome set. The outcome set for a single operation is essentially the same as in classical probability theory. The main difference here is that many physical operations for testing the system are included while in classical probability theory only one operation at a time is considered. Outcome sets corresponding to different physical operations may overlap. If two different experiments are performed, it is possible that a portion of their outcomes give physically equivalent results, in which case the outcomes are identified.

A manual of physical operations refers to a collection or catalogue of experiments that usually satisfies certain mild completeness and normative conditions. Since we can represent a physical operation by its outcome set, a manual of physical operations will be represented by a collection of nonempty sets.

We now give precise definitions for the relevant concepts. A *quasimanual* $\mathcal{A}$ is a nonempty collection of nonempty sets, and any $E \in \mathcal{A}$ is called an $\mathcal{A}$-*operation*. An $\mathcal{A}$-*outcome* is an element $a \in E$, where $E \in \mathcal{A}$. We call $X = \cup \mathcal{A}$ the *outcome set* for $\mathcal{A}$. If $A \subseteq E \in \mathcal{A}$, we call A an $\mathcal{A}$-*event* and denote the set of $\mathcal{A}$-events by $\mathcal{E}(\mathcal{A})$. In the sequel, $\mathcal{A}$ will denote a quasimanual with outcome set $X = X(\mathcal{A})$. If $A, B \in \mathcal{E}(\mathcal{A})$ satisfy $A \cap B = \emptyset$ and $A \cup B \in \mathcal{E}(\mathcal{A})$, we say that A, B are *orthogonal* and write $A \perp B$. If $A \perp B$ and $A \cup B \in \mathcal{A}$, we say that A and B are *operational complements* and write $A \, oc \, B$. If $A, B \in \mathcal{E}(\mathcal{A})$ share a common operational complement $C \in \mathcal{E}(\mathcal{A})$, then we say that A and B are *op-*

BASIC DEFINITIONS 73

erationally perspective with *axis* C and write $A\,op\,B$. Notice that op is a symmetric, reflexive relation. We now form the transitive completion of op to obtain an equivalence relation. If there exists a finite sequence of $\mathcal{A}$-events $D_0, D_1, \ldots, D_n$ such that $A = D_0$, $B = D_n$ and $D_{i-1}\,op\,D_i$, $i = 1, 2, \ldots, n$, then we say that A, B are *weakly perspective* and write $A\,wp\,B$. We denote by $p(A)$ the equivalence class of all $B \in \mathcal{E}(\mathcal{A})$ such that $B\,wp\,A$.

We now explain the above terminology. As mentioned earlier, an $\mathcal{A}$-operation corresponds to a physical experiment, so an $\mathcal{A}$-event represents a subset of the possible outcomes of an experiment. If $A \subseteq E \in \mathcal{A}$, we call E a *test operation* for A. If a test operation for A is performed, then A is said to *occur* or to *nonoccur* according to whether the resulting outcome belongs or does not belong to A. Now $A \perp B$ means that A, B can be simultaneously tested (that is, have a common test operation E) and when E is executed, if one occurs the other nonoccurs. In particular, if $A\,oc\,B$ and $E = A \cup B$ is executed then A occurs if and only if B nonoccurs. However, unlike classical probability theory, if $A \perp B$ and we know that A occurs, we cannot conclude that B nonoccurs. This is because there may be test operations for A which are not test operations for B. This is one of the reasons why the present theory is more general than classical probability theory and this increased generality enables us to include quantum mechanics within the present framework.

Suppose $A\,oc\,B$ and A occurs. Although we can not say that B nonoccurs, there is some weak sense in which B fails to occur. In this case we say that B is *refuted*. Similarly, if A nonoccurs we say that B is *confirmed*. Extending this idea, if $A\,oc\,B$ and A is confirmed (refuted) we say that B is *refuted (confirmed)*. In this way A is confirmed if and only if B is refuted. Thus the trichotomy occurs-nonoccurs-not tested has a corresponding coarser dichotomy confirm-refute. Notice that when A occurs if and only if B occurs, we have $A = B$. On the other hand, if $A\,wp\,B$, then A is confirmed if and only if B is confirmed. Thus elements in the equivalence class $p(A)$ are operationally equivalent to A in the sense of confirm-refute.

If $B\,oc\,A$ and $C\,oc\,A$, then $B\,op\,C$ so $p(B) = p(C)$. For $A \in \mathcal{E}(\mathcal{A})$ we define the *negation* $p(A)'$ of $p(A)$ by $p(A)' = p(B)$ where B is any $\mathcal{A}$-event such that $B\,oc\,A$. Our previous observation shows that $p(A)'$ is well-defined. The term negation stems from the fact that any element of $p(A)'$ is confirmed if and only if A is refuted. A test opertation for $A \in \mathcal{E}(\mathcal{A})$ will also be referred to as a test operation for $p(A)$. If such a test operation is performed, we say that $p(A)$ is *confirmed* (*refuted*) according to whether A is confirmed (refuted). Of course, $p(A)$ is confirmed if and only if $p(A)'$ is refuted. Since $p(A)$ is either confirmed or refuted when it is tested, we call $p(A)$ an *operational proposition*, $A \in \mathcal{E}(\mathcal{A})$.

We now define various additional properties that a quasimanual $\mathcal{A}$ may possess. We say that $\mathcal{A}$ is *irredundant* if $E, F \in \mathcal{A}$ with $E \subseteq F$ implies $E = F$. (Foulis and Randall include irredundance in their definition of a quasimanual.) For $A \in \mathcal{E}(\mathcal{A})$ define

$$A^\perp = \{x \in X(\mathcal{A}) : \{x\} \perp A\}$$

and define $\mathcal{A}$ to be *coherent* if $A, B \in \mathcal{E}(\mathcal{A})$ with $A \subseteq B^\perp$ imply $A \perp B$. We say that $\mathcal{A}$ is a *manual* if $A, B \in \mathcal{E}(\mathcal{A})$ with $B \in p(A)'$ imply $B \, oc \, A$. Notice that this is equivalent to $B \, wp \, C$ with $C \, oc \, A$ imply $B \, oc \, A$. It is easy to show that a manual $\mathcal{A}$ is irredundant. Indeed, suppose $E, F \in \mathcal{A}$ with $E \subseteq F$. Since $\emptyset \, oc \, E$ and $\emptyset \, oc \, F$, $F \, op \, E$. Since $E \, oc \, (F \setminus E)$, we have $F \, oc \, (F \setminus E)$. But this implies that $F \setminus E = \emptyset$ so $E = F$. A *premanual* is a quasimanual that is a subset of a manual. That is, $\mathcal{A}$ is a premanual if it is either a manual or can be enlarged to a manual by adjoining additional operations. The following theorem shows that a premanual has a unique smallest extension to a manual.

Theorem 3.1. *If $\mathcal{A}$ is a premanual, then there is a unique smallest manual $\langle \mathcal{A} \rangle$ containing $\mathcal{A}$. Moreover, $X(\mathcal{A}) = X(\langle \mathcal{A} \rangle)$.*

Proof. Let $\langle \mathcal{A} \rangle$ be the intersection of the set of manuals

$$\hat{\mathcal{B}} = \{\mathcal{B} : \mathcal{B} \text{ is a manual}, \mathcal{A} \subseteq \mathcal{B}\}.$$

To show that $\langle \mathcal{A} \rangle$ is a manual, let $A, B, C \in \mathcal{E}(\langle \mathcal{A} \rangle)$ with $B \, wp_{\langle \mathcal{A} \rangle} C$ where $C \, oc_{\langle \mathcal{A} \rangle} A$. Then for every $\mathcal{B} \in \hat{\mathcal{B}}$ we have $A, B, C \in \mathcal{E}(\mathcal{B})$ with $B \, wp_\mathcal{B} C$ and $C \, oc_\mathcal{B} A$. Hence, $B \, oc_\mathcal{B} A$ for every $\mathcal{B} \in \hat{\mathcal{B}}$ so $B \, oc_{\langle \mathcal{A} \rangle} A$. It now easily follows that $\langle \mathcal{A} \rangle$ is the unique smallest manual containing $\mathcal{A}$. To show that $X(\mathcal{A}) = X(\langle \mathcal{A} \rangle)$, suppose $\mathcal{A} \subseteq \mathcal{B} \in \hat{\mathcal{B}}$. Assume that $x \in X(\mathcal{B}) \setminus X(\mathcal{A})$. Let $\mathcal{B}'$ be the quasimanual defined by

$$\mathcal{B}' = \{E \in \mathcal{B} : x \notin E\}.$$

Since $x \notin E$ for every $E \in \mathcal{A}$, we have $\mathcal{A} \subseteq \mathcal{B}'$. To show that $\mathcal{B}'$ is a manual, suppose $A, B, C \in \mathcal{E}(\mathcal{B}')$ with $B \, wp_{\mathcal{B}'} C$, $C \, oc_{\mathcal{B}'} A$. It follows that

OPERATIONAL LOGICS   75

$B\,wp_{\mathcal{B}}C$ and $C\,oc_{\mathcal{B}}A$. Since $\mathcal{B} \in \hat{\mathcal{B}}$ we have $B\,oc_{\mathcal{B}}A$. Since $x \notin A, x \notin B$, we have $x \notin A \cup B \in \mathcal{B}$. Hence, $A \cup B \in \mathcal{B}'$ so $B\,oc_{\mathcal{B}'}A$. Since $\mathcal{B}' \in \hat{\mathcal{B}}$ and $x \notin X(\mathcal{B}')$, $x \notin X(\langle \mathcal{A} \rangle)$. Therefore, $X(\mathcal{A}) = X(\langle \mathcal{A} \rangle)$. □

We call $\langle \mathcal{A} \rangle$ the manual *generated* by $\mathcal{A}$.

3.2. Operational Logics

This section shows that the set $\Pi(\mathcal{A})$ of operational propositions has a structure which generalizes a Boolean algebra. Since a Boolean algebra forms the framework for the traditional predicate calculus of propositions, we see that $\Pi(\mathcal{A})$ generalizes the traditional mathematical logic framework.

If S is a nonempty set, a *partial binary operation* on S is a function $f: T \to S$ where $T \subseteq S \times S$. We call T the *domain* of f and write $T = D(f)$. An *associative ortho-algebra* (AOA) is a system $(L, +, 0, 1)$ where L is a set, $0, 1 \in L$, and $+$ is a partial binary operation with domain $D(+)$ satisfying the following four conditions:

(i) If $(p, q) \in D(+)$, then $(q, p) \in D(+)$ and $p + q = q + p$.

(ii) If $(p, q), (p + q, r) \in D(+)$, then $(q, r), (p, q + r) \in D(+)$ and $(p + q) + r = p + (q + r)$.

(iii) For each $p \in L$ there exists a unique $q \in L$ such that $(p, q) \in D(+)$ and $p + q = 1$.

(iv) If $(p, p) \in D(+)$, then $p = 0$.

Example 3.1. Let S be a nonempty set, L the power set of S, $0 = \emptyset, 1 = S$. For $A, B \in L$, let $(A, B) \in D(+)$ if and only if $A \cap B = \emptyset$ and if $(A, B) \in D(+)$, then $A + B = A \cup B$. It is easy to check that $(L, +, 0, 1)$ is an AOA. □

Example 3.2. Let $\mathcal{H}$ be a Hilbert space, $L = L(\mathcal{H})$, 0 the zero operator, $1 = I$. For $P_1, P_2 \in L$, let $(P_1, P_2) \in D(+)$ if and only if $P_1 \perp P_2$ and if $(P_1, P_2) \in D(+)$, then $P_1 + P_2$ is the operator sum. It is easy to check that $(L, +, 0, 1)$ is an AOA. □

We now show the relationship between an AOA and other mathematical structures used in the foundations of quantum mechanics. In particular, we are concerned with the framework of the quantum logic approach (see preface for references). If $(P, \leq)$ is a partially ordered set (poset), we call $0, 1 \in P$ *first* and *last* elements of P, respectively, if $0 \leq p \leq 1$

for all $p \in P$. We denote the least upper bound and greatest lower bound of $p,q \in P$, if they exist, by $p \vee q$, $p \wedge q$, respectively. A *minimal upper bound* for $p,q \in P$, if it exists, is an element $r \in P$ satisfying $p,q \leq r$ and if $p,q \leq s \leq r$, then $s = r$. If $(P, \leq, 0, 1)$ is a poset with 0 and 1, an *orthocomplementation* on P is a map $' : P \to P$ satisfying

(a) $p'' = p$ for all $p \in P$;
(b) if $p \leq q$, then $q' \leq p'$;
(c) $p \vee p' = 1$ for all $p \in P$.

We then call $(P, \leq, ', 0, 1)$ an *orthocomplemented* poset. In this case, if $p,q \in P$ satisfy $p \leq q'$ we say that p and q are *orthogonal* and write $p \perp q$. An orthocomplemented poset P is *orthomodular* if $p \perp q$ implies $p \vee q$ exists and if $p \leq q$ implies $q = p \vee (q \wedge p')$.

The quantum logic approach takes the set of experimental propositions P as axiomatic elements and usually comes to the conclusion that P forms an orthomodular poset. As we shall later see, operational statistics comes to a similar conclusion from a more basic level.

Now let $(L, +, 0, 1)$ be an AOA. For $p,q \in L$ we define $p \leq q$ if there exists $r \in L$ such that $(p,r) \in D(+)$ and $p + r = q$. For $p \in L$ we define p' to be the unique element of L satisfying $(p, p') \in D(+)$ and $p + p' = 1$.

Theorem 3.2. *Let $(L, +, 0, 1)$ be an AOA.*

(a) $(L, \leq, ', 0, 1)$ *is an orthocomplemented poset.*

(b) $p \perp q$ *if and only if* $(p, q) \in D(+)$.

(c) *If $p \leq q$, then* $(p, q'), (p, (p+q')') \in D(+)$ *and* $q = p + (p + q')'$.

(d) *If $(p, q) \in D(+)$, then $p+q$ is a minimal upper bound for p, q and if $p \vee q$ exists, then $p \vee q = p + q$.*

Proof. (a) Since $1 + 1' = 1$, it follows that $(1', 1') \in D(+)$ so $1' = 0$. Since $p + p' = 1$, we have $p'' = p$ for all $p \in L$. Moreover, we have

$$1 = 1 + 0 = (p' + p) + 0 = p' + (p + 0)$$

so $(p, 0) \in D(+)$ and $p + 0 = p$ for all $p \in L$. Hence, $0 \leq p \leq 1$ for all $p \in L$ so $0, 1$ are first and last elements, respectively. We now show that $\leq$ is a partial order relation. Since $p + 0 = p$, $p \leq p$ for all $p \in L$. If $p \leq q$ and $q \leq p$, then there exist $r, s \in L$ such that $p + r = q$ and $q + s = p$. Hence,

$$p + (r + s) = (p + r) + s = q + s = p$$

so we have
$$1 + (r + s) = (p' + p) + (r + s) = p' + p = 1.$$

Hence, $r + s = 0$. This implies that $(r, r) \in D(+)$ so $r = 0$. Thus, $p = q$. If $p \leq q$ and $q \leq r$, then there exists $s, t \in L$ with $p + s = q$ and $q + t = r$. Hence,
$$p + (s + t) = (p + s) + t = q + t = r$$

so $p \leq r$. To show that $'$ is an orthocomplementation, we already have $p'' = p$ for all $p \in L$. Now suppose that $p \leq q$. Then there is an $r \in L$ with $p + r = q$. This implies
$$p + (r + q') = (p + r) + q' = q + q' = 1.$$

Hence, $r + q' = p'$ so $q' \leq p'$. To show that $p \vee p' = 1$, we have from the above that $p, p' \leq 1$. Now suppose that $p, p' \leq q$. Then $q' \leq p \leq q$. Hence, there is an $r \in L$ with $q' + r = q$. Since $(q, q') \in D(+)$, it follows that $(q', q') \in D(+)$. Thus, $q' = 0$ so $q = q'' = 1$.
(b) If $p \perp q$, then $p \leq q'$ so there exists an $r \in L$ such that $q' = p + r$. Since $(q', q) \in D(+)$ we conclude that $(p, q) \in D(+)$. Conversely, if $(p, q) \in D(+)$, then since
$$q + [p + (p + q')] = (p + q) + (p + q)' = 1$$

we have $q' = p + (p + q)'$. Hence, $p \leq q'$ so $p \perp q$.
(c) If $p \leq q = q''$, then $p \perp q'$. By (b) we have $(p+q'), (p, (p+q')') \in D(+)$ and $q = p + (p + q')'$.
(d) If $(p, q) \in D(+)$, then clearly $p, q \leq p+q$ so $p+q$ is an upper bound for p, q. Now suppose that $r \in L$ satisfies $p, q \leq r \leq p + q$. Then there exist $s, t \in L$ with $r + s = p + q$ and $p + t = r$. Hence,
$$p + (t + s) = (p + t) + s = r + s = p + q.$$

Then by (b) we have
$$q' + (t + s) = [p + (p + q)'] + (t + s) = (p + q) + (p + q)' = 1.$$

Hence, $t + s = q$ so $s \leq q$. But since $s \perp r$ we have $s \leq r' \leq q'$. Thus, $s \leq q \wedge q' = 0$ so $s = 0$. It follows that $r = p + q$ so $p + q$ is a minimal upper bound. The last statement of (d) now easily follows. □

Corollary 3.3. *Let $(L, +, 0, 1)$ be an AOA.*

(a) *For every $p \in L$, $(p, 0) \in D(+)$ and $p + 0 = p$.*

(b) *If $(p, q), (p, r) \in D(+)$ and $p + q = p + r$, the $q = r$.*

(c) *If $(p, 1) \in D(+)$, then $p = 0$.*

Proof. (a) This is proved in Theorem 3.2a. (b) If p+q=p+r, then by Theorem 3.2b, $p \leq r'$. Applying Theorem 3.2c gives

$$r' + q = [p + (p + r)'] + q = (p + q) + (p + r)' = (p + r) + (p + r)' = 1.$$

Hence, $q = r'' = r$. (c) If $(p, 1) \in D(+)$, we have $(p, p') \in D(+)$ and $(p, p + p') \in D(+)$. Hence, $(p, p) \in D(+)$ and $p = 0$. □

We have seen that if $(L, +, 0, 1)$ is an AOA, then $(L, \leq, ', 0, 1)$ is an orthocomplemented poset. In general, L need not be orthomodular. That is, if $p \perp q$, then $p \vee q$ need not exist and if $p \leq q$ we need not have

$$q = p \vee (q \wedge p') = p \vee (p \vee q')'.$$

However, we do have a weakened form of orthomodularity in which $\vee$ is replaced by $+$. Indeed, from Theorem 3.2b and (c), if $p \perp q$ then $p + q$ exists and if $p \leq q$ then $q = p + (p + q')'$. We now give a simple characterization of orthomodularity. We say that an AOA $(L, +, 0, 1)$ is *orthocoherent* if $(p, q), (p, r), (q, r) \in D(+)$ imply $(p + q, r) \in D(+)$.

Corollary 3.4. *If $(L, +, 0, 1)$ is an AOA, then the following statements are equivalent.*

(a) $(L, \leq, ', 0, 1)$ *is an orthomodular poset.*

(b) $(L, +, 0, 1)$ *is orthocoherent.*

(c) *If $p \perp q$, then $p \vee q$ exists.*

Proof. (a)⇒(b) Suppose $(L, \leq, ', 0, 1)$ is an orthomodular poset and $(p, q), (p, r), (q, r) \in D(+)$. Then $p, q \leq r'$ so $p + q = p \vee q \leq r'$. Hence, $(p + q) \perp r$ so $(p + q, r) \in D(+)$.
(b)⇒(c) Suppose $(L, +, 0, 1)$ is orthocoherent and $p \perp q$. If $p, q \leq s$, then $(p, q), (p, s'), (q, s') \in D(+)$. Hence, $(p + q, s') \in D(+)$ so $p + q \leq s$. It follows that $p \vee q$ exists and $p \vee q = p + q$.
(c)⇒(a) This follows from Theorem 3.2b, c and d. □

The next result shows that orthocoherent AOA's are equivalent to orthomodular posets.

Theorem 3.5. *Let $(L, \leq,', 0, 1)$ be an orthomodular poset. Define*

$$D(+) = \{(p,q) \in L \times L : p \perp q\}$$

and for $(p,q) \in D(+)$ define $p + q = p \vee q$. Then $(L, +, 0, 1)$ is an orthocoherent AOA.

Proof. We check the four axioms (i)–(iv). (i) clearly holds. (ii) If $(p,q), (p+q, r) \in D(+)$, then $p \perp q$ and $(p \vee q) \perp r$. Then $q \leq p \vee q \leq r'$ so $(q,r) \in D(+)$. Since $p \leq p \vee q \leq r'$ we have $r \leq p'$ and $q \leq p'$. Hence, $q \vee r \leq p'$ so $(p, q+r) \in D(+)$. Moreover,

$$(p+q) + r = (p \vee q) \vee r = p \vee (q \vee r) = p + (q+r).$$

(iii) For $p \in L$ we have $(p, p') \in D(+)$ and $p + p' = p \vee p' = 1$. For uniqueness, suppose $(p,q) \in D(+)$ and $p + q = p \vee q = 1$. Since $q \leq p'$, by orthomodularity

$$q = q \vee 0 = q \vee 1' = q \vee (p \vee q)' = q \vee (p' \wedge q') = p'.$$

(iv) If $(p,p) \in D(+)$, then $p \leq p'$. Hence,

$$p \leq p' \wedge p = (p \vee p')' = 1' = 0$$

so that $p = 0$. The proof of orthocoherence is the same as in Corollary 3.4 (a)$\Rightarrow$(b). □

Let $\mathcal{A}$ be a manual and let $\Pi(\mathcal{A}) = \{p(A) : A \in \mathcal{E}(\mathcal{A})\}$ be its set of operational propositions. Suppose that $A, B \in \mathcal{E}(\mathcal{A})$ and $A \perp B$. Let $A_1 \in p(A)$, $B_1 \in p(B)$. Since $A \perp B$ there exists a $C \in \mathcal{E}(\mathcal{A})$ such that $A \cup C \, oc \, B$. Hence, $A \cup C \in p(B)' = p(B_1)'$. Since $\mathcal{A}$ is a manual, we have $A \cup C \, oc \, B_1$. It follows that $A \perp B_1$. Repeating this argument we have $A_1 \perp B_1$. Hence, if $A \perp B$ we can write $p(A) \perp p(B)$ and the above argument shows that this is well-defined. If $p(A) \perp p(B)$ we define $p(A) + p(B) = p(A \cup B)$. Again, it is not hard to show that $p(A) + p(B)$ is well-defined. We define $0 = p(\emptyset)$ and $1 = 0'$ and we call $\{\Pi(\mathcal{A}), +, 0, 1\}$ the *operational logic* of $\mathcal{A}$. If one wishes to study confirmation and refutation, as opposed to the more stringent requirements entailed by occurrence and nonoccurrence, then one naturally identifies operationally perspective events and analyzes only the operational logic $\Pi(\mathcal{A})$. One must then pay the cost of discarding information that is lost by this identification and this lost information is sometimes crucial.

We say that a manual $\mathcal{A}$ is *orthocoherent* if $\Pi(\mathcal{A})$ is orthocoherent. It is easy to show that coherent manuals are orthocoherent. Although the converse does not hold, in general, it does hold for manuals possessing only finite operations or only a finite number of operations.

Our next result shows that operational logics and AOA's are equivalent concepts. Two AOA's L_1, L_2 are *isomorphic* if there exists a bijection $J: L_1 \to L_2$ such that $(J \times J)D(+) = D(+)$, $J(0) = 0$, $J(1) = 1$, and $J(p + q) = Jp + Jq$.

Theorem 3.6. *The operational logic of a manual is an AOA. Conversely, if L is an AOA, then there is a manual $\mathcal{A}$ such that $\Pi(\mathcal{A})$ is isomorphic to L.*

Proof. Let $\mathcal{A}$ be a manual and let $(\Pi(\mathcal{A}), +, 0, 1)$ be its operational logic. To show that $(\Pi(\mathcal{A}), +, 0, 1)$ is an AOA we check the four conditions (i)–(iv). Condition (i) is clear. (ii) Suppose that $(p(A), p(B)), (p(A) + p(B), p(C)) \in D(+)$. Then $A \perp B$ and $p(A \cup B) \perp p(C)$. Hence, $(A \cup B) \perp C$ so $B \perp C$ which implies $(p(B), p(C)) \in D(+)$. Also, $A \perp (B \cup C)$ so $(p(A), p(B) + p(C)) \in D(+)$. Moreover,

$$[p(A) + p(B)] + p(C) = p(A \cup B) + p(C)$$
$$= p(A \cup B \cup C)$$
$$= p(A) + p(B \cup C)$$
$$= p(A) + [p(B) + p(C)].$$

(iii) If $B \, oc \, A$, then $p(A) + p(B) = p(A \cup B) = 1$. For uniqueness, suppose $(p(A), p(C)) \in D(+)$ and $p(A) + p(C) = 1$. Then $p(A \cup C) = 1$ so $C \in P(A)' = p(B)$. Hence, $p(C) = p(B)$.
(iv) If $(p(A), p(A)) \in D(+)$, then $p(A) \perp p(A)$. Hence, $A \perp A$ so $A = \emptyset$. Thus, $p(A) = p(\emptyset) = 0$.

Conversely, let $(L, +, 0, 1)$ be an AOA. We call a finite nonempty subset $\{p_1, \ldots, p_n\}$ of L with $p_i \neq 0$, $i = 1, \ldots, n$, a *partition* of L if $p_1 + \cdots + p_n$ exists and equals 1. Let $\mathcal{A}$ be the set of partitions of L. Notice that $\{1\} \in \mathcal{A}$ and $\{p, p'\} \in \mathcal{A}$ for every $p \in L$ with $p \neq 0, 1$. Now $\mathcal{A}$ is a quasimanual. For $A = \{p_1, \ldots, p_n\} \in \mathcal{E}(\mathcal{A})$ we use the notation $\Sigma(A) = p_1 + \cdots + p_n$, $\Sigma(\emptyset) = 0$. If $A, B \in \mathcal{E}(\mathcal{A})$ and $\Sigma(A) = \Sigma(B)$, then $[\Sigma(A)]' \, oc \, A$ and $[\Sigma(A)]' \, oc \, B$ so

$A \, op \, B$. Conversely, if $A \, op \, B$, then there is a $C \in \mathcal{E}(\mathcal{A})$ with $C \, oc \, A$ and $C \, oc \, B$. Then
$$\Sigma(C) + \Sigma(A) = \Sigma(C) + \Sigma(B) = 1$$
so by Corollary 3.3(b), $\Sigma(A) = \Sigma(B)$. It follows that for arbitrary $A, B \in \mathcal{E}(\mathcal{A})$ we have $A \, wp \, B$ if and only if $\Sigma(A) = \Sigma(B)$. Hence,
$$p(A) = \{B \in \mathcal{E}(\mathcal{A}) : \Sigma(B) = \Sigma(A)\}$$
and
$$p(A)' = \{B \in \mathcal{E}(\mathcal{A}) : \Sigma(B) = [\Sigma(A)]'\}.$$

To show that $\mathcal{A}$ is a manual, suppose $A, B \in \mathcal{E}(\mathcal{A})$ with $B \in p(A)'$. Then $\Sigma(B) = [\Sigma(A)]'$ so $\Sigma(A) + \Sigma(B) = 1$. It follows that $B \, oc \, A$. Define $J: \Pi(\mathcal{A}) \to L$ by $Jp(A) = \Sigma(A)$. Clearly, J is injective. Also, J is surjective since $J0 = Jp(\emptyset) = 0$ and if $q \in L$, with $q \neq 0$, then $Jp(\{q\}) = q$. It is clear that $J(1) = 1$. If $p(A) \perp p(B)$, then $A \perp B$ so $\Sigma(A) \perp \Sigma(B)$. Conversely, if $\Sigma(A) \perp \Sigma(B)$, then $A \perp B$ so $(J \times J)D(+) = D(+)$. Finally, if $A \perp B$, then
$$J[p(A) + p(B)] = Jp(A \cup B)$$
$$= \Sigma(A \cup B)$$
$$= \Sigma(A) + \Sigma(B)$$
$$= Jp(A) + Jp(B). \quad \square$$

It follows from Corollary 3.4 and Theorem 3.6 that the operational logic of an orthocoherent manual forms an orthomodular poset. One mild additional condition provides the framework for the quantum logic approach. An orthomodular poset L is said to be *σ-orthocomplete* if the least upper bound of any sequence of mutually orthogonal elements of L exists. A σ-orthocomplete, orthomodular poset is called a *quantum logic*.

The quantum logic approach postulates that the set of experimental propositions for a physical system forms a quantum logic $(L, \leq, ', 0, 1)$. An experimental proposition is interpreted as an experiment with at most two outcomes, $\leq$ corresponds to an experimental implication, and $'$ to a negation. As we have stressed earlier, operational statistics is more general and more basic. Moreover, information can be lost while going to the operational logic (or quantum logic) level. Nevertheless, there are some important, physically applicable concepts that can be defined at this level. We now briefly discuss some of these and refer the reader to the literature for more details [Gudder, 1979; Varadarajan, 1968].

Let $(L, \leq,', 0, 1)$ be a quantum logic. Two propositions $p, q \in L$ are *compatible* (written $q \leftrightarrow p$) if there exist mutually orthogonal elements p_1, q_1, and r such that $p = p_1 \vee r$, $q = q_1 \vee r$. Compatible propositions correspond to propositions that are simultaneously testable. It can be shown that $q \leftrightarrow p$ if and only if q and p are contained in a Boolean subalgebra of L. An *observable* on L is a σ-homomorphism $x: B(\mathbf{R}) \to L$; that is,

(a) $x(\mathbf{R}) = 1$,
(b) if $A \cap B = \emptyset$, then $x(A) \perp x(B)$,
(c) if $A_i \in B(\mathbf{R})$ is a sequence of mutually disjoint sets, then $x(\cup A_i) = \vee x(A_i)$.

Two observables x, y are *compatible* if $x(A) \leftrightarrow y(B)$ for every $A, B \in B(\mathbf{R})$. Compatible observables correspond to noninterfering measurements. It can be shown that x, y are compatible if and only if x and y are both functions of a single observable. A *normal state* on L is a map $m: L \to [0, 1] \subseteq \mathbf{R}$ satisfying $m(1) = 1$, $m(\vee q_i) = \Sigma m(q_i)$ if $q_i \perp q_j$, $i \neq j$. Normal states correspond to probability measure on L. If x is an observable and $A \in B(\mathbf{R})$, then $x(A)$ is the proposition that is verified when x has a value in A. Moreover, for a normal state m, $m[x(A)]$ is the probability of verification when the system is in the state m. The probability measure $A \mapsto m[x(A)]$ on $B(\mathbf{R})$ is called the *distribution* of x in the state m and

$$E_m(x) = \int \lambda m[x(d\lambda)]$$

is the *expectation* of x in the state m. The *spectrum* of an observable x is the smallest closed subset A of $\mathbf{R}$ such that $x(A) = 1$. The spectrum of x corresponds to the allowable values of x; that is, the values that x may attain (or approach arbitrarily closely).

Example 3.3. If (X, Σ) is a measurable space, it is easy to check that $(\Sigma, \subseteq,^c, \emptyset, X)$ is a quantum logic. It can be shown that x is an observable if and only if there exists a measurable function (random variable) $f: X \to \mathbf{R}$ such that $x(A) = f^{-1}(A)$ for every $A \in B(\mathbf{R})$. The spectrum of x is the closure of the range of f. Normal states are just probability measures and the distribution and expectation of x reduce to those for f. We have thus arrived at classical probability theory. All propositions and observables are compatible so there is no possibility of interference. □

Example 3.4. If $\mathcal{H}$ is a complex Hilbert space and $P_1, P_2 \in L(\mathcal{H})$ we define $P_1 \leq P_2$ if $P_1 P_2 = P_1$. Then it is easy to check that $(L(\mathcal{H}), \leq, \perp, 0, I)$ is a quantum logic. By the spectral theorem, observables can be identified with self-adjoint operators on $\mathcal{H}$. Applying Gleason's theorem (if $(\dim \mathcal{H} \geq 3)$), normal states can be identified with density operators. The spectrum, distribution, and expectation of an observable reduce to their usual operator counterparts. It is easy to show that two propositions or observables are compatible if and only if their corresponding operators commute. We have thus arrived at Hilbert space probability theory. □

3.3 States

Until now we have mainly studied the algebraic properties of physical experiments and their corresponding operational propositions. We now consider probabilistic aspects of this framework by introducing probability measures.

Let $\mathcal{A}$ be a quasimanual with outcome set $X = X(\mathcal{A})$. An $\mathcal{A}$-*state* is a function $\omega \colon X \to [0,1] \subseteq \mathbf{R}$ such that $\Sigma_{x \in E} \omega(x) = 1$ for every $E \in \mathcal{A}$. The set of all $\mathcal{A}$-states is denoted by $\Omega = \Omega(\mathcal{A})$. An $\mathcal{A}$-state may be thought of as a probability weight on the outcomes. The defining relation for an $\mathcal{A}$-state means that if an operation is tested, then one of its outcomes is certain to result. For $\omega \in \Omega$ and $A \in \mathcal{E}(\mathcal{A})$, we define

$$\omega(A) = \sum_{x \in A} \omega(x).$$

The resulting map $\omega \colon \mathcal{E}(\mathcal{A}) \to [0,1]$ is called a *regular $\mathcal{E}(\mathcal{A})$-state*. Notice that for $\omega \in \Omega$, and $A, B \in \mathcal{E}(\mathcal{A})$ with $A \perp B$ we have

$$\omega(A \cup B) = \omega(A) + \omega(B).$$

Frequently, ω is interpreted as corresponding to a preparation procedure and $\omega(A)$ gives the probability that A occurs when A is tested after the ω-preparation procedure is performed. Notice that if ω is restricted to the events of a single operation, then ω becomes a (countably additive) probability measure on this set of events. More generally, a map $\alpha \colon \mathcal{E}(\mathcal{A}) \to [0,1]$ such that $\alpha(E) = 1$ for every $E \in \mathcal{A}$ and

$$\alpha(A \cup B) = \alpha(A) + \alpha(B)$$

whenever $A \perp B$ is called an $\mathcal{E}(\mathcal{A})$-*state*. In general, an $\mathcal{E}(\mathcal{A})$-state need not be regular.

If $(L, +, 0, 1)$ is an AOA, a *state* on L is a map $\alpha \colon L \to [0, 1] \subseteq \mathbb{R}$ such that $\alpha(1) = 1$ and $\alpha(p+q) = \alpha(p) + \alpha(q)$ whenever $(p, q) \in D(+)$. Similarly, a *state* α on an orthomodular poset $(L, \leq, ', 0, 1]$ is defined to be a state on L considered as an AOA. That is, $\alpha \colon L \to [0, 1]$ satisfies $\alpha(1) = 1$ and

$$\alpha(p \vee q) = \alpha(p) + \alpha(q)$$

whenever $p \perp q$. We now show that if $\mathcal{A}$ is a manual, then $\mathcal{E}(\mathcal{A})$-states can be lifted to states on the AOA $\Pi(\mathcal{A})$. If ω is an $\mathcal{E}(\mathcal{A})$-state and $A \operatorname{oc} B$, then $\omega(B) = 1 - \omega(A)$. It follows that if $A \operatorname{wp} B$, then $\omega(A) = \omega(B)$. Hence, $\omega(B) = \omega(A)$ for all $B \in p(A)$ so we can define $\omega[p(A)] = \omega(A)$. Thus, $\omega \colon \Pi(\mathcal{A}) \to [0, 1]$ and we now show that ω is a state on $\Pi(\mathcal{A})$. Since $1 = p(E)$ for any $E \in \mathcal{A}$ we have

$$\omega(1) = \omega[p(E)] = \omega(E) = 1.$$

Moreover, if $(p(A), p(B)) \in D(+)$, then $A \perp B$ so

$$\omega[p(A) + p(B)] = \omega[p(A \cup B)]$$
$$= \omega(A \cup B) = \omega(A) + \omega(B)$$
$$= \omega[p(A)] + \omega[p(B)].$$

In this way $\mathcal{E}(\mathcal{A})$-states can be identified with states on the operational logic $\Pi(\mathcal{A})$.

If ω is an $\mathcal{A}$-state, since the ω-preparation procedure can be regarded as part of the instructions for the manual of procedures, we can interpret ω as a complete stochastic model for the physical system described by $\mathcal{A}$. This is usually understood in the following empirical sense. For every $A \in \mathcal{E}(\mathcal{A})$, $\omega(A)$ is interpreted as the "long run relative frequency" with which the event A occurs as a consequence of executions of test operations. Hence, $\omega[p(A)]$ is the frequency with which $p(A)$ is confirmed when tested. If $\mathcal{A}$ is a manual, this provides an interpretation for the regular $\mathcal{E}(\mathcal{A})$-states on the operational logic $\Pi(\mathcal{A})$. However, there is no way, in general, to tell just which states on an AOA are regular unless the underlying manual is known. This indicates how vital information may be lost in passing from a manual to its operational logic.

If $\mathcal{A}$ and $\mathcal{B}$ are quasimanuals, a *morphism* from $\mathcal{A}$ to $\mathcal{B}$ is a map

$$\phi: X(\mathcal{A}) \to \mathcal{E}(\mathcal{B})$$

satisfying:

(a) for $E \in \mathcal{A}$, $\cup_{x \in E} \phi(x) \in \mathcal{B}$
(b) if $x_1, x_2 \in E \in \mathcal{A}$ with $x_1 \neq x_2$, then $\phi(x_1) \cap \phi(x_2) = \emptyset$.

We say that ϕ is *positive* if $\phi(x) \neq \emptyset$ for all $x \in X(\mathcal{A})$. We call ϕ *outcome preserving* if $\phi(x) \in X(\mathcal{B})$ for all $x \in X(\mathcal{A})$. Corresponding to a morphism $\phi: X(\mathcal{A}) \to \mathcal{E}(\mathcal{B})$ we have a map

$$\hat{\phi}: \mathcal{E}(\mathcal{A}) \to \mathcal{E}(\mathcal{B})$$

defined by

$$\hat{\phi}(A) = \cup_{x \in A} \phi(x).$$

If ϕ is an outcome preserving, bijective morphism and $\hat{\phi}$ is surjective, we call ϕ an *isomorphism*. If $\phi: X(\mathcal{A}) \to \mathcal{E}(\mathcal{B})$ is a morphism and $A, B \in \mathcal{E}(\mathcal{A})$ with $A \perp B$, then clearly $\hat{\phi}(A) \perp \hat{\phi}(B)$. Moreover, if $A \, oc \, B$, then $\hat{\phi}(A) \, oc \, \hat{\phi}(B)$ and hence, if $A \, wp \, B$, then $\hat{\phi}(A) \, wp \, \hat{\phi}(B)$. Also notice that if $\omega \in \Omega(\mathcal{B})$, then $\omega \circ \phi \in \omega(\mathcal{A})$. We then define the *conjugate map*

$$\phi^+: \Omega(\mathcal{B}) \to \Omega(\mathcal{A})$$

by $\phi^+(\omega) = \omega \circ \phi$ for all $\omega \in \Omega(\mathcal{B})$.

A simple example of a morphism can be obtained from a premanual $\mathcal{A}$. Let

$$\phi: X(\mathcal{A}) \to X(\langle \mathcal{A} \rangle)$$

be the imbedding identity map. Then ϕ is an outcome preserving morphism. Although ϕ is bijective, $\hat{\phi}$ is surjective if and only if $\mathcal{A}$ is a manual. Hence, if $\mathcal{A}$ is not a manual, ϕ is not an isomorphism. Our next result shows that $\Omega(\langle \mathcal{A} \rangle) = \Omega(\mathcal{A})$, Hence, $\hat{\phi}$ is bijective.

Theorem 3.7. *If $\mathcal{A}$ is a premanual, then $\Omega(\langle \mathcal{A} \rangle) = \Omega(\mathcal{A})$.*

Proof. Since $\mathcal{A}$ is a premanual, there exists a manual $\mathcal{A}'$ such that $\mathcal{A} \subseteq \mathcal{A}'$. Let $\omega \in \Omega(\mathcal{A})$ and define

$$\mathcal{B} = \left\{ E \in \mathcal{A}' : \sum_{x \in E} \omega(x) = 1 \right\}$$

Then $\mathcal{B}$ is a quasimanual and $\mathcal{A} \subseteq \mathcal{B} \subseteq \mathcal{A}'$. We now show that $\mathcal{B}$ is a manual. Suppose $A, B, C \in \mathcal{E}(\mathcal{B})$ with $A\,wp_\mathcal{B}\,C$ and $C\,oc_\mathcal{B}\,B$. From a previous observation we have $\omega \in \Omega(\mathcal{B})$ and

$$\omega(A) = \omega(C) = 1 - \omega(B) \tag{3.1}$$

Since $\mathcal{B} \subseteq \mathcal{A}'$, $A\,oc_{\mathcal{A}'}\,B$. Hence, $A \cap B = \emptyset$ and from (3.1) we have

$$\sum_{x \in A \cup B} \omega(x) = 1.$$

Thus, $A \cup B \in \mathcal{B}$ so $A\,oc_\mathcal{B}\,B$. Since, $\mathcal{B}$ is a manual, $\langle \mathcal{A} \rangle \subseteq \mathcal{B}$ and $\omega \in \Omega(\langle \mathcal{A} \rangle)$. It follows that $\Omega(\mathcal{A}) \subseteq \Omega(\langle \mathcal{A} \rangle)$. Since the reverse inclusion is clear, the result follows. □

It is frequently useful in studying the state space $\Omega(\mathcal{A})$ to consider its convexity structure. Notice that if $\omega_1, \omega_2 \in \Omega(\mathcal{A})$ and $\lambda \in \mathbb{R}$ with $0 < \lambda < 1$, then the map $\omega\colon X(\mathcal{A}) \to [0, 1]$ defined by

$$\omega(x) = \lambda \omega_1(x) + (1 - \lambda)\omega_2(x)$$

is also an $\mathcal{A}$-state. We call ω a *convex combination* of ω_1 and ω_2. Moreover, if we define $L^1(X)$ to be the real linear space of all $\phi\colon X \to \mathbb{R}$ such that $\sum_{x \in E} |f(x)| < \infty$ for all $E \in \mathcal{A}$, then $\Omega(\mathcal{A}) \subseteq L^1(X)$ and $\Omega(\mathcal{A})$ is closed under convex combinations. Hence, $\Omega(\mathcal{A})$ is a convex set. We say that an $\mathcal{A}$-state is *pure* or *extremal* if it is not a convex combination of other $\mathcal{A}$-states. Also, an $\mathcal{A}$-state is *dispersion-free* if it assumes only the values 0 and 1. Notice that a dispersion-free state is pure. Of course, the coverse need not hold.

If $X = X(\mathcal{A})$ is a finite set, then $L^1(X)$ is a finite dimensional linear space with $\dim L^1(X) = |X|$. In this case, if $\Omega \neq \emptyset$, we define $\dim \Omega$ to be the dimension of the smallest flat (linear variety) containing Ω. A set of outcomes $X_1 \subseteq X$ is called Ω-*dense* if for every $\omega_1, \omega_2 \in \Omega$ with $\omega_1 \neq \omega_2$ there exists an $x \in X_1$ such that $\omega_1(x) \neq \omega_2(x)$. An Ω-dense set is *minimal* if it does not contain a strictly smaller Ω-dense set. It can be shown that if $|X| < \infty$, then $\dim \Omega = |X_1|$ where X_1 is any minimal Ω-dense subset [Gudder, Kläy, and Rüttimann, 1986].

It is frequently important to know that a quasimanual has a sufficiently rich supply of states. Two ways in which this can be described are the following. We say that $\Delta \subseteq \Omega(\mathcal{A})$ is *unital* if for every $x \in X(\mathcal{A})$ there exists an $\omega \in \Delta$ such that $\omega(x) = 1$. We call $\Delta \subseteq \Omega(\mathcal{A})$ *separating* if $x, y \in X(\mathcal{A})$ with $x \neq y$ implies there is an $\omega \in \Delta$ such that $\omega(x) \neq \omega(y)$. The following result illustrates the usefulness of unitality.

Theorem 3.8. *If a quasimanual $\mathcal{A}$ has a unital set of $\mathcal{A}$-states Δ, then $\mathcal{A}$ is a premanual.*

Proof. Let
$$\mathcal{B} = \left\{ E \subseteq X(\mathcal{A}) : \sum_{x \in E} \omega(x) = 1 \text{ for all } \omega \in \Delta \right\}$$

Then $\mathcal{B}$ is a quasimanual, $\mathcal{A} \subseteq \mathcal{B}$ and $\Delta \subseteq \Omega(\mathcal{B})$. To show that $\mathcal{B}$ is a manual, let $A, B, C \in \mathcal{E}(\mathcal{B})$ and suppose that $A\,wp\,C$ where $C\,oc\,B$. As in Equation 3.1 we have
$$\omega(A) = \omega(C) = 1 - \omega(B)$$
for all $\omega \in \Delta$. Now suppose that $A \cap B \neq \emptyset$ and let $x \in A \cap B$. Since Δ is unital, there is an $\omega \in \Delta$ such that $\omega(x) = 1$. Hence, $\omega(A) = \omega(B) = 1$, but this contradicts (3.1). Thus $A \cap B = \emptyset$ and from (3.1) we have
$$\sum_{x \in A \cup B} \omega(x) = 1$$
for all $\omega \in \Delta$. It follows that $A \cup B \in \mathcal{B}$ so $A\,oc\,B$. Hence, $\mathcal{B}$ is a manual so $\mathcal{A}$ is a premanual. $\square$

Example 3.5. The simplest example of a premanual that is not a manual is the *hook* $\mathcal{K} = \{\{a, b\}, \{b, c\}, \{c, d\}\}$. Now $\mathcal{K}$ is not a manual since $\{d\}\,wp\,\{b\}$ where $\{b\}\,oc\,\{a\}$, yet $\{d\}\,o\!\!\!/\,\{a\}$. (See Figure 3.1a). Notice that
$$\langle \mathcal{K} \rangle = \{\{a, b\}, \{b, c\}, \{c, d\}, \{d, a\}\}$$
as shown in Figure 3.1b. In these figures, straight lines represent operations. In this case $\Pi(\langle \mathcal{K} \rangle)$ is the Boolean algebra with four elements $p(\emptyset), p(\{a\}), p(\{b\}), p(1)$. Since any single element of $X(\mathcal{K})$ is minimal Ω-dense, $\dim \Omega(\mathcal{K}) = \dim \Omega(\langle \mathcal{K} \rangle) = 1$. Notice that $\Omega(\mathcal{K})$ contains a unital separating set of dispersion-free states. $\square$

Example 3.6. The simplest example of a manual that is not coherent is the *Wright triangle*
$$\mathcal{W} = \{\{a, b, c\}, \{c, d, e\}, \{e, f, a\}\}$$

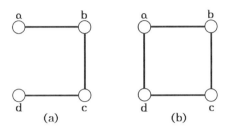

Figure 3.1. The Hook.

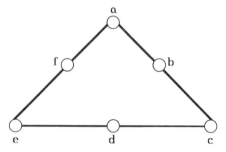

Figure 3.2. Wright Triangle.

shown in Figure 3.2. To see that $\mathcal{W}$ is not coherent, note that $\{e\} \subseteq \{a,c\}^{\perp}$, yet $\{e\} \not\perp \{a,c\}$. The operational logic $\Pi(\mathcal{W})$ consists of fourteen elements $p(\emptyset)$, $p(1)$, $p(\{a\})$, $p(\{b\})$, $p(\{c\})$, $p(\{d\})$, $p(\{e\})$, $p(\{f\})$, $p(\{a,f\})$, $p(\{a,b\})$, $p(\{b,c\})$, $p(\{a,c\})$, $p(\{e,c\})$, $p(\{a,e\})$. This can be thought of as three eight-element Boolean algebras "pasted together" in a loop. By Theorem 3.5, $\Pi(\mathcal{W})$ is not an orthomodular poset. Since $\{a,b,f\}$ is a minimal Ω-dense set, $\dim \Omega(\mathcal{W}) = 3$. Notice that there exists a unital, separating set of dispersion-free states. Also, the state that assigns value $\frac{1}{2}$ to the outcomes a,c,e is pure but is not dispersion-free. □

In classical probability theory, classical mechanics and statistical mechanics, the basic structure is a measurable space (X, Σ). A *countable Σ-partition* of X is a sequence of nonempty sets $A_i \in \Sigma$ such that $A_i \cap A_j = \emptyset$, $i \neq j$, and $\cup A_i = X$. We define a *Borel manual* $\mathcal{B} = \mathcal{B}(X, \Sigma)$ to be the set of all countable Σ-partitions of X. We regard $\mathcal{B}$-operations as "coarsenings" of a (possibly) idealized physical operation with outcome set X. The manual $\mathcal{B}$ is coherent and $\Pi(\mathcal{B})$ is canonically isomorphic to the σ-complete Boolean algebra Σ. (More details of this are given in the next section.) The $\mathcal{B}$-states are precisely the countably additive probability measures on (X, Σ). Moreover, the $\mathcal{E}(\mathcal{B})$-states that are not necessary regular are the finitely additive probability measures on (X, Σ).

In traditional quantum mechanics, the basic structure is a Hilbert space $\mathcal{H}$. Corresponding to $\mathcal{H}$ we define the associated *frame manual* $\mathcal{F} = \mathcal{F}(\mathcal{H})$ to be the set of all orthonormal bases (frames) for $\mathcal{H}$. Let us show that $\mathcal{F}$ is indeed a manual. If $A, B \in \mathcal{E}(\mathcal{F})$, then it is clear that $A \, op \, B$ if and only if $P_A = P_B$ where P_A is the orthogonal projection onto the closed subspace generated by A (and similarly for P_B). It follows that $p(A) = p(B)$ if and only if $P_A = P_B$. Hence, if $B \in p(A)'$, then $P_B = P_A^\perp = I - P_A$ so $B \, oc \, A$. It is also easy to show that $\mathcal{F}$ is coherent. In this way $\Pi(\mathcal{F})$ is canonically isomorphic to the complete orthomodular lattice $L(\mathcal{H})$ of projection operators on $\mathcal{H}$. Due to Gleason's theorem, if $\dim \mathcal{H} \geq 3$, the $\mathcal{F}$-states are in one-to-one correspondence with the density operators on $\mathcal{H}$. It follows from this and the previous paragraph that operational statistics generalizes both classical and Hilbert space probability theory. (The same is true of quantum logic.)

3.4 Hidden Variables

Let $\mathcal{A}$ be a quasimanual and let ω be an $\mathcal{A}$-state. Then ω corresponds to a stochastic model of a physical system. If ω is dispersion-free, then the model is in fact deterministic. This is because for any operation $E \in \mathcal{A}$ there exists a unique outcome $x \in E$ such that $\omega(x) = 1$ and $\omega(y) = 0$ for all $y \in E \setminus \{x\}$. Thus, if the operation E is executed, then the outcome x will result with certainty. In this way the outcome of any operation can be predicted with certainty. If ω is not dispersion-free, then it might be possible to refine our model by replacing ω by a state with less dispersion. That is, we might be able to gain more information about the system in order to provide probabilities close to 1 for certain outcomes.

As a simple example, suppose we flip a fair coin. Without further information, we can only say that a head will result with probability $\frac{1}{2}$. However, if we had more accurate information about the nature of the flip, we could provide probabilities closer to 1 or 0. If we had information concerning the initial position, momentum, angular momentum, air resistence, etc., we could with almost certainty predict the result of this experiment. We would then obtain a refined model in which our original state is replaced by one which is dispersion-free (or almost so). This additional information (initial position, momentum, etc.) could be thought of as a space of hidden parameters, a knowledge of which enables us to determine the outcome of the experiment with certainty.

If $\mathcal{A}$ has a rich enough supply of dispersion-free $\mathcal{A}$-states, then deterministic refinements are, in principle, possible. For example, in classical probability theory such a supply is always available. In this case we have a Borel manual $\mathcal{B}(X, \Sigma)$ and we can always construct a probability measure concentrated at any outcome in X. However, there are quasimanuals far more general than Borel manuals for which this is possible. The present section is devoted to these.

As usual, a *partition* of a nonempty set S is a collection of mutually disjoint nonempty subsets $P_\alpha \subseteq S$ such that $\cup P_\alpha = S$. A *partition quasimanual* $\mathcal{P} = \mathcal{P}(S)$ is a nonempty set of partitions for some set S. Notice that $\mathcal{P}$ is a premanual since the set of all partitions of S is a manual. If $A, B \in \mathcal{E}(\mathcal{P})$, then it is clear that $A\, op\, B$ if and only if $\cup A = \cup B$ where $\cup A$ means $\cup P_\alpha, P_\alpha \in A$. Thus, $p(A) = p(B)$ if and only if $\cup A = \cup B$. Moreover, it is easy to show that $B \in p(A)'$ if and only if $\cup A = (\cup B)^c$ so

$$p(A)' = \{B \in \mathcal{E}(\mathcal{P}) : \cup B = (\cup A)^c\}.$$

Denoting the power set of S by $P(S)$, the map $J : \Pi(\mathcal{P}) \to P(S)$ defined by $Jp(A) = \cup A$ becomes an injection satisfying $J[p(A)'] = [Jp(A)]^c$ and $A \perp B$ implies $Jp(A) \cap Jp(B) = \emptyset$. Denote the range of J by $R(J)$.

A collection of subsets Λ of a nonempty set S is called an *additive class* if

(i) $S \in \Lambda$
(ii) $A \in \Lambda$ implies $A^c \in \Lambda$
(iii) $A, B \in \Lambda$ with $A \cap B = \emptyset$ implies $A \cup B \in \Lambda$.

We call Λ a *σ-additive class* if Λ satisfies (i), (ii) and

(iii') $A_i \in \Lambda$, $i = 1, 2, \ldots$, with $A_i \cap A_j = \emptyset$, $i \neq j$, implies $\cup A_i \in \Lambda$.

Notice that a σ-additive class is a generalization of a σ-algebra.

We have seen that if $\mathcal{P} = \mathcal{P}(S)$ is a manual then $\Pi(\mathcal{P})$ can be represented by sets via the map J. Moreover, $Jp(1) = S$, $J[p(A)'] = [Jp(A)]^c$, and

HIDDEN VARIABLES 91

$p(A) \perp p(B)$ implies $Jp(A) \cap Jp(B) = \emptyset$. Also, if $p(A) \perp p(B)$ then

$$J[p(A) + p(B)] = Jp(A \cup B)$$
$$= \cup(A \cup B)$$
$$= Jp(A) \cup Jp(B).$$

Thus, $J: \Pi(\mathcal{P}) \to R(J)$ preserves the AOA structure of $\Pi(\mathcal{P})$. In fact, J would be an isomorphism if $R(J)$ formed an AOA. Now in order for $R(J)$ to form an AOA, $R(J)$ must be closed under disjoint unions. Since $R(J)$ satisfies (i), (ii) this is equivalent to $R(J)$ being an additive class. This indicates that in order to completely preserve the AOA structure of $\Pi(\mathcal{P})$ in its representation space $P(S)$ we should consider partitions whose sets belong to an additive class.

Let Λ be an additive class of subsets of S. Then a Λ-*partition* of S is a finite partition of S whose elements are in Λ. We call the set $\mathcal{P}(S, L)$ of all Λ-partitions of S an *additive partition manual* (the next theorem shows that $\mathcal{P}(S, \Lambda)$ is indeed a manual). Similarly, if Λ is a σ-additive class, a *countable Λ-partition* is a countable partition of S whose elements are in Λ. Moreover, the set $\mathcal{P}(S, \Lambda)$ of all countable Λ-partitions of S is a σ-*additive partition manual*.

Theorem 3.9. *Let Λ be an additive class on S.*

(i) $\mathcal{P}(S, \Lambda)$ *is a coherent manual.*

(ii) $\Lambda = (\Lambda, \subseteq, ^c, \emptyset, S)$ *is an orthomodular poset and* $J: \Pi[\mathcal{P}(S, \Lambda)] \to \Lambda$ *is an isomorphism.*

Proof. (i) Let $A, B \in \mathcal{E}(\mathcal{P})$ and suppose $B \in p(A)'$. Then $\cup B = (\cup A)^c$. It follows that $A \cup B \in \mathcal{P}$. Hence, $B \, oc \, A$ and $\mathcal{P}(S, \Lambda)$ is a manual. To show that $\mathcal{P}(S, \Lambda)$ is coherent, suppose $A, B \in \mathcal{E}(\mathcal{P})$ with $A \subseteq B^\perp$. It follows that $(\cup A) \cap (\cup B) = \emptyset$. Since $\cup A, \cup B \in \Lambda$, we have $C = (\cup A) \cup (\cup B) \in \Lambda$. Then $C^c \in \Lambda$ and the subsets in A and B together with the set C gives a Λ-partition of S. Hence, $A \cup B \in \mathcal{E}(\mathcal{P})$ and $A \perp B$.

(ii) It is clear that Λ is an orthocomplemented poset. To show orthomodularity, suppose $A, B \in \Lambda$ with $A \subseteq B$. Then $A \cap B^c = \emptyset$ so $A \cup B^c \in \Lambda$. Hence, $A^c \cap B \in \Lambda$ and moreover $A \cap (A^c \cap B) = \emptyset$. Hence, $A \cup (A^c \cap B) \in \Lambda$ and by distributivity $B = A \cup (A^c \cap B)$. To show that J is an isomorphism, we have already shown that J is a bijective homomorphism, so it suffices to show that J^{-1} preserves order. Suppose $A, B \in \Lambda$ with $A \subseteq B$. Then $\{A, B \cap A^c, B^c\} \in \mathcal{P}$ so $\{A\}, \{A, B \cap A^c\} \in \mathcal{E}(\mathcal{P})$. Now $J^{-1}(A) = p(\{A\})$, $J^{-1}(B) = p(\{A, B \cap A^c\})$ and $p(\{A\}) \leq p(\{A, B \cap A^c\})$. □

Corollary 3.10. *Let Λ be a σ-additive class on S.*

(i) $\mathcal{P}(S, \Lambda)$ *is a coherent manual.*

(ii) $\Lambda = (\Lambda, \subseteq, ^c, \emptyset, S)$ *is a quantum logic and $J: \Pi[\mathcal{P}(S, \Lambda)] \to \Lambda$ is a σ-isomorphism.*

A set of states Δ on an orthomodular poset Λ is *order determining* if for $p, q \in \Lambda$, $\alpha(p) \leq \alpha(q)$ for all $\alpha \in \Delta$ implies $p \leq q$. A quasimanual $\mathcal{A}$ *admits hidden variables* if $\mathcal{A}$ has a unital, separating set of dispersion-free $\mathcal{A}$-states. An orthomodular poset L *admits hidden variables* if L has an order-determining set of dispersion-free states.

Theorem 3.11. (i) *A quasimanual $\mathcal{A}$ admits hidden varibales if and only if $\mathcal{A}$ is isomorphic to a partition quasimanual.*
(ii) *An orthomodular poset L admits hidden variables if and only if L is isomorphic to an additive class.*

Proof. (i) Suppose $\mathcal{A}$ is isomorphic to a partition quasimanual $\mathcal{P}(S, \Lambda)$. For $s \in S$, define the $\mathcal{P}$-state α_s by $\alpha_s(A) = 1$ if $s \in A$ and $\alpha_s(A) = 0$ if $s \notin A$. Let $\Delta = \{\alpha_s : s \in S\}$. Clearly Δ is a set of dispersion-free $\mathcal{P}$-states. For $A \in X(\mathcal{P})$, if $s \in A$, then $\alpha_s(A) = 1$ so Δ is unital. For $A, B \in X(\mathcal{P})$, if $\alpha_s(A) = \alpha_s(B)$ for every $s \in S$, then $A = B$ so Δ is separating. If the isomorphism is given by $\phi: X(\mathcal{A}) \to X(\mathcal{P})$ then $\phi^+(\Delta)$ is a unital, separating set of dispersion-free $\mathcal{A}$-states. Conversely, suppose $\mathcal{A}$ admits a unital, separating set S of dispersion-free $\mathcal{A}$-states. Define $\phi: X(\mathcal{A}) \to P(S)$ by $\phi(x) = \{s \in S : s(x) = 1\}$ and let $X(\mathcal{P}) = \{\phi(x) : x \in X(\mathcal{A})\}$. Then $\phi(x) \neq \emptyset$ for all $x \in X(\mathcal{A})$ since S is unital. To show that ϕ is injective, suppose $x \neq y \in X(\mathcal{A})$. Since S is separating, there is an $s \in S$ such that $s(x) = 1$ and $s(y) = 0$. Hence, $\phi(x) \neq \phi(y)$. It follows that $\phi: X(\mathcal{A}) \to X(\mathcal{P})$ is bijective. Let $x, y \in E \in \mathcal{A}$ with $x \neq y$, and suppose that $\phi(x) \cap \phi(y) \neq \emptyset$. Then there is an $s \in S$ such that $s(x) = s(y) = 1$. This contradicts the fact that $\Sigma_{z \in E} s(z) = 1$. Hence, $\phi(x) \cap \phi(y) = \emptyset$. Let $\mathcal{P} = \{\hat{\phi}(E) : E \in \mathcal{A}\}$. Let $P = \hat{\phi}(E)$, $E \in \mathcal{A}$, and suppose $s \in S$. If $s(z) = 0$ for all $z \in E$, then we contradict the fact that $\Sigma_{z \in E} s(z) = 1$. Hence, there is an $x \in E$ such that $s(x) = 1$.

Thus, $s \in \phi(x)$ and $\cup P = S$. It follows that P is a partition of S. Hence, $\mathcal{P} = \mathcal{P}(S)$ is a partition quasimanual that is isomorphic to $\mathcal{A}$.

(ii) The proof is similar to that of (i). If Λ is an additive class on S, define the set of states $\Delta = \{\alpha_s : s \in S\}$ as in (i). Suppose $A, B \in \Lambda$ with $\alpha(A) \leq \alpha(B)$ for all $\alpha \in \Delta$. If $s \in A$, then $\alpha_s(A) = 1$, so $\alpha_s(B) = 1$. Hence, $s \in B$ and $A \subseteq B$. Thus, Δ is an order determining set of dispersion-free states on Λ. Sufficiency now easily follows. For necessity, suppose L has an order determining set S of dispersion-free states. As in (i) define $\phi: L \to P(S)$ by $\phi(p) = \{s \in S : s(p) = 1\}$, and let $\Lambda = \{\phi(p) : p \in \Lambda\}$. Clearly, $\phi(1) = S$ so $S \in \Lambda$. If $A = \phi(p) \in \Lambda$, then

$$\phi(p') = \phi(p)^c = A^c$$

so $A^c \in \Lambda$. Let $A, B \in \Lambda$ with $A \cap B = \emptyset$ and suppose $\phi(p) = A, \phi(q) = B$. If $s \in S$ with $s(p) = 1$, then $s \in A$. Hence, $s \notin B$ so $s(q) = 0$. Then $s(q') = 1$ and it follows that $s(p) \leq s(q')$ for all $s \in S$. Since S is order determining, $p \perp q$. It then follows that $A \cup B = \phi(p \vee q) \in \Lambda$, so Λ is an additive class. It is now straightforward to show that $\phi: L \to \Lambda$ is an isomorphism. □

Corollary 3.12. *A quantum logic L admits hidden variables if and only if L is σ-isomorphic to a σ-additive class.*

We thus see that partition quasimanuals are precisely the quasimanuals that admit hidden variables and σ-additive classes are precisely the quantum logics that admit hidden variables. Moreover, if L is a quantum logic that admits hidden variables then L is the operational logic of a σ-additive partition manual $\mathcal{P}(S, \Lambda)$ where Λ is a σ-additive class on S.

3.5 Products of Quasimanuals

In order to represent a combination or composite of two physical systems a suitable definition of a product of quasimanuals is needed. Now there are various ways of defining products and this section considers three: the cartesian, operational, and tensor products. We shall show that the three products are used to represent bilateral, unilateral, and no influence (or interference), respectively, between the two quasimanuals involved.

In the sequel, $\mathcal{A}$ and $\mathcal{B}$ will denote quasimanuals with outcome sets X and Y, respectively. If $x \in X$, $y \in Y$ we write the ordered pair (x, y) simply as xy. Similarly, if $A \subseteq X$, $B \subseteq Y$, we write the cartesian product $A \times B$ as AB. For simplicity xB will mean $\{x\}B$. We define the *cartesian product* of $\mathcal{A}$ and $\mathcal{B}$ to be the quasimanual

$$\mathcal{A} \times \mathcal{B} = \{EF \colon E \in \mathcal{A}, F \in \mathcal{B}\}.$$

An execution of an operation EF in $\mathcal{A} \times \mathcal{B}$ would proceed as follows: operation E in $\mathcal{A}$ is executed to secure an outcome, say $x \in E$, operation F in $\mathcal{B}$ is executed to secure an outcome, say $y \in F$, and the outcome of this execution of the compound operation EF is recorded as xy.

It is easy to show that the cartesian product of two manuals is again a manual; similarly for premanuals, coherent manuals, or orthocoherent manuals. If $\mathcal{A}$ and $\mathcal{B}$ are manuals, it is not possible in general to determine the structure of the operational logic $\Pi(\mathcal{A} \times \mathcal{B})$ from the structure of $\Pi(\mathcal{A})$ and $\Pi(\mathcal{B})$. Again, vital information is lost in the passage from manuals to their operational logics.

If $E \in \mathcal{A}$ and $F_x \in \mathcal{B}$ for every $x \in E$, we call $\cup_{x \in E} x F_x$ a *two-stage operation*. The quasimanual consisting of all such two-stage operations is called the *forward operational product* of $\mathcal{A}$ and $\mathcal{B}$ and is denoted by $\overrightarrow{\mathcal{AB}}$. An execution of a two-stage operation $\cup_{x \in E} x F_x$ in $\overrightarrow{\mathcal{AB}}$ would proceed as follows: operation E in $\mathcal{A}$ is executed to secure an outcome, say $x \in E$, then operation F_x in $\mathcal{B}$ is executed to secure an outcome, say $y \in F_x$ and the outcome of this execution of the two-stage operation is recorded as xy.

If $S \subseteq YX$, we define $S^{-1} = \{xy \colon yx \in S\}$. The *backward operational product* of $\mathcal{A}$ and $\mathcal{B}$ is defined to be the quasimanual

$$\overleftarrow{\mathcal{AB}} = \{S^{-1} \colon S \in \overrightarrow{\mathcal{BA}}\}.$$

To execute an operation S^{-1} in $\overleftarrow{\mathcal{AB}}$ we would execute the two-stage operation S in $\overrightarrow{\mathcal{BA}}$ to secure an outcome, say yx, but we would record this as the $\overleftarrow{\mathcal{AB}}$ outcome xy. It is not hard to show that the forward or backward operational product of two manuals is again a manual, and similarly for premanuals, coherent manuals or orthocoherent manuals. If $\mathcal{A}$ and $\mathcal{B}$ are manuals, the structure of $\Pi(\overrightarrow{\mathcal{AB}})$ depends on the detailed structure of $\mathcal{A}$ but only the structure of $\Pi(\mathcal{B})$.

Notice that all three quasimanuals $\mathcal{A} \times \mathcal{B}$, $\overrightarrow{\mathcal{AB}}$, and $\overleftarrow{\mathcal{AB}}$ have the outcome set XY. Moreover, it is easy to show that $\mathcal{A} \times \mathcal{B} = \overrightarrow{\mathcal{AB}} \cap \overleftarrow{\mathcal{AB}}$. We now define the quasimanual $\overline{\mathcal{AB}} = \overrightarrow{\mathcal{AB}} \cup \overleftarrow{\mathcal{AB}}$. Even if $\mathcal{A}$ and $\mathcal{B}$ are manuals, $\overline{\mathcal{AB}}$ is usually not a manual although it is frequently a premanual. For example, if $\Omega(\mathcal{A})$ and $\Omega(\mathcal{B})$ are unital, it follows from Theorem 3.7 that $\overline{\mathcal{AB}}$ is a premanual.

Two manuals, $\mathcal{A}, \mathcal{B}$ are called *tensorial* if $\overline{\mathcal{AB}}$ is a premanual, in which case the *tensor product* of $\mathcal{A}$ and $\mathcal{B}$ is defined to be the manual $\mathcal{A} \otimes \mathcal{B} = \langle \overline{\mathcal{AB}} \rangle$. Notice that the outcome set for $\mathcal{A} \otimes \mathcal{B}$ is again XY.

If $\alpha \in \Omega(\mathcal{A})$, $\beta \in \Omega(\mathcal{B})$, we define the $\mathcal{A} \times \mathcal{B}$-*product state* $\alpha\beta \in \Omega(\mathcal{A} \times \mathcal{B})$ by $\alpha\beta(xy) = \alpha(x)\beta(y)$. Notice that $\mathcal{A} \times \mathcal{B}$-product states exhibit a statistical independence of operations in $\mathcal{A}$ from the operations in $\mathcal{B}$. We denote the set of $\mathcal{A} \times \mathcal{B}$-product states by $\Omega(\mathcal{A})\Omega(\mathcal{B})$. Let $\mathcal{A}, \mathcal{B}$ be tensorial manuals. Since $\Omega(\langle \overline{\mathcal{AB}} \rangle) = \Omega(\overline{\mathcal{AB}})$, it is easy to show that the following containments hold

$$\Omega(\mathcal{A})\Omega(\mathcal{B}) \subseteq \Omega(\mathcal{A} \otimes \mathcal{B}) = \Omega(\overline{\mathcal{AB}})$$
$$= \Omega(\overrightarrow{\mathcal{AB}}) \cap \Omega(\overleftarrow{\mathcal{AB}})$$
$$\subseteq \Omega(\overrightarrow{\mathcal{AB}}) \cup \Omega(\overleftarrow{\mathcal{AB}}) \subseteq \Omega(\mathcal{A} \times \mathcal{B}).$$

There are examples of orthocoherent tensorial manuals whose tensor product is not orthocoherent [Randall and Foulis, 1981]. Using such examples, it can be shown that there is no hope of obtaining a general "tensor product" $P \otimes Q$ of orthomodular posets with the following desirable properties [Randall and Foulis, 1981].

(i) $P \otimes Q$ is an orthomodular poset.

(ii) $\otimes$ is a map from $P \times Q$ to $P \otimes Q$ such that $p_1 \otimes q_1 \perp p_2 \otimes q_2$ if either $p_1 \perp p_2$ or $q_1 \perp q_2$.

(iii) If α, β are states for P, Q, respectively, then there exists a state γ for $P \otimes Q$ such that

$$\gamma(p \otimes q) = \alpha(p)\beta(q) \text{ for all } p \in P, q \in Q.$$

Nevertheless, counterparts of these three properties hold for the tensor product of tensorial manuals. Thus, we must work at the operational level of manuals if we are to obtain a reasonable definition of a tensor product.

Although $\mathcal{A} \times \mathcal{B}$ is set-theoretically contained in $\overrightarrow{\mathcal{AB}}$, the instructions for performing an operation in $\mathcal{A} \times \mathcal{B}$ are different from those in $\overrightarrow{\mathcal{AB}}$ since for

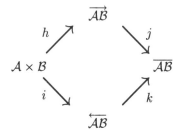

Figure 3.3. Natural Morphisms.

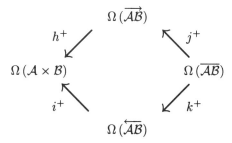

Figure 3.4. Conjugate Maps.

the latter a definite temporal order is explicit. However, there is the natural morphism h from $\mathcal{A} \times \mathcal{B}$ to $\overrightarrow{\mathcal{A}\mathcal{B}}$ given by the identity map $h(xy) = xy$ for all $x \in X$, $y \in Y$. Similarly, we have the natural morphisms i, j, and k shown in Figure 3.3. We also have the corresponding injective conjugate maps shown in Figure 3.4. If $\mathcal{A}$ and $\mathcal{B}$ are tensorial manuals, we can replace $\overrightarrow{\mathcal{A}\mathcal{B}}$ by $\mathcal{A} \otimes \mathcal{B}$ in Figure 3.3 and 3.4.

Let $\omega \in \Omega(\mathcal{A} \times \mathcal{B})$ be fixed. For any $E \in \mathcal{A}$, we define $_E\omega \in \Omega(\mathcal{B})$ by $_E\omega(y) = \omega(Ey)$ for all $y \in Y$. It is clear that $_E\omega(y)$ gives the probability that y is secured if an operation $EF \in \mathcal{A} \times \mathcal{B}$ is executed with $y \in F$. We call $_E\omega$ the state ω *preconditioned* by the operation E. Similarly, we define $\omega_F \in \Omega(\mathcal{A})$, the state ω *postconditioned* by the operation $F \in \mathcal{B}$, by $\omega_F(x) = \omega(xF)$ for all $x \in X$. If $E, G \in \mathcal{A}$ satisfy $_E\omega = {}_G\omega$, we say that E and G have the same *influence* on $\mathcal{B}$ relative to ω. If every pair of $\mathcal{A}$-operations have the same influence on $\mathcal{B}$ relative to ω, we say that $\mathcal{A}$ has *no influence* on $\mathcal{B}$ relative to ω.

Lemma 3.13. *Let $\omega \in \Omega(\mathcal{A} \times \mathcal{B})$.*

(a) *$\omega \in \Omega(\overleftarrow{\mathcal{AB}})$ if and only if $_E\omega = {_G\omega}$ for every $E, G \in \mathcal{A}$.*

(b) *$\omega \in \Omega(\overrightarrow{\mathcal{AB}})$ if and only if $\omega_F = \omega_H$ for every $F, H \in \mathcal{B}$.*

Proof. (a) Suppose $\omega \in \Omega(\overleftarrow{\mathcal{AB}})$ and $E, G \in \mathcal{A}$. If $y \in F \in \mathcal{B}$, form the operations $F_1, F_2 \in \overleftarrow{\mathcal{AB}}$ defined by

$$F_1 = \cup_{y' \in F} Ey', \quad F_2 = Gy \cup (\cup_{y' \in F \setminus \{y\}} Ey').$$

Since, $\omega(F_1) = \omega(F_2) = 1$, we have

$$\sum_{y' \in F} \omega(Ey') = \omega(Gy) + \sum_{y' \in F \setminus \{y\}} \omega(Ey'). \quad (3.2)$$

It follows from (3.2) that

$$_E\omega(y) = \omega(Ey) = \omega(Gy) = {_G\omega}(y).$$

Conversely, suppose $_E\omega = {_G\omega}$ for every $E, G \in \mathcal{A}$. An operation in $\overleftarrow{\mathcal{AB}}$ has the form $\cup_{y \in F} H_y y$, where $F \in \mathcal{B}$ and $H_y \in \mathcal{A}$ for every $y \in F$. If $E_0 \in \mathcal{A}$, then $\omega(H_y y) = \omega(E_0 y)$ for every $y \in F$. Hence,

$$\omega(\cup_{y \in F} H_y y) = \sum_{y \in F} \omega(E_y y)$$

$$= \sum_{y \in F} \omega(E_0 y)$$

$$= \omega(E_0 F) = 1$$

so $\omega \in \Omega(\overleftarrow{\mathcal{AB}})$. The proof of (b) is similar. □

As a consequence of Lemma 3.13, if $\mathcal{A}$ and $\mathcal{B}$ are tensorial manuals, then $\omega \in \Omega(\mathcal{A} \otimes \mathcal{B})$ if and only if neither $\mathcal{A}$ or $\mathcal{B}$ has any influence on the other relative to ω. Moreover, in view of Lemma 3.13, the following are well-defined. For $\omega \in \Omega(\overrightarrow{\mathcal{AB}})$ we define $\omega_1 \in \Omega(\mathcal{A})$ by $\omega_1(x) = \omega(xF)$ for every $F \in \mathcal{B}$ and $x \in X$. For $\omega \in \Omega(\overleftarrow{\mathcal{AB}})$ we define $\omega_2(y) = \omega(Ey)$ for every $E \in \mathcal{A}$ and $y \in Y$. In physical terminology, ω_1 and ω_2 are called *reduced* or *component* states for $\mathcal{A}$ and $\mathcal{B}$, while in probability terminology they are called *marginal* states. Notice, if $A \in \mathcal{E}(\mathcal{A}), B \in \mathcal{E}(\mathcal{B})$, then we have $\omega_1(A) = \omega(AF)$ for all $F \in \mathcal{B}$ and $\omega_2(B) = \omega(EB)$ for all $E \in \mathcal{A}$.

Suppose that a known operation $EF \in \mathcal{A} \times \mathcal{B}$ has been executed and it is known that the event $A \subseteq E$ occurred. Then evidently the probability that $y \in F$ is secured is $\omega(Ay)/\omega(AF)$. However, suppose that some unknown operation $EF \in \mathcal{A} \times \mathcal{B}$ has been executed and that it is known that the event $A \subseteq E$ occurred and that $y \in F$. In this case we may not be able to give a probability that y is secured since we cannot account for the possible influence of the unknown operation F. However, if there is no influence of such operations on $\mathcal{A}$, then the probability that y is secured is clearly $\omega(Ay)/\omega_1(A)$. We are then led to the following definitions.

Let $\omega \in \Omega(\overrightarrow{\mathcal{AB}})$ and let $A \in \mathcal{E}(\mathcal{A})$ with $\omega_1(A) \neq 0$. Then we define $_A\omega \in \Omega(\mathcal{B})$ by $_A\omega(y) = \omega(Ay)/\omega_1(A)$ for all $y \in Y$. We refer to $_A\omega$ as ω *preconditioned* by A. Similarly, let $\omega \in \Omega(\overleftarrow{\mathcal{AB}})$ and let $B \in \mathcal{E}(\mathcal{B})$ with $\omega_2(B) \neq 0$. Then we define $\omega_B \in \Omega(\mathcal{A})$ by $\omega_B(x) = \omega(xB)/\omega_2(B)$ for all $x \in X$, and ω_B is referred to as ω *postconditioned* by B. Notice that

$$_A\omega(B) = \frac{\omega(AB)}{\omega(AF)}$$

$$\omega_B(A) = \frac{\omega(AB)}{\omega(EB)}$$

$E \in \mathcal{A}$, $F \in \mathcal{B}$, when the denominators are nonvanishing. We use the notation $_x\omega$ and ω_y for $_{\{x\}}\omega$ and $\omega_{\{y\}}$, respectively. The next results give operational versions of the classical law of total probability and Bayes' rule, respectively.

Lemma 3.14. *Let* $\omega \in \Omega(\overrightarrow{\mathcal{AB}})$, $E \in \mathcal{A}$, $F \in \mathcal{B}$. *Then*
(i) $\omega_2 = \sum_{x \in E} \omega_1(x) {}_x\omega$ *and*
(ii) $\omega_1 = \sum_{y \in F} \omega_2(y) \omega_y$
(where terms for which $_x\omega$ or ω_y are undefined are omitted).

Proof. (i) If $y \in Y$, then

$$\sum_{x \in E} \omega_1(x) {}_x\omega(y) = \sum_{x \in E} \omega_1(x) \frac{\omega(xy)}{\omega_1(x)}$$

$$= \sum_{x \in E} \omega(xy) = \omega(Ey) = \omega_2(y).$$

PRODUCTS OF QUASIMANUALS   99

The proof of (ii) is similar. □

Lemma 3.15. *Let $\omega \in \Omega(\overrightarrow{AB})$, $A \in \mathcal{E}(\mathcal{A})$, $B \in \mathcal{E}(\mathcal{B})$, $\omega_1(A) \neq 0$, and $\omega_2(B) \neq 0$. Then*
$$_A\omega(B) = \omega_2(B)\frac{\omega_B(A)}{\omega_1(A)}.$$

Proof. The following equalities are easily verified.
$$_A\omega(B) = \sum_{y \in B} \frac{\omega(Ay)}{\omega_1(A)} = \frac{\omega(AB)}{\omega_1(A)}$$
$$= \frac{1}{\omega_1(A)} \sum_{x \in A} \omega(xB)$$
$$= \frac{\omega_2(B)}{\omega_1(A)} \sum_{x \in A} \frac{\omega(xB)}{\omega_2(B)}$$
$$= \omega_2(B)\frac{\omega_B(A)}{\omega_1(A)}. \quad \square$$

Combining the results of Lemmas 3.14(ii) and 3.15 gives
$$_A\omega(B) = \frac{\omega_2(B)\omega_B(A)}{\sum_{y \in F} \omega_2(y)\omega_y(A)} \tag{3.3}$$
which is the analog of the classical Bayes' rule.

Let $\omega \in \Omega(\overrightarrow{AB})$ and let $A, C \in \mathcal{E}(\mathcal{A})$ with $\omega_1(A) \neq 0$, and $\omega_1(C) \neq 0$. In general, we have $_A\omega \neq {}_C\omega$ even if $A\,op\,C$. If $p(A) = p(C)$ and $_A\omega \neq {}_C\omega$, we say that ω exhibits *interference* between A and C. A similar definition applies for $\omega \in \Omega(\overleftarrow{AB})$. The next result shows that $\omega \in \Omega(\overleftrightarrow{AB})$ if and only if ω exhibits no interference whatsoever.

Theorem 3.16. *Let $\omega \in \Omega(\overrightarrow{AB})$. Then $\omega \in \Omega(\overleftrightarrow{AB})$ if and only if $A, C \in \mathcal{E}(\mathcal{A})$ with $p(A) = p(C)$ and $\omega_1(A) \neq 0$ imply $_A\omega = {}_C\omega$.*

Proof. Suppose $\omega \in \Omega(\overleftrightarrow{AB})$, $A, C \in \mathcal{E}(\mathcal{A})$, $p(A) = p(C)$ and $\omega_1(A) \neq 0$. Since $\alpha(A) = \alpha(C)$ for every $\alpha \in \Omega(\mathcal{A})$, applying Lemma 3.15 gives
$$_A\omega(y) = \frac{\omega_2(y)\omega_y(A)}{\omega_1(A)}$$
$$= \frac{\omega_2(y)\omega_y(C)}{\omega_1(C)}$$
$$= {}_C\omega(y).$$

Conversely, if $E, G \in \mathcal{A}$, then $p(E) = p(G)$ so by hypothesis $_E\omega = {}_G\omega$. It follows from Lemma 3.13 that $\omega \in \Omega(\overleftrightarrow{AB})$ so $\omega \in \Omega(\overline{AB})$. □

Because of the possible interference between operationally perspective events, it is not possible in general to condition a state by an operational proposition. However, applying Theorem 3.16, if $\omega \in \Omega(\overline{AB})$, $A \in \mathcal{E}(\mathcal{A})$ and $\omega_1(A) \neq 0$, we can and do define $_{p(A)}\omega = {}_A\omega$. In fact, the usual von Neumann formula given by Equation 2.22 conditions Hilbert space states in this way by propositions. Even if $\omega \in \Omega(\overline{AB})$, there will, in general, exist events $A, C \in \mathcal{E}(\mathcal{A})$ such that $_A\omega \neq {}_C\omega$. In this case, we say that ω exhibits *correlations* between A and C. Thus, a state may not exhibit interference, yet still exhibit correlations. However, a product state $\omega = \alpha\beta$ never exhibits correlations since $_A\omega = \beta$ and $\omega_B = \alpha$ for every $A \in \mathcal{E}(\mathcal{A})$, $B \in \mathcal{E}(\mathcal{B})$, whenever they are defined. Moreover, the converse of this statement also holds. Indeed, let $\omega \in \Omega(\overrightarrow{AB})$ and suppose ω never exhibits correlations between events in $\mathcal{E}(\mathcal{A})$. Fix $A \in \mathcal{E}(\mathcal{A})$ such that $\omega_1(A) \neq 0$. Then for every $xy \in XY$ we have

$$_A\omega(y) = \frac{\omega(Ay)}{\omega_1(A)}$$

and if $\omega_1(x) \neq 0$ we have $_A\omega = {}_x\omega$. Hence,

$$\omega(xy) = \omega_1(x)_x\omega(y)$$
$$= \omega_1(x)_A\omega(y)$$

so ω is the product state $\omega = \omega_1{}_A\omega$.

We now consider the special case of traditional quantum mechanics. Suppose that two quantum systems are represented by Hilbert spaces $\mathcal{H}_1$ and $\mathcal{H}_2$. As mentioned in Chapter 2, the combined system is usually represented by the Hilbert space tensor product $\mathcal{H}_1 \otimes \mathcal{H}_2$. In operational statistics these systems are represented by the frame manuals $\mathcal{F}_1 = \mathcal{F}(\mathcal{H}_1)$, $\mathcal{F}_2 = \mathcal{F}(\mathcal{H}_2)$, and $\mathcal{F} = \mathcal{F}(\mathcal{H}_1 \otimes \mathcal{H}_2)$, respectively. All of these frame manuals are coherent, carry unital, separating sets of states, and $\mathcal{F}_1, \mathcal{F}_2$ are tensorial. Since in operational statistics, the combined system is represented by $\mathcal{F}_1 \otimes \mathcal{F}_2$, it is natural to ask how $\mathcal{F}_1 \otimes \mathcal{F}_2$ and $\mathcal{F}$ are related.

To begin with, if $\dim \mathcal{H}_1, \dim \mathcal{H}_2 \geq 3$, then $\mathcal{F}_1 \otimes \mathcal{F}_2$ is not coherent. Indeed, let $x_i, y_i, z_i \in \mathcal{H}_i$, $i = 1, 2$, be distinct unit vectors satisfying:

$$x_1 \perp y_1, \quad x_1 \perp z_1, \quad y_1 \not\perp z_1, \quad y_2 \perp z_2, \quad x_2 \not\perp z_2.$$

Letting $A = \{(z_1, z_2)\}$, $B = \{(x_1, x_2), (y_1, y_2)\}$ we have $A, B \in \mathcal{E}(\mathcal{F}_1 \otimes \mathcal{F}_2)$ and $A \subseteq B^\perp$. However, it is easy to check that $A \not\perp B$. Thus, since $\mathcal{F}$ is coherent, $\mathcal{F}$ and $\mathcal{F}_1 \otimes \mathcal{F}_2$ are not isomorphic. Since the outcome sets for $\mathcal{F}_1, \mathcal{F}_2$ and $\mathcal{F}$ are the unit spheres S_1, S_2, and S of $\mathcal{H}_1$, $\mathcal{H}_2$, and $\mathcal{H}_1 \otimes \mathcal{H}_2$, respectively; the outcome set of $\mathcal{F}_1 \otimes \mathcal{F}_2$ is $S_1 S_2$. The natural map $\phi: S_1 S_2 \to S$ given by $\phi(xy) = x \otimes y$ defines a positive morphism from $\mathcal{F}_1 \otimes \mathcal{F}_2$ to $\mathcal{F}$. However, ϕ is not injective since $\phi((e^{i\theta}x)(e^{-i\theta}y)) = \phi(xy)$. Thus, it appears that $\mathcal{F}_1 \otimes \mathcal{F}_2$ carries phase information that is lost in $\mathcal{F}$.

The conjugate map

$$\phi^+: \Omega(\mathcal{F}) \to \Omega(\mathcal{F}_1 \otimes \mathcal{F}_2)$$

when restricted to pure states, is injective. Indeed, a pure state ω in $\Omega(\mathcal{F})$ can be represented by a unit vector $t \in S$ and the corresponding state

$$\phi^+(\omega) \in \Omega(\mathcal{F}_1 \otimes \mathcal{F}_2)$$

is evaluated according to the formula

$$[\phi^+(\omega)](xy) = |\langle t, x \otimes y \rangle|^2.$$

If $\omega_1 \in \Omega(\mathcal{F})$ is another pure state in $\Omega(\mathcal{F})$ that is represented by $t_1 \in S$, then

$$\phi^+(\omega) = \phi^+(\omega_1)$$

implies that

$$|\langle t, x \otimes y \rangle| = |\langle t_1, x \otimes y \rangle|$$

for every $xy \in S_1 S_2$. It follows that $t^+ = t_1^+$ and hence t and t_1 differ by a scalar multiple of modulus 1.

From this last result we conclude that, as far as the physical (pure) states are concerned, $\mathcal{F}_1 \otimes \mathcal{F}_2$ contains at least as much information as does $\mathcal{F}(\mathcal{H}_1 \otimes \mathcal{H}_2)$. Whether ϕ^+ is surjective is at present an open question. If it is not it would be interesting to know if the extra states in $\mathcal{F}_1 \otimes \mathcal{F}_2$ are physically realizable.

3.6 Simpson's Paradox

Operational statistics is not only useful as a language for discussing quantum probability, it also has applications in classical statistics. One can have incompatible operations even in classical problems. When this happens, the methods of classical statistics break down and we must apply the more general theory of operational statistics. This section uses this more general theory to resolve a classical problem called Simpson's paradox. The presentation is based on the work of M. Kläy, C. Randall, and D. Foulis [1988].

We present Simpson's paradox by giving a typical simplified example in which it occurs. Suppose that a patient suffering from a severe disease can be treated either by a standard medical procedure or by a new one, but not both. The treatments are administered to a group of patients in two cities and it is recorded whether each patient survives or dies within one year after receiving the treatment. The results are shown in Table 3.1. Adding the numbers for both cities we get the totals shown in Table 3.2.

	First City		Second City	
	Standard	New	Standard	New
Number of Survivals	50	1000	5000	95
Number of Deaths	950	9000	5000	5

Table 3.1 (Totals by City)

	Standard	New
Number of Survivals	5050	1095
Number of Deaths	5950	9005

Table 3.2 (Totals)

Using classical statistics for Table 3.1 we see that a patient from the first city has a probability $50/(950 + 50) = 0.05$ of surviving after the standard treatment and a probability $1000/(9000 + 1000) = 0.10$ of surviving after the new treatment. Thus the new treatment is better for patients in the first city. In the second city, the probability of surviving after the standard

SIMPSON'S PARADOX

treatment is 0.50 while the probability of surviving after the new treatment is 0.95. Again, the new treatment appears to be better. However, using Table 3.2, the probability of survival after the standard treatment is 0.4591 while the probability of survival after the new treatment is 0.1084. We must now reverse our judgement and say that overall the standard treatment is better. This is an illustration of Simpson's paradox.

This paradox has serious consequences. How does a patient or doctor decide which treatment to use? How does a govenment funding agency decide whether to grant money to further develop the new treatment? If the agency only sees Table 3.2 it will make the wrong decision since it is clear that the new treatment is better. This gives an example of how statistics can lie. A party with vested interests can apply the statistics to draw any conclusion it wants.

Of course, the paradox results from the unequal sample sizes from the cities and treatments and from the fact that patients from the first city appear to have a later stage of the disease and thus a lower survival rate. Such problems can be eliminated by taking equal sample sizes but this is sometimes impossible in practice. For example, retroactive studies may be involved or there may be inherent imbalances in studies concerning factors such as race, religion, or sex.

There is a deeper reason for the emergence of Simpson's paradox. This reason is that the two treatments are incompatible in the sense that once a patient is administered one treatment, there is no way to determine the outcome of the other treatment for this same patient. There is no definite outcome corresponding to the performance of the two treatments simultaneously. This is similar to the incompatibility of position and momentum measurements in quantum mechanics. The way to overcome these difficulties is to apply operational statistics.

Since there is a clear sense of direction for the given data, we treat the situation as a two-stage experiment. In the first stage, a patient is selected from either the first or second city. In the second stage, the patient is administered either the standard or new treatment. The outcomes of these experiments are as follows:

c = Patient comes from first city
$\bar{c}$ = Patient comes from second city
s = Patient survives after standard treatment
$\bar{s}$ = Patient dies after standard treatment
n = Patient survives after new treatment

$\bar{n}$ = Patient dies after new treatment.

The outcome set for the first stage is $X = \{c, \bar{c}\}$ and the corresponding manual is $\mathcal{A} = \{X\}$. The outcome set for the second stage is $Y = \{s, \bar{s}, n, \bar{n}\}$. There are two operations for the second stage, namely $S = \{s, \bar{s}\}$ in which the standard treatment is given and $N = \{n, \bar{n}\}$ in which the new treatment is given. Notice that other subsets of Y cannot be considered as operations since there are no experiments whose outcomes would result as such subsets. The manual for the second stage becomes $\mathcal{B} = \{S, N\}$. The two stage experiment is described by the forward operational product $\overrightarrow{\mathcal{AB}}$. The outcome set of $\overrightarrow{\mathcal{AB}}$ is XY and the operations in $\overrightarrow{\mathcal{AB}}$ are $E = XS$, $F = XN$, $G = (cS) \cup (\bar{c}N)$, $H = (cN) \cup (\bar{c}S)$. Thus, $\overrightarrow{\mathcal{AB}} = \{E, F, G, H\}$.

To perform a statistical analysis, we must choose a state (or probability measure) that best describes the situation. The actual probability distribution for the population under study is in general unknown (which is the reason that a sample of the population is taken) and has to be estimated from the data. The most common way to do this is by means of the maximum likelihood estimator. We first proceed in the classical framework and then compare the results to those obtained using operational statistics. Let $\mathcal{M}$ denote the collection of all probability measures on the classical sample space XY. Denote by N_{xy} the number of trials resulting in the outcome $xy \in XY$. Then assuming stochastic independence of the trials, the maximum likelihood estimator is the $\hat{\mu} \in \mathcal{M}$ that maximizes the expression

$$P = \prod_{xy \in XY} \mu(xy)^{N_{xy}}$$

among all possible $\mu \in \mathcal{M}$. Denoting the total number of trials by

$$N = \sum_{xy \in XY} N_{xy}$$

it is well known that $\hat{\mu}(xy) = N_{xy}/N$ is the solution of this problem.

In our example, Table 3.1 contains the numbers N_{xy} and with $N = 21,000$ we obtain the following maximum likelihood estimator

$\hat{\mu}(cs) = 0.0024$, $\hat{\mu}(c\bar{s}) = 0.0450$
$\hat{\mu}(cn) = 0.0474$, $\hat{\mu}(c\bar{n}) = 0.4265$
$\hat{\mu}(\bar{c}s) = \hat{\mu}(\bar{c}\bar{s}) = 0.2370$
$\hat{\mu}(\bar{c}n) = 0.0045$, $\hat{\mu}(\bar{c}\bar{n}) = 0.0002$.

SIMPSON'S PARADOX

Notice that $\hat{\mu}(cs) < \hat{\mu}(cn)$ but $\hat{\mu}(\bar{c}s) > \hat{\mu}(\bar{c}n)$ so that the maximum likelihood estimator gives different judgements. The probabilities that we computed earlier are conditional probabilities using $\hat{\mu}$. For example, the probability that a patient survives given that the patient is from the first city and is administered the standard treatment is

$$\hat{\mu}(\{cs, cn, \bar{c}s, \bar{c}n\} \mid \{cs, c\bar{s}\}) = \frac{\hat{\mu}(\{cs\})}{\hat{\mu}(\{cs, c\bar{s}\})} = 0.05.$$

We now solve the estimator problem for the manual $\overrightarrow{\mathcal{AB}}$. Again, let N_{xy} be the number of trials resulting in outcome xy and let $\mathcal{N}$ be the collection of all states on $\overrightarrow{\mathcal{AB}}$. We must find a state $\hat{\nu} \in \mathcal{N}$ which maximizes

$$P' = \prod_{xy \in XY} \nu(xy)^{N_{xy}}$$

among all states $\nu \in \mathcal{N}$. The maximum likelihood estimator $\hat{\nu}$ can be found using Lagrange multipliers [Kläy, Randall, and Foulis, 1988]. We shall just summarize the results here. Denote the "column totals" by

$$N_{cS} = N_{cs} + N_{c\bar{s}}, \quad N_{cN} = N_{cn} + N_{c\bar{n}}$$
$$N_{\bar{c}S} = N_{\bar{c}s} + N_{\bar{c}\bar{s}}, \quad N_{\bar{c}N} = N_{\bar{c}n} + N_{\bar{c}\bar{n}}$$

the "city totals" by

$$N_c = N_{cS} + N_{cN}, \quad N_{\bar{c}} = N_{\bar{c}S} + N_{\bar{c}N}$$

and the total number of outcomes by $N = N_c + N_{\bar{c}}$. Then $\hat{\nu}$ is given by

$\hat{\nu}(cs) = N_{cs}N_c/N_{cS}N = 0.0261$
$\hat{\nu}(c\bar{s}) = N_{c\bar{s}}N_c/N_{cS}N = 0.4953$
$\hat{\nu}(cn) = N_{cn}N_c/N_{cN}N = 0.0521$
$\hat{\nu}(c\bar{n}) = N_{c\bar{n}}N_c/N_{cN}N = 0.4692$
$\hat{\nu}(\bar{c}s) = N_{\bar{c}s}N_{\bar{c}}/N_{\bar{c}S}N = 0.2393$
$\hat{\nu}(\bar{c}\bar{s}) = N_{\bar{c}\bar{s}}N_{\bar{c}}/N_{\bar{c}S}N = 0.2393$
$\hat{\nu}(\bar{c}n) = N_{\bar{c}n}N_{\bar{c}}/N_{\bar{c}N}N = 0.4547$
$\hat{\nu}(\bar{c}\bar{n}) = N_{\bar{c}\bar{n}}N_{\bar{c}}/N_{\bar{c}N}N = 0.0239.$

Thus, the judgement based directly on $\hat{\nu}$ is consistent: $\hat{\nu}(cs) < \hat{\nu}(cn)$ and $\hat{\nu}(\bar{c}s) < \hat{\nu}(\bar{c}n)$. Moreover, if we compute the marginal probabilities

$\hat{\nu}_X(s) = \hat{\nu}(cs) + \hat{\nu}(\bar{c}s) = 0.2654$
$\hat{\nu}_X(n) = \hat{\nu}(cn) + \hat{\nu}(\bar{c}n) = 0.5068$

we again obtain consistency: $\hat{\nu}_X(s) < \hat{\nu}_X(n)$.

Similar analyses can be applied to other problems involving Simpson's paradox. We conclude that Simson's paradox dissolves when operational statistics is used. For a further discussion and more details we refer the reader to [Kläy, Randall, and Foulis, 1988].

3.7 Spin Manuals

In the previous sections of this chapter we have presented a formalism for studying the set of experiments (operations) corresponding to a physical system. However, as we have stressed earlier, this does not yet capture the essence of quantum mechanics since the distinguishing feature of quantum mechanics is that probabilities are computed in terms of amplitudes.

This section considers some of the simplest manuals that have physical significance, namely manuals that describe the spin of a particle. These manuals show in a clear way how amplitudes are used to compute probabilities. The situation is simpler than the two-hole experiment considered in Chapter 2 since only a finite number of outcomes are involved. In this section we follow the work of R. Wright [1978].

Recall that a spin-s particle has $2s + 1$ possible spin values, $s = 0, \frac{1}{2}, 1, \frac{3}{2}, \ldots$. A useful tool for testing the spin values of a particle is a Stern-Gerlach experiment. Such an experiment is composed of two magnets which create an inhomogeneous magnetic field and thereby splits a beam of spin-s particles into $n = 2s + 1$ subbeams depending on the component of the spin along the gradient of the magnetic field. Figure 3.5 illustrates a beam of spin-1 particles passing through a Stern-Gerlach apparatus. In this figure, the source emits one particle at a time and the particle will travel along one of three paths depending upon whether its spin value is $1, 0$, or -1. A small conducting loop is placed around each of the three paths. We are assuming that the particles are charged so that a current is induced in the conducting loop through which the particle passes. Moreover, the loops are attached to small lights that flash when a current is induced in a loop. In this way we can observe the spin value of the particles.

Let $\mathcal{S}$ be the quasimanual corresponding to the Stern-Gerlach apparatus that we assume is fixed in space. Each orientation of the source will result in a different experiment or operation in $\mathcal{S}$ and each operation will have $n = 2s + 1$ outcomes. Thus, there is a bijection between the set of proper

SPIN MANUALS

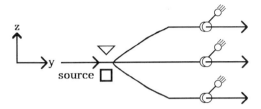

Figure 3.5. Stern-Gerlach Experiment.

rotations in $\mathbb{R}^3$ and operations in $\mathcal{S}$. Let $E(R)$ be the operation corresponding to the rotation R in $SO(3)$ where $SO(3)$ is the special orthogonal group. Thus,

$$\mathcal{S} = \{E(R): R \in SO(3)\}.$$

We shall label the $2s+1$ outcomes of $E(R)$ by $-s, -s+1, s-1, \ldots, s$, so that

$$E(R) = \{(R, m): -s \leq m \leq s\}.$$

Also, let $S = \cup \mathcal{S}$ be the outcome set for $\mathcal{S}$. Thus,

$$S = \{(R, m): R \in SO(3), -s \leq m \leq s\}.$$

It is not hard to show that $\mathcal{S}$ is a coherent manual.

Not every $\mathcal{S}$-state is physically realizable within the present interpretation of $\mathcal{S}$ (that is, by a Stern-Gerlach apparatus). For example, $\mathcal{S}$ has dispersion-free $\mathcal{S}$-states and these would not correspond to any physical source. It must be remembered, however, that other experimental arrangements might be represented by a manual with the same structure as $\mathcal{S}$. For example, $\mathcal{S}$ might just as well represent an uncountable collection of n-sided dice. What is needed is a way to decide which $\mathcal{S}$-states are relevant and which are not within each given context. Thus, within $\Omega(\mathcal{S})$ there is a set $\Delta(\mathcal{S})$ of meaningful $\mathcal{S}$-states. The composition of $\Delta(\mathcal{S})$ depends on the physics of the situation.

The traditional model for $\mathcal{S}$ is the frame manual $\mathcal{F}(\mathbb{C}^n)$. By using a unitary (projective) representation of $SO(3)$ on $\mathbb{C}^n$ one can construct a

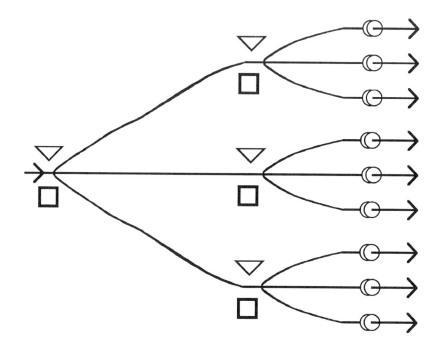

Figure 3.6. Compound Experiment.

morphism from $\mathcal{S}$ to $\mathcal{F}(\mathbb{C}^n)$ [Wright, 1978] but we shall not go into that here.

If we are only concerned with the collection of one-stage experiments, then there is really no need to study amplitude functions, since the corresponding $\mathcal{S}$-states carry all the necessary information. However, as we have seen earlier, interference becomes important for combined or compound experiments. It is in the study of such interference effects that amplitude functions play a crucial role.

The operational product $\overrightarrow{\mathcal{S}\mathcal{S}}$ describes a compound two-stage Stern-Gerlach experiment such as that illustrated in Figure 3.6 for a spin-1 particle. Denoting $\overrightarrow{\mathcal{S}\mathcal{S}}$ by $\mathcal{S}_2$, we can model three-stage operations by $\mathcal{S}_3 = \overrightarrow{\mathcal{S}_2\mathcal{S}}$, and m-stage operations by $\mathcal{S}_m = \overrightarrow{\mathcal{S}_{m-1}\mathcal{S}}$.

Another concept that we shall need is that of coarsening. Let $E \in \mathcal{S}$ and let $\mathcal{D}$ be a partition of the set E. We shall define two types of coarsenings corresponding to $\mathcal{D}$. In the first type, for each $D \in \mathcal{D}$, we disconnect

SPIN MANUALS

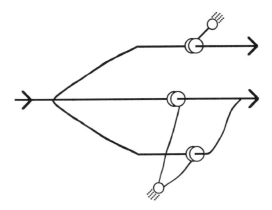

Figure 3.7. Incoherent Coarsening.

the wires between the conducting loops and the lights on those subbeams corresponding to outcomes in D, and hook up all the loops to a single light. We now perform a new experiment as follows. Send a particle into the apparatus and see which light flashes; if the light flashes corresponding to the outcomes of D, then record x_D as the outcome secured. This corresponds to the case in the two-hole experiment in which we "watch" to see through which hole the particle passed. Let us further suppose that past the detecting loops there are more Stern-Gerlach magnets (with reversed polarities as necessary) which, for each $D \in \mathcal{D}$ bend the subbeams corresponding to D back into a single beam. We do this so that we can perform further experiments on these recombined beams. Since each recombined beam is an incoherent superposition (or mixture) of the component beams, we call the set $\{x_D : D \in \mathcal{D}\}$ an *incoherent coarsening* of the operation E. An incoherent coarsening is illustrated in Figure 3.7.

For the second type of coarsening, again let $E \in \mathcal{S}$ and let $\mathcal{D}$ be a partition of E. We remove all the detecting loops entirely and for each $D \in \mathcal{D}$, we use further Stern-Gerlach magnets to recombine those subbeams corresponding to outcomes in D into a single beam. At the end of the experiment we place detecting loops around each of the recombined partial beams, each loop being connected to its own light. This setup is illustrated in Figure 3.8. The experiment consists in sending a particle into the apparatus and seeing which light flashes. If the light flashes corresponding to the recombination of the D-subbeams, we record y_D as the

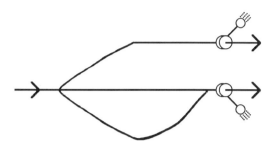

Figure 3.8. Coherent Coarsening.

outcome secured. This corresponds to the case in the double-hole experiment in which we do not "watch" to see through which hole the particle passes. In this type of coarsening, we obtain a coherent superposition of the component subbeams. We call the set $\{y_D : D \in \mathcal{D}\}$ a *coherent coarsening* of E. We shall also allow the possibility that part of an operation is incoherently coarsened and part is coherently coarsened. For example, if $\{D_1, D_2\}$ is a partition of $E \in \mathcal{S}$, then $\{x_{D_1}, y_{D_2}\}$ is a viable coarsening of E.

We can form a manual from the set of all coarsenings. Let $P(E)$ denote the set of all partitions of $E \in \mathcal{S}$, and let

$$\mathcal{S}^* = \{\{D : D \in \mathcal{D}\} : \mathcal{D} \in P(E),\ E \in \mathcal{S}\}.$$

Then $\mathcal{S}^*$ is a manual whose outcomes are nonempty events of $\mathcal{S}$ and whose operations are partitions of $\mathcal{S}$-operations. Of course, an operation $\{D : D \in \mathcal{D}\}$, $\mathcal{D} \in P(E)$, in $\mathcal{S}^*$ can be identified with a coarsening $\{x_D : D \in \mathcal{D}\}$ or $\{y_D : D \in \mathcal{D}\}$ or a combination of both types of coarsenings. In order to include these possibilities, we need two labeled copies of $\mathcal{S}^*$. Let

$$\bar{\mathcal{S}} = \{\{(D_i, z_i) : \{D_i\} \in P(E)\} : E \in \mathcal{S},\ z_i = x \text{ or } y\}.$$

Then $\bar{\mathcal{S}}$ is a manual whose operations are labeled partitions of $\mathcal{S}$-operations and whose outcome set is

$$\bar{\mathcal{S}} = \big(\mathcal{E}(\mathcal{S}) \setminus \{\emptyset\}\big) \times \{x, y\}.$$

The outcomes (D, x) and (D, y) are written x_D and y_D, repsectively. If d is an outcome of $\mathcal{S}$, then for both coherent and incoherent coarsenings, the outcome $\{d\}$ is physically equivalent to d. Hence, we identify the outcomes

SPIN MANUALS   111

$x_{\{d\}}$ and $y_{\{d\}}$ in $\bar{S}$ with the outcome d in S so that S becomes a submanual of $\bar{S}$.

It is easy to check that $(D_1, z_1)\, op\, (D_2, z_2)$ in $\bar{S}$ if and only if $D_1\, op\, D_2$ in S. It follows that $\Pi(S), \Pi(S^*)$, and $\Pi(\bar{S})$ are isomorphic. In particular, x_D and y_D are logically equivalent since they correspond to the same proposition in the operational logic. Since we know that x_D and y_D are physically different we have again lost information in going to the operational logic. A consequence of the logical equivalence of x_D and y_D is their statistical equivalence; for every $\omega \in \Omega(\bar{S})$, $\omega(x_D) = \omega(y_D)$. In fact, there is a canonical isomorphism of $\Omega(S)$ onto $\Omega(\bar{S})$ given by $\omega \mapsto \bar{\omega}$ where

$$\bar{\omega}(D, z) = \sum_{d \in D} \omega(d).$$

Thus, statistically equivalent outcomes need not be physically equivalent. As we shall show later, we can distinguish x_D and y_D statistically, but only if we consider compound experiments.

We now consider a compound coarsened experiment and its relationship to interference. Suppose we have a spin-$\frac{1}{2}$ particle with manual S. Let $E, F \in S$ be two Stern-Gerlach experiments whose outcomes we denote by $E = \{(1, \frac{1}{2}), (1, -\frac{1}{2})\}$ and $F = \{(2, \frac{1}{2}), (2, -\frac{1}{2})\}$. We now consider three compound experiments, EF, $\{x_E\}F$ and $\{y_E\}F$ where $\{x_E\}$ and $\{y_E\}$ are the incoherent and coherent coarsenings of E, respectively. These three experiments are illustrated in Figure 3.9. For a given source, each outcome has a certain probability, given by some state ω. As we saw in Section 3.5, interference can be described by relationships among these probabilities.

Classical probability theory would lead us to the following analysis. Both x_E and y_E mean that either $(1, \frac{1}{2})$ or $(1, -\frac{1}{2})$ occurred, but we don't know which. Hence, the probability of $x_E(2, m)$ or $y_E(2, m)$ is the sum of the probabilities of $(1, \frac{1}{2})(2, m)$ and $(1, -\frac{1}{2})(2, m)$. For the state ω we have for $m = \pm\frac{1}{2}$

$$\omega(x_E(2, m)) = \omega(y_E(2, m))$$
$$= \omega\left(\left(1, \frac{1}{2}\right)(2, m)\right) + \omega\left(\left(1, -\frac{1}{2}\right)(2, m)\right). \tag{3.4}$$

Now we can test (3.4) using our three experiments. According to (3.4) the relative number of particles leaving channel a in the last two experiments equals the sum of the relative number of particles leaving channels a_1 and

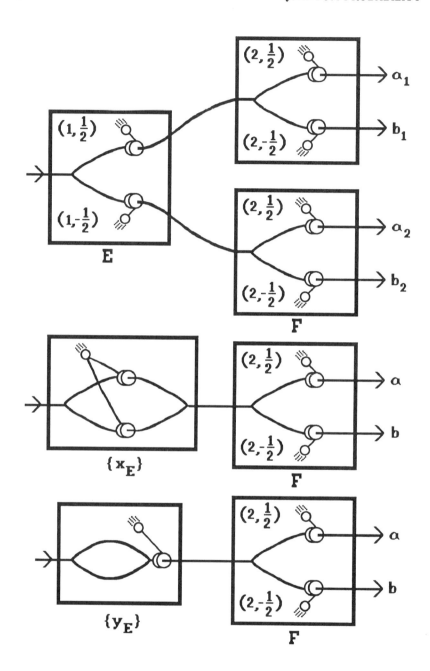

Figure 3.9. Three Experiments.

SPIN MANUALS   113

a_2 in the first experiment, and similarly for b and b_1, b_2. One finds experimentally that $\omega(x_E(2,m))$ satisfies (3.4) but $\omega(y_E(2,m))$ does not. In the terminology of Section 3.5, ω exhibits interference between the events x_E and y_E. It is probably more accurate to say that in the experiment $\{y_E\}F$ there is interference between the outcomes of E while in the experiment $\{x_E\}F$ there is none. Of course, the correct formula for $\omega(y_E(2,m))$ is obtained by introducing amplitudes α_1, α_2 where $|\alpha_1|^2 = \omega((1, \frac{1}{2})(2,m))$, $|\alpha_2|^2 = \omega((1, -\frac{1}{2})(2,m))$ and $\omega(y_E(2,m)) = |\alpha_1 + \alpha_2|^2$.

In order to introduce these amplitudes correctly, we follow the three rules of Feynman and Hibbs [1965].

(1) The probability that a particle will arrive at x, after emission from a source s, can be represented quantitatively by the absolute square of a complex number called the probability amplitude.

(2a) When an event can occur in several indistinguishable ways, the probability amplitude for the event is the sum of the probability amplitudes for each way considered separately (there is interference).

(2b) When an experiment is performed which is capable of determining whether one or another alternatives is actually taken, the probability of an event is the sum of the probabilities for each alternative (interference is lost).

(3) When a particle goes by some particular route, the amplitude for that route can be written as the product of the amplitude to go part way with the amplitude to go the rest of the way.

Although these rules were intended for describing a particle moving in space, they can also be applied to spin. Notice that (2b) gives the prescription for finding $\omega(x_E(2,m))$ according to (3.4) while (2a) gives the prescription for finding $\omega(y_E(2,m))$. Following Rule 1, we define a (normalized) *amplitude function* on the manual S to be a function $f: S \to \mathbb{C}$ such that $\sum_{x \in E} |f(x)|^2 = 1$ for every $E \in \mathcal{S}$. Now suppose we have a two-stage compound experiment. Then the output of the first stage can be considered as the source of the second stage. Following Rule 1, for every $x \in S$ there is an amplitude function $f_x: S \to \mathbb{C}$ such that $f_x(y)$ is the amplitude of going from outcome x in the first stage to outcome y in the second stage. Moreover, we have an amplitude function f_s where $f_s(x)$ is the amplitude of the outcome x in the first stage. In the next chapter we shall call the function $g: S \times S \to \mathbb{C}$ defined by $g(x,y) = f_x(y)$ a *transition amplitude*.

Now suppose we want to describe a k-stage experiment. Then by Rule 3, the amplitude that the particle goes from the source s to the outcome x in k-stages by way of the outcomes $x_1, x_2, \ldots, x_{k-1}$ becomes

$$f(x_1 x_2 \ldots x_{k-1} x) = f_s(x_1) f_{x_1}(x_2) f_{x_2}(x_3) \ldots f_{x_{k-1}}(x). \tag{3.5}$$

Applying Rule 2a and Equation 3.5, the amplitude of a coherent coarsening is given by

$$f(y_{D_1} y_{D_2} \ldots y_{D_{k-1}} x) = \sum_{x_i \in D_i} f_s(x_1) f_{x_1}(x_2) \ldots f_{x_{k-1}}(x). \tag{3.6}$$

Applying Rule 1 and Equation 3.6, the probability of $y_{D_1} y_{D_2} \ldots y_{D_{k-1}} x$ becomes

$$\begin{aligned} \omega(y_{D_1} y_{D_2} \ldots y_{D_{k-1}} x) &= \left| f(y_{D_1} y_{D_2} \ldots y_{D_{k-1}} x) \right|^2 \\ &= \left| \sum_{x_i \in D_i} f_s(x_1) f_{x_1}(x_2) \ldots f_{x_{k-1}}(x) \right|^2. \end{aligned} \tag{3.7}$$

Clearly, it is the cross terms in (3.7) that describe the interference. Rule 2b tells how to compute probabilities for incoherent coarsenings:

$$\begin{aligned} \omega(x_{D_1} x_{D_2} \ldots x_{D_{k-1}} x) &= \sum_{x_i \in D_i} \omega(x_1 x_2 \ldots x_{k-1} x) \\ &= \sum_{x_i \in D_i} |f_s(x_1) f_{x_1}(x_2) \cdots f_{x_{k-1}}(x)|^2 \\ &= \sum_{x_i \in D_i} |f_s(x_1)|^2 |f_{x_1}(x_2)|^2 \cdots |f_{x_{k-1}}(x)|^2 \\ &= \sum_{x_i \in D_i} {}_s\omega(x_1)_{x_1}\omega(x_2) \cdots {}_{x_{k-1}}\omega(x). \end{aligned}$$

In summary, the physics of the situation is given by the structure of the manual $\mathcal{S}$, the form of the amplitude function $f: \mathcal{S} \to \mathbb{C}$ and the rules for obtaining compound amplitudes from f. Combining these ingredients we obtain a quantum probability theory. The next chapter develops the axiomatics for such a theory.

3.8 Notes and References

Sections 3.1, 3.2, 3.3, and 3.5 follow the work of Foulis and Randall closely [Foulis and Randall, 1981; Randall and Foulis, 1981, 1983]. We have done this for two reasons. First, we want to be faithful to their interpretation and do not want to cause misunderstandings in what they mean. Second, they have stated their case clearly and succinctly and we can not improve on it. Operational statistics has developed in various directions and we refer the reader to the literature for more details [Foulis and Randall, 1985; Foulis, Piron, and Randall, 1983; Cook, 1985; Fischer and Rüttimann, 1978]. Section 3.2 overlaps with the work of G. Rüttimann [1987] and Section 3.7 follows the interesting paper of R. Wright [1978].

Some of the results in Section 3.4 are contained in [Gudder, 1979]. The subject of hidden variables has a vast literature going back to the beginnings of quantum mechanics and we shall return to this subject in later chapters. A sample of the references in this subject are [Belifante, 1973; Bell, 1966; Bohm and Bub, 1966a, 1966b; Bohr, 1935; Cohen, 1966; Einstein, Podolsky, and Rosen, 1935; Gudder, 1968a, 1970; Gudder and Armstrong, 1985; Jauch and Piron, 1965; Kochen and Specker, 1967; Pitowski, 1982, 1983a, 1983b; von Neumann, 1955]. For the past sixty years there has been an ongoing debate among physicists, mathematicians, and philosophers concerning hidden variables and the debate still continues today. One must be careful when reading articles on this subject since there are various definitions and interpretations of hidden variables. There are also proofs and experiments purporting that (local) hidden variables cannot exist. However, these are always based on classical probability theory, while our approach relies on quantum probability theory. We shall return to this subject in Chapters 5 and 6 (specifically Sections 5.2 and 6.4).

As we have seen, there is a close connection between operational statistics and quantum logic although the former is more basic. Further details on the quantum logic approach can be found in [Beltrametti and Cassinelli, 1981; Birkhoff and von Neumann, 1936; Bodiou, 1964; Gudder, 1979; Jauch, 1968; Mackey, 1963; Piron, 1976; Varadarajan, 1968, 1970].

4

Amplitudes and Transition Amplitudes

The previous chapter presented the "language" of operational statistics. The basic structure consists of a quasimanual $\mathcal{A}$ where the elements of $\mathcal{A}$ correspond to physical experiments or operations and subsets of an operation correspond to physical events that can be tested by executing the operation. A statistical model can be constructed by giving an $\mathcal{A}$-state ω. In this model, $\omega(x)$ gives the probability that an outcome x is secured when tested. Moreover, the probability that an event A occurs, when tested, becomes $\omega(A) = \sum_{x \in A} \omega(x)$. However, the essence of quantum mechanics is not captured unless amplitudes are introduced. This was done in Section 3.7 for the simple case of spin manuals.

The present chapter considers amplitudes and transition amplitude functions together with the structure of spaces of these functions. We show that such spaces are closely related to Hilbert spaces. In this way we see how Hilbert spaces come into play in quantum mechanics as a derived concept. This gives a deeper understanding of the role of Hilbert space than is possible in traditional quantum mechanics where one starts with Hilbert space as the basic axiomatic structure. Moreover, it allows us to begin with amplitudes as the primitive axiomatic elements of a quantum probability theory. We first introduce amplitude functions on a quasimanual and study the resulting consequences. Later we begin with the amplitudes and transition amplitudes themselves and derive the corresponding quasimanual.

4.1 Partial Hilbert Spaces

This section treats a generalization of Hilbert space, called a partial Hilbert space, that will be useful in the sequel. As we shall see in the next section, the set of amplitude functions on a quasimanual forms a partial Hilbert space. In our opinion, this is the correct and natural place at which Hilbert space enters quantum mechanics. It shall also become evident that the superposition principle, first stressed by Dirac (see Section 2.7), is responsible for this entrance.

Let H be a nonempty set and let $s \subseteq H \times H$ be a reflexive, symmetric relation on H. If $(f,g) \in s$, we write fsg. For $A \subseteq H$ we define

$$A^s = \{f \in H : fsg \text{ for all } g \in A\}.$$

We call $A \subseteq H$ an *s-set* if $A \subseteq A^s$. Thus, A is an s-set if and only if fsg for all $f, g \in A$. It is clear that singleton sets are s-sets and hence every $f \in H$ is contained in an s-set. Moreover, by Zorn's lemma, every s-set is contained in a maximal s-set. Denote the collection of s-sets in H by $s(H)$ and the collection of maximal s-sets by $m(H)$. For an s-set A we define

$$\hat{A} = \cap \{B \in m(H) : A \subseteq B\}.$$

Lemma 4.1. (a) *For any $A \subseteq H$ we have $A \subseteq A^{ss}$ and $A^{sss} = A^s$.*
(b) *If $A \subseteq B \subseteq H$, then $B^s \subseteq A^s$.* (c) *$A \in s(H)$ if and only if $A^{ss} \in s(H)$.*
(d) *If $A \in s(H)$, then $A \subseteq A^{ss} \subseteq A^s$.* (e) *If $A \in s(H)$, then $A^s \in s(H)$ if and only if $A^s = A^{ss}$.* (f) *$A = A^s$ if and only if $A \in m(H)$.* (g) *If $A \in m(H)$, then $A = A^{ss} = A^s$.* (h) *If $A \in s(H)$, then $\hat{A} = A^{ss}$.*

Proof. The proofs of (a)–(g) are straightforward. To prove (h), let $B \in m(H)$ with $A \subseteq B$. Then by (b) and (g) we have, $A^{ss} \subseteq B^{ss} = B$. Hence, $A^{ss} \subseteq \hat{A}$. Conversely, suppose $f \in \hat{A}$ and $g \in A^s$. Then there exists a $B \in m(H)$ such that $A \cup \{g\} \subseteq B$. Since $f \in B$, we have fsg. Hence, $f \in A^{ss}$ so $\hat{A} \subseteq A^{ss}$. □

We call (H, s) a *partial Hilbert space* if it satisfies
(1) every $A \in m(H)$ is a complex Hilbert space;
(2) $0_A = 0_B$ for all $A, B \in m(H)$ where 0_A is the zero in A;
(3) if $A, B \in m(H)$ and $f, g \in A \cap B$, $\alpha, \beta \in \mathbb{C}$ then

$$(\alpha f + \beta g)_A = (\alpha f + \beta g)_B \text{ and } \langle f, g \rangle_A = \langle f, g \rangle_B.$$

PARTIAL HILBERT SPACES 119

Of course, in (3) we mean by $(\alpha f + \beta g)_A$ the element $\alpha f + \beta g$ considered as a member of A and $\langle f, g \rangle_A$ is the inner product in A. If (H, s) is a partial Hilbert space, we say that $A \subseteq H$ is a *closed subspace* of H if $(A, s \mid A)$ is a partial Hilbert space with the linear structures and inner products inherited from H. It is easy to show that $A \subseteq H$ is a closed subspace of H if and only if every $B \in m(A)$ is a Hilbert space with linear structure and inner product inherited from H.

Let (H, s) be a partial Hilbert space and let $A \in s(H)$. Then there exists a $B \in m(H)$ such that $A \subseteq B$. We define $[A]$ to be the closed linear span of A in the Hilbert space B. It is clear that $[A]$ is independent of the $B \in m(H)$ used to define it. The following lemma summarizes basic relationships for A and $[A]$.

Lemma 4.2. *Let (H, s) be a partial Hilbert space and let $A \in s(H)$. Then $[A], A^{ss}, \hat{A}$ are Hilbert spaces in H and*

$$A \subseteq [A] \subseteq A^{ss} = \hat{A} \subseteq A^s = [A]^s.$$

Proof. All of the equalities except the last are immediate consequences of Lemma 4.1. To show that $A^s \subseteq [A]^s$ let $f \in A^s$. Then there exists a $B \in m(H)$ such that $\{f\} \cup A \subseteq B$. Moreover, since $A \subseteq B$, $[A] \subseteq B$. It follows that $f \in [A]^s$. □

If (H, s) is a partial Hilbert space and $f \in H$, we write $\|f\| = \|f\|_A$ where $f \in A \in m(H)$ and A is arbitrary. If $f, g \in H$, fsg and $\alpha, \beta \in \mathbb{C}$, we write $\alpha f + \beta g = (\alpha f + \beta g)_A$ and $\langle f, g \rangle = \langle f, g \rangle_A$ where $f, g \in A \in m(H)$ and A is arbitrary. Two partial Hilbert spaces H, K are *isomorphic* if there exists a bijection $\phi \colon H \to K$ satisfying

(1) fsg in H if and only if $\phi(f) s \phi(g)$ in K;
(2) if fsg in H, $\alpha, \beta \in \mathbb{C}$, then $\phi(\alpha f + \beta g) = \alpha \phi(f) + \beta \phi(g)$ and $\langle \phi(f), \phi(g) \rangle_K = \langle f, g \rangle_H$.

If H_0 is a Hilbert space and we define s to be the relation $H_0 \times H_0$, then clearly (H_0, s) is a partial Hilbert space. Moreover, the above definitions of closed subspace and isomorphic reduce to the usual definitions on a Hilbert space. For a less trivial example of a partial Hilbert space, let $S_\alpha, \alpha \in \Delta$, be a collection of closed subspaces of H_0 satisfying $S_\alpha \cap S_\beta = S$ for $\alpha \neq \beta$, where $S \neq S_\alpha$ for any $\alpha \in \Delta$. Let $H = \cup_{\alpha \in \Delta} S_\alpha$ and for $f, g \in H$ define fsg if $f, g \in S_\alpha$ for some $\alpha \in \Delta$. We then have the following result.

Lemma 4.3. *With the above definitions, (H,s) is a partial Hilbert space and $m(H) = \{S_\alpha : \alpha \in \Delta\}$.*

Proof. Clearly, $S_\alpha \in s(H)$ for every $\alpha \in \Delta$. To show that $S_\alpha \in m(H)$, suppose $g \in H \setminus S_\alpha$ and $g \in S_\alpha^s$. Let $f \in S_\alpha$. Since gsf we must have $f, g \in S_\beta$ for some $\beta \in \Delta$. Since $g \notin S_\alpha$, $\beta \neq \alpha$. We conclude that

$$S_\alpha \subseteq \cup\{S_\beta : \beta \neq \alpha\}.$$

But then

$$S_\alpha \subseteq S_\alpha \cap (\cup_{\beta \neq \alpha} S_\beta) = \cup_{\beta \neq \alpha}(S_\alpha \cap S_\beta) = S$$

so $S_\alpha = S$ which is a contradiction. Conversely, let $A \in m(H)$. Since $S \subseteq H^s$, we must have $S \subseteq A$. Since S is a proper subset of S_α for $\alpha \in \Delta$, $S \notin m(H)$. Hence, there exists an $f \in A \setminus S$. Assume $f \in S_\alpha$. Now suppose that $g \in A \setminus S$. Since fsg, we must have $f, g \in A_\gamma$ for some $\gamma \in \Delta$. If $\gamma \neq \alpha$, then $S_\alpha \cap S_\gamma = S$ which contradicts the fact that $f \in S_\alpha \cap S_\gamma$. Hence, $g \in S_\alpha$. It follows that $A \subseteq S_\alpha$ and by maximality $A = S_\alpha$. The proof is now easily completed. □

The next example shows that we can not generalize the previous work to an arbitrary collection of closed subspaces S_α of H_0 even if the S_α are required to be maximal and proper.

Example 4.1. Let $H_0 = \mathbb{C}^3$ and let f_1, f_2, f_3 be the standard orthonormal basis for H_0. Define the three subspaces $S_1 = \text{span}\{f_2, f_3\}$, $S_2 = \text{span}\{f_1, f_3\}$, $S_3 = \text{span}\{f_1, f_2\}$ and let $H = S_1 \cup S_2 \cup S_3$. For $f, g \in H$, define fsg if $f, g \in S_i$ for some $i = 1, 2, 3$. Then (H, s) is not a partial Hilbert space under the induced linear structure and inner product. Indeed, it is easy to show that

$$S = \text{span}\{f_1\} \cup \text{span}\{f_2\} \cup \text{span}\{f_3\} \in m(H)$$

and yet S is not a linear space. □

Example 4.2. Let X be a nonempty set and let $(X_\alpha, \Sigma_\alpha, \mu_\alpha)$ be a collection of measure spaces with $X_\alpha \subseteq X$. Let H be the set of functions $f: X \to \mathbb{C}$ such that $f \mid X_\alpha$ is measurable for all α and

$$\int_{X_\alpha} |f|^2 d\mu_\alpha = \int_{X_\beta} |f|^2 d\mu_\beta < \infty$$

for all α and β. For $f, g \in H$, define fsg if

$$\int_{X_\alpha} f\bar{g} d\mu_\alpha = \int_{X_\beta} f\bar{g} d\mu_\beta$$

for all α and β. It can be shown using methods which are similar to those developed in the next section that (H, s) is a partial Hilbert space. $\square$

We denote the topological dual of a Hilbert space H_0 by H_0^*. The *dual* H^* of a partial Hilbert space H is the set of functionals $F: H \to \mathbb{C}$ such that $F \mid A \in A^*$ for all $A \in m(H)$. By the Riesz lemma, if $F \in H^*$ then for any $A \in m(H)$ there exists a unique $f_F^A \in A$ such that $F(g) = \langle g, f_F^A \rangle$ for all $g \in A$. Moreover, if $B, C \in m(H)$ and $f_F^B, f_F^C \in B \cap C$, then $f_F^B = f_F^C$. We define the *strong dual* H^σ of H to be

$$H^\sigma = \{F \in H^*: \|F \mid A\| = \|F \mid B\| \text{ for all } A, B \in m(H)\}.$$

It follows that an $F \in H^*$ is in H^σ if and only if $\|f_F^A\| = \|f_F^B\|$ for all $A, B \in m(H)$. For $F, G \in H^\sigma$ we write FsG if

$$\langle f_F^A, f_G^A \rangle = \langle f_F^B, f_G^B \rangle$$

for all $A, B \in m(H)$.

Theorem 4.4. *If (H, s) is a partial Hilbert space, then (H^σ, s) is a partial Hilbert space under the usual linear structure.*

Proof. First, it is clear that if $F, G \in H^*$, $\alpha, \beta \in \mathbb{C}$, then $\alpha F + \beta G \in H^*$ and for any $A \in m(H)$,

$$f_{\alpha F + \beta G}^A = \bar{\alpha} f_F^A + \bar{\beta} f_G^A$$

It is also clear that if $F \in H^\sigma$, $\alpha \in \mathbb{C}$, then $\alpha F \in H^\sigma$. Now suppose that $F, G \in H^\sigma$ and FsG. If $A, B \in m(H)$, we have

$$\|f_{F+G}^A\|^2 = \langle f_F^A + f_G^A, f_F^A + f_G^A \rangle$$
$$= \|f_F^A\|^2 + \|f_G^A\|^2 + 2\operatorname{Re}\langle f_F^A, f_G^A \rangle$$
$$= \|f_F^B\|^2 + \|f_G^B\|^2 + 2\operatorname{Re}\langle f_F^B, f_G^B \rangle = \|f_{F+G}^B\|^2.$$

Hence, $F + G \in H^\sigma$. Moreover, if $\alpha, \beta \in \mathbb{C}$, then clearly $\alpha F s \beta G$ so $\alpha F + \beta G \in H^\sigma$. For $F \in H^\sigma$ we write $\|F\| = \|f_F^A\|$ and if FsG we write

$$\langle F, G \rangle = \langle f_G^A, f_F^A \rangle$$

where $A \in m(H)$ is arbitrary. Now let $\tilde{A} \in m(H^\sigma)$. If $F, G \in \tilde{A}$, $\alpha, \beta \in \mathbb{C}$, then it is easy to show that $\alpha F + \beta G \in \tilde{A}^s$ so $\alpha F + \beta G \in \tilde{A}$. Also, it is clear that $\langle \cdot, \cdot \rangle$ is an inner product on $\tilde{A}$ so $\tilde{A}$ is an inner product space. To show that $\tilde{A}$ is complete, suppose $F_n \in \tilde{A}$ is a Cauchy sequence. If $f \in H$, then $f \in A$ for some $A \in m(H)$. We then have

$$|F_n(f) - F_m(f)| = |(F_n - F_m)(f)| \leq \|F_n - F_m\| \|f\| \to 0$$

as $n, m \to \infty$. Hence, $F_n(f)$ converges. Define $F: H \to \mathbb{C}$ by $F(f) = \lim F_n(f)$. It is straightforward to show that $F \in H^\sigma$, and for any $A \in m(H)$, $F_n \mid A \to F \mid A$ in the norm topology. It follows from the Riesz lemma that $f_{F_n}^A \to f_F^A$ for every $A \in m(H)$. Now let $G \in \tilde{A}$ and let $A, B \in m(H)$. Then

$$\langle f_F^A, f_G^A \rangle = \lim \langle f_{F_n}^A, f_G^A \rangle$$
$$= \lim \langle f_{F_n}^B, f_G^B \rangle = \langle f_F^B, f_G^B \rangle.$$

Hence, FsG and since $\tilde{A} \in m(H^\sigma)$, $F \in \tilde{A}$. Thus, $\tilde{A}$ is a Hilbert space. Conditions 2 and 3 in the definition of a partial Hilbert space are clearly satisfied for H^σ so the proof is complete. □

4.2 Amplitude Spaces

Operational statistics provides a language for discussing physical theories and in order to capture the way that quantum probabilities are computed, we must introduce amplitude functions into this language. The present section studies the structure of the space of amplitude functions on a quasimanual. Throughout this section, $\mathcal{A}$ will denote a quasimanual with outcome space X.

A function $\mu: X \to [0, \infty) \subseteq \mathbb{R}$ is called a *positive measure* if there exists a constant $\hat{\mu} \in \mathbb{R}$ such that $\sum_{x \in E} \mu(x) = \hat{\mu}$ for every $E \in \mathcal{A}$. We denote the set of positive measures on $\mathcal{A}$ by $M_+(\mathcal{A})$. As before, we denote the set of $\mathcal{A}$-states by $\Omega(\mathcal{A})$. Of course, for $\mu \in M_+(\mathcal{A})$ we have $\mu \in \Omega(\mathcal{A})$ if and only if $\hat{\mu} = 1$. A function $f: X \to \mathbb{C}$ is called an *amplitude function* if the function $\mu_f(x) = |f(x)|^2$ is in $M_+(\mathcal{A})$. An amplitude function f is *normal* if $\hat{\mu}_f = 1$. The *amplitude space* on $\mathcal{A}$ is the set $H = H(\mathcal{A})$ of all amplitude functions on $\mathcal{A}$. For $f \in H$ we use the notation $\|f\| = (\hat{\mu}_f)^{\frac{1}{2}}$, and we define $\text{supp}(f) = X \setminus \ker f$ where $\ker f = \{x \in X : f(x) = 0\}$.

AMPLITUDE SPACES   123

Let $G = \{\chi: X \to \mathbb{C}: |\chi(x)| = 1 \text{ for all } x \in X\}$. Then G is a group under pointwise multiplication and its elements correspond to "phase" transformations. If $f \in H$, then clearly $\chi f \in H$ for any $\chi \in G$. We call $f \in H$ *pure* if $\text{supp}(g) \subseteq \text{supp}(f)$ for $g \in H$ implies that $g = \alpha \chi f$ for some $\alpha \in \mathbb{C}$, $\chi \in G$. The next result gives the relationship between $H(\mathcal{A})$ and $\Omega(\mathcal{A})$.

Lemma 4.5. (a) *If $r \geq 0$, $\chi \in G$, and $\omega \in \Omega(\mathcal{A})$, then*

$$f = r\chi\omega^{\frac{1}{2}} \in H(\mathcal{A}).$$

Conversely, if $f \in H(\mathcal{A})$, then f admits a representation $f = r\chi\omega^{1/2}$, $r \geq 0$, $\chi \in G$, $\omega \in \Omega(\mathcal{A})$. Moreover, if $f \neq 0$, then $r = \|f\|$, $\omega = \mu_f/\|f\|^2$, $\chi(x) = f(x)/|f(x)|$ for all $x \in \text{supp}(f)$ and χ is unique on $\text{supp}(f)$. (b) *If $0 \neq f \in H$ is pure, then $\mu_f/\|f\|^2$, is extremal in the convex set $\Omega(\mathcal{A})$.*

Proof.   (a) The first statement clearly holds. Conversely, if $f \in H$, define $\chi(x) = 1$ for $x \in \ker f$ and $\chi(x) = f(x)/|f(x)|$ for $x \in \text{supp}(f)$. Then $\chi \in G$ and if $f \neq 0$ we have

$$f(x) = \frac{\|f\|\chi(x)|f(x)|}{\|f\|}$$

where

$$\frac{|f(x)|^2}{\|f\|^2} = \omega \in \Omega(\mathcal{A}).$$

For uniqueness, suppose $0 \neq f \in H(\mathcal{A})$ has the form $f = r\chi\omega^{1/2}$, $r \geq 0$, $\chi \in G$, and $\omega \in \Omega(\mathcal{A})$. For $E \in \mathcal{A}$ we have

$$\|f\|^2 = \sum_{x \subseteq E} |f(x)|^2 = r^2 \hat{\omega} = r^2.$$

Hence, $r = \|f\|$, and

$$\omega(x) = \frac{|f(x)|^2}{\|f\|^2} = \frac{\mu_f(x)}{\|f\|^2}.$$

The result now easily follows. (b) Suppose $0 \neq f \in H$ is pure and let $\omega = \mu_f/\|f\|^2$. To show that ω is extremal, assume that

$$\omega = \lambda\omega_1 + (1-\lambda)\omega_2, \quad 0 < \lambda < 1$$

where $\omega_1, \omega_2 \in \Omega(\mathcal{A})$. Then $\text{supp}(\omega_1) \subseteq \text{supp}(\omega)$. Define $g(x) = \omega_1(x)^{1/2}$ for all $x \in X$. Then $g \in H$ and $\text{supp}(g) \subseteq \text{supp}(f)$. Since f is pure, $g = \alpha \chi f$ for some $\alpha \in \mathbb{C}$, $\chi \in G$. For $x \in X$, we obtain

$$\omega_1(x) = |g(x)|^2 = |\alpha|^2 |f(x)|^2$$
$$= |\alpha|^2 \mu_f(x)$$
$$= |\alpha|^2 \|f\|^2 \omega(x).$$

Since $\omega_1, \omega \in \Omega(\mathcal{A})$, we have $\omega_1 = \omega$. Hence, $\omega_1 = \omega_2$ so ω is extremal. □

If $f \in H(\mathcal{A})$, then clearly $\alpha f \in H(\mathcal{A})$ for all $\alpha \in \mathbb{C}$. However, if $f, g \in H(\mathcal{A})$, then the following example shows that $f + g$ need not be in $H(\mathcal{A})$.

Example 4.3. Let $\mathcal{A} = \{E_1, E_2\}$ where $E_1 = \{x_1, x_2\}$ and $E_2 = \{x_2, x_3, x_4\}$. Then $\mathcal{A}$ is a quasimanual with outcome set

$$X = X(\mathcal{A}) = \{x_1, x_2, x_3, x_4\}.$$

Define the functions $f, g \colon X \to \mathbb{C}$ as follows
$f(x_1) = f(x_2) = 1, \quad f(x_3) = f(x_4) = 1/\sqrt{2}$
$g(x_1) = g(x_3) = 1, \quad g(x_2) = \sqrt{2}, \quad g(x_4) = 0$.
It is easy to check that $f, g \in H(\mathcal{A})$ but $f + g \notin H(\mathcal{A})$. □

We now define a relation on $H(\mathcal{A})$ which characterizes closure under linear combinations. For $f, g \in H(\mathcal{A})$, we write fsg if

$$\sum_{x \in E} f(x) \bar{g}(x) = \sum_{x \in F} f(x) \bar{g}(x)$$

for all $E, F \in \mathcal{A}$. It is clear that s is reflexive and symmetric. However, s need not be transitive. For instance, in Example 4.3, $fs0$, $0sg$ but f is not s related to g. If fsg, we write

$$\langle f, g \rangle = \sum_{x \in E} f(x) \bar{g}(x), \quad E \in \mathcal{A}.$$

Notice that $\langle f, f \rangle = \|f\|^2$.

AMPLITUDE SPACES

Lemma 4.6. *If $f, g \in H(\mathcal{A})$, then fsg if and only if $f+g, f+ig \in H(\mathcal{A})$.*

Proof. For any $E \in \mathcal{A}$ we have

$$\sum_{x \in E}[f(x) + g(x)][\bar{f}(x) + \bar{g}(x)] = \|f\|^2 + \|g\|^2 + 2\operatorname{Re} \sum_{x \in E} f(x)\bar{g}(x). \quad (4.1)$$

Hence, if fsg, then $f + g \in H(\mathcal{A})$. Moreover, $fsig$ so $f + ig \in H(\mathcal{A})$. Conversely, if $f + g \in H(\mathcal{A})$, then from (4.1) we have

$$\operatorname{Re} \sum_{x \in E} f(x)\bar{g}(x) = \operatorname{Re} \sum_{x \in F} f(x)\bar{g}(x)$$

for any $E, F \in \mathcal{A}$. If in addition, $f + ig \in H(\mathcal{A})$, then

$$\sum_{x \in E}[f(x) + ig(x)][\bar{f}(x) - i\bar{g}(x)] = \|f\|^2 + \|g\|^2 + 2\operatorname{Im} \sum_{x \in E} f(x)\bar{g}(x). \quad (4.2)$$

Applying (4.2) gives

$$\operatorname{Im} \sum_{x \in E} f(x)\bar{g}(x) = \operatorname{Im} \sum_{x \in F} f(x)\bar{g}(x)$$

for any $E, F \in \mathcal{A}$. It follows that fsg. □

It follows from Lemma 4.6 that fsg if and only if $\alpha f + \beta g \in H(\mathcal{A})$ for all $\alpha, \beta \in \mathbb{C}$. Since the latter condition corresponds to a Dirac superposition (Section 2.7), we may think of s as the "superposition" relation. That is, fsg if and only if superpositions of f and g are allowed. Moreover, if $A \in m(H)$ we may think of A as a maximal superposition set. Our next result shows that $(H(\mathcal{A}), s)$ is a partial Hilbert space so A is a Hilbert space for any $A \in m(H)$. Thus, Dirac's original idea of a superposition principle leads, in a natural way, to Hilbert spaces as a derived concept. In our opinion this gives the correct role of Hilbert spaces in the theory.

For later convenience, we introduce the following notation. For $E \in \mathcal{A}$, we define the Hilbert space

$$\ell^2(E) = \left\{f : E \to \mathbb{C} : \sum_{x \in E} |f(x)|^2 < \infty\right\}$$

where the inner product is given by

$$\langle f, g \rangle_E = \sum_{x \in E} f(x)\bar{g}(x).$$

Notice that $f \in H(\mathcal{A})$ if and only if $f \mid E \in \ell^2(E)$ for all $E \in \mathcal{A}$ and $\|f \mid E\|_E = \|f \mid F\|_F$ for all $E, F \in \mathcal{A}$.

Theorem 4.7. $(H(\mathcal{A}), s)$ *is a partial Hilbert space.*

Proof. Suppose $A \in m(H(\mathcal{A}))$. If $f \in A$, $\alpha \in \mathbb{C}$, then $\alpha f s g$ for all $g \in A$. Since A is maximal, $\alpha f \in A$. If $f, g \in A$, then by Lemma 4.6, $f + g \in H(\mathcal{A})$. Also, it is clear that $(f + g)sh$ for all $h \in A$. Again by maximality, $f + g \in A$. Hence, A is a linear space. It is straightforward to show that $\langle \cdot, \cdot \rangle$ is an inner product on A. For completeness, let $f_n \in A$ be a Cauchy sequence. Let $x_0 \in X$ and suppose $x_0 \in E \in \mathcal{A}$. Then

$$|f_n(x_0) - f_m(x_0)|^2 \leq \sum_{x \in E} |f_n(x) - f_m(x)|^2 = \|f_n - f_m\|^2.$$

Hence, $f_n(x_0)$ is Cauchy in $\mathbb{C}$ so $f_n(x_0)$ converges. Define $f \colon X \to \mathbb{C}$ by $f(x) = \lim f_n(x)$. A standard argument shows that $f \mid E \in \ell^2(E)$ for all $E \in \mathcal{A}$ and that $\|f \mid E\|_E = \lim \|f_n \mid E\|_E$. But since $\|f_n \mid E\|_E = \|f_n \mid F\|_F$ for all $E, F \in \mathcal{A}$ we have $\|f \mid E\|_E = \|f \mid F\|_F$ for all $E, F \in \mathcal{A}$. Hence, $f \in H(\mathcal{A})$. If $g \in A$ and $E, F \in \mathcal{A}$, then a standard argument gives

$$\langle f \mid E, g \mid E \rangle_E = \lim \langle f_n \mid E, g \mid E \rangle_E$$
$$= \lim \langle f_n \mid F, g \mid F \rangle_F$$
$$= \langle f \mid F, g \mid F \rangle_F.$$

Hence, fsg and by maximality $f \in A$. It easily follows that $\lim \|f_n - f\| = 0$ so A is complete. Conditions 2 and 3 for a partial Hilbert space clearly hold so the proof is complete. □

A set $A \subseteq H(\mathcal{A})$ is *unital* if for any $x \in X(\mathcal{A})$ there exists a $0 \neq f \in A$ such that $|f(x)| = \|f\|$.

Theorem 4.8. *If $A \in s(H(\mathcal{A}))$ is unital, then $[A]$ is the unique maximal s-set containing A and*

$$A \subseteq [A] = A^{ss} = \hat{A} = A^s = [A]^s$$

Proof. Let $B \in m(H(\mathcal{A}))$ with $A \subseteq B$. Then clearly $[A] \subseteq B$. Let $f \in B$ and suppose $f \perp A$. If $x \in X(\mathcal{A})$, then $x \in E$ for some $E \in \mathcal{A}$. Since A is unital, there exists a $g \in A$ such that $g(x) \neq 0$ and $g(y) = 0$ for all $y \in E \setminus \{x\}$. Then

$$0 = \sum_{z \in E} f(z)\bar{g}(z) = f(x)\bar{g}(x).$$

Hence, $f(x) = 0$. It follows that $f = 0$ and hence $B = [A]$. The result now follows from Lemmas 4.1 and 4.2. □

Corollary 4.9. *Let $A \in s(H(\mathcal{A}))$ be unital. (a) If A is a linear space, the its closure $\bar{A}$ is the unique maximal s-set containing A. (b) If A is a Hilbert space, then $A \in m(H(\mathcal{A}))$.*

A set of functions $Y \subseteq \mathbb{C}^S$ is *separating* for a set S if $x \neq y \in S$ implies that there exists an $f \in Y$ such that $f(x) \neq f(y)$. We denote the power set of a set S by $P(S)$. The next result characterizes closed subspaces of $H(\mathcal{A})$ up to an isomorphism.

Theorem 4.10. *A partial Hilbert space H is isomorphic to a closed subspace of $H(\mathcal{A})$ for some quasimanual $\mathcal{A}$ if and only if there exists a $Y \subseteq P(H^*)$ satisfying*

(1) *$\cup Y$ is separating for H;*

(2) *if fsg in H, then $\sum_{F \in E} F(f)\bar{F}(g) = \langle f, g \rangle$ for every $E \in Y$;*

(3) *if $\sum_{F \in E_1} F(f)\bar{F}(g) = \sum_{F \in E_2} F(f)\bar{F}(g)$ for every $E_1, E_2 \in Y$, then fsg.*

Proof. Suppose $\phi: H \to K$ is an isomorphism, where K is a closed subspace of $H(\mathcal{A})$ for some quasimanual $\mathcal{A}$. For $x \in X(\mathcal{A})$, let $F_x: K \to \mathbb{C}$ be defined by $F_x(f) = f(x)$. We first show that $F_x \in K^*$. Let $A \in m(K)$. It is clear that $F_x \mid A$ is a linear functional. To show that $F_x \mid A$ is continuous, suppose $f_n \in A$ and $f_n \to f \in A$ in the norm topology of A. If $x \in E \in \mathcal{A}$, we have

$$|F_x(f_n) - F_x(f)|^2 = |f_n(x) - f(x)|^2$$
$$\leq \sum_{y \in E} |f_n(y) - f(y)|^2 = \|f_n - f\|^2 \to 0.$$

Hence, $F_x \mid A \in A^*$. For $F \in K^*$, it is easy to show that $F \circ \phi \in H^*$. Let

$$Y = \{\{F_x \circ \phi : x \in E\} : E \in \mathcal{A}\} \subseteq P(H^*).$$

To show that $\cup Y$ is separating for H, let $f \neq g \in H$. Then $\phi(f) \neq \phi(g) \in K$ so there exists an $x \in X(\mathcal{A})$ such that $\phi(f)(x) \neq \phi(g)(x)$. Now $F_x \circ \phi \in \cup Y$ and

$$(F_x \circ \phi)f = \phi(f)(x) \neq \phi(g)(x) = (F_x \circ \phi)g.$$

To show that (2) holds, let fsg in $H, E \in \mathcal{A}$, and

$$E' = \{F_x \circ \phi : x \in E\} \in Y.$$

Then $\phi(f)s\phi(g)$ in K and

$$\sum_{F\in E'} F(f)\bar{F}(g) = \sum_{x\in E} \phi(f)(x)\bar{\phi}(g)(x)$$
$$= \langle \phi(f), \phi(g)\rangle_K = \langle f, g\rangle_H.$$

Finally, suppose the hypothesis in (3) holds for $f, g \in H$. For $E_1, E_2 \in \mathcal{A}$ we have

$$E_1' = \{F_x \circ \phi : x \in E_1\} \in Y$$
$$E_2' = \{F_x \circ \phi : x \in E_2\} \in Y$$

and

$$\sum_{x\in E_1} \phi(f)(x)\bar{\phi}(g)(x) = \sum_{F\in E_1'} F(f)\bar{F}(g)$$
$$= \sum_{F\in E_2'} F(f)\bar{F}(g)$$
$$= \sum_{x\in E_2} \phi(f)(x)\bar{\phi}(g)(x).$$

Hence, $\phi(f)s\phi(g)$ in K and since ϕ is an isomorphism, we have fsg.

Conversely, suppose there exists a $Y \subseteq P(H^*)$ satisfying (1), (2), and (3). If $\mathcal{A} = Y$, then $X = X(\mathcal{A}) = \cup Y$ and $\mathcal{A}$ is a quasimanual. For $f \in H$, define $\phi(f): X \to \mathbb{C}$ by $\phi(f)(x) = x(f)$. To show that $\phi(f) \in H(\mathcal{A})$, since fsf, applying Condition 2 we have for every $E \in Y$

$$\sum_{x\in E} |\phi(f)(x)|^2 = \sum_{x\in E} |x(f)|^2 = \|f\|^2.$$

It follows that ϕ is a map from H into $H(\mathcal{A})$. To show that ϕ is injective, suppose $f \neq g \in H$. Since X is separating for H, there exists an $x \in X$ such that $x(f) \neq x(g)$. Hence, $\phi(f)(x) \neq \phi(g)(x)$ so $\phi(f) \neq \phi(g)$. Suppose fsg in H. If $E_1, E_2 \in \mathcal{A}$, then applying Condition 2 gives

$$\sum_{x\in E_1} \phi(f)(x)\bar{\phi}(g)(x) = \sum_{x\in E_1} x(f)\bar{x}(g) = \langle f, g\rangle$$
$$= \sum_{x\in E_2} \phi(f)(x)\bar{\phi}(g)(x).$$

Hence, $\phi(f)s\phi(g)$ and $\langle \phi(f), \phi(g)\rangle_{H(\mathcal{A})} = \langle f, g\rangle_H$. Moreover, it is clear that $\phi(\alpha f + \beta g) = \alpha\phi(f) + \beta\phi(g)$ for any $\alpha, \beta \in \mathbb{C}$. Also, it easily follows

AMPLITUDE SPACES 129

from (3) that if $\phi(f)s\phi(g)$, then fsg. We now show that $\phi(H)$ is a closed subspace of $H(\mathcal{A})$. Let $A \in m(\phi(H))$. It is clear that $\phi^{-1}(A) \in s(H)$. Suppose $f \in H$ and $f \in \phi^{-1}(A)^s$. Then $\phi(f) \in A^s$ and since A is maximal in $\phi(H)$, $\phi(f) \in A$. Hence, $f \in \phi^{-1}(A)$ so $\phi^{-1}(A) \in m(H)$ and $\phi^{-1}(A)$ is a Hilbert space. Since $\|\phi(f)\| = \|f\|$ for all $f \in H$, it follows that A is a Hilbert space in $\phi(H)$. We conclude that $\phi(H)$ is a closed subspace of $H(\mathcal{A})$ and that H and $\phi(H)$ are isomorphic. □

Theorem 4.10 characterizes partial Hilbert spaces which are isomorphic to a closed subspace of $H(\mathcal{A})$ for some quasimanual $\mathcal{A}$. The next result shows that every strong dual H^σ has this property.

Theorem 4.11. *If H is a partial Hilbert space, then H^σ is isomorphic to a closed subspace of $H(\mathcal{A})$ for some coherent, irredundant quasimanual $\mathcal{A}$.*

Proof. For each $A \in m(H)$, let E_A be the collection of all orthonormal bases for the Hilbert space A. Let $\mathcal{A}$ be the quasimanual $\{E : E \in E_A, A \in m(H)\}$ and let $X = X(\mathcal{A}) = \cup \mathcal{A}$. For $F \in H^\sigma$, define $\phi(F) = F \mid X$. To show that $\phi(F) \in H(\mathcal{A})$, let $E \in \mathcal{A}$. Then $E \in E_A$ for some $A \in m(H)$ and we have

$$\sum_{x \in E} |\phi(f)(x)|^2 = \sum_{x \in E} |F(x)|^2$$
$$= \sum_{x \in E} |\langle x, f_F^A \rangle|^2$$
$$= \|f_F^A\|^2 = \|F\|^2.$$

Hence, $\phi : H^\sigma \to H(\mathcal{A})$. To show that f is injective, suppose $\phi(F) = \phi(G)$ so $F \mid X = G \mid X$. For $f \in H$, we have $f \in A \in m(H)$ for some A. If $f = 0$, then clearly $F(f) = G(f)$. Otherwise, $f/\|f\| \in E$ for some $E \in E_A$. Hence,

$$F(f) = \|f\| F\left(\frac{f}{\|f\|}\right)$$
$$= \|f\| G\left(\frac{f}{\|f\|}\right)$$
$$= G(f).$$

Thus, $F = G$ and ϕ is injective. For $F, G \in H^\sigma$, $A \in m(H)$ we have for any $E \in E_A$

$$\langle f_F^A, f_G^A \rangle = \sum_{x \in E} \langle f_F^A, x \rangle \langle x, f_G^A \rangle$$
$$= \sum_{x \in E} G(x)\bar{F}(x)$$
$$= \sum_{x \in E} \phi(G)(x)\bar{\phi}(F)(x).$$

It follows that FsG in H^σ if and only if $\phi(f)s\phi(g)$ in $H(\mathcal{A})$. Moreover, if FsG in H^σ we have $\langle G, F \rangle = \langle \phi(G), \phi(F) \rangle$. It is also clear that if FsG in H^σ and $\alpha, \beta \in \mathbb{C}$, then $\phi(\alpha F + \beta G) = \alpha\phi(F) + \beta\phi(G)$. It is straightforward to show that $\phi(H)$ is a closed subspace of $H(\mathcal{A})$. To show that $\mathcal{A}$ is irredundant, suppose $E_1, E_2 \in \mathcal{A}$ with $E_1 \subseteq E_2$. If $E_1, E_2 \in E_A$ for some $A \in m(H)$, then clearly $E_1 = E_2$. Otherwise, $E_1 \in E_A$, $E_2 \in E_B$ for $A, B \in m(H)$. Applying Lemmas 4.1 and 4.2 we have

$$B = B^s = [E_2]^s = E_2^s \subseteq E_1^s = [E_1]^s = A^s = A.$$

Hence, $A = B$ and again $E_1 = E_2$. To show that $\mathcal{A}$ is coherent, let $A, B \in \mathcal{E}(\mathcal{A})$ with $A \subseteq B^\perp$. It follows that fsg for every $f \in A$, $g \in B$. Hence, there exists a $C \in m(H)$ such that $A \cup B \subseteq C$. We can then extend $A \cup B$ to an orthonormal basis for C. Thus, $A \cup B \subseteq E \in E_C$ so $A \cup B \in \mathcal{E}(\mathcal{A})$. □

Although our previous discussion of amplitude functions on a quasimanual has been useful in clarifying the role of superposition and the Hilbert space structure in quantum mechanics, it has not yet reached the heart of the matter. The main importance of amplitude functions is that when they are summed over an event and the resulting modulus squared is taken, the corresponding probabilities describe quantum interferences. This phenomena is not apparent in one-stage experiments. As pointed out in Section 3.7, the above sum rule is only used for compound (or multi-stage) experiments. Such experiments are described by operational or tensor products of quasimanuals. We now briefly consider tensor products of amplitude functions. Further examples and applications will be given in later sections.

Let $\mathcal{A}_1, \mathcal{A}_2$ be quasimanuals with $X_1 = X(\mathcal{A}_1)$, $X_2 = X(\mathcal{A}_2)$. Recall that the cartesian product $\mathcal{A}_1 \times \mathcal{A}_2$ is a quasimanual with outcome set $X_1 \times X_2$. For $f_1 \in H(\mathcal{A}_1)$, $f_2 \in H(\mathcal{A}_2)$, define $f_1 f_2 : X_1 \times X_2 \to \mathbb{C}$ by $f_1 f_2(x_1 x_2) = f_1(x_1) f_2(x_2)$. Then $f_1 f_2 \in H(\mathcal{A}_1 \times \mathcal{A}_2)$. Indeed, let $E, F \in$

$\mathcal{A}_1 \times \mathcal{A}_2$ where $E = E_1 E_2$, $F = F_1 F_2$. Then

$$\sum_{x_1 x_2 \in E} |f_1 f_2(x_1 x_2)|^2 = \sum_{x_1 \in E_1} |f_1(x_1)|^2 \sum_{x_2 \in E_2} |f_2(x_2)|^2$$
$$= \sum_{x_1 \in F_1} |f_1(x_1)|^2 \sum_{x_2 \in F_2} |f_2(x_2)|^2 \qquad (4.3)$$
$$= \sum_{x_1 x_2 \in F} |f_1 f_2(x_1 x_2)|^2.$$

We also see from (4.3) that $\|f_1 f_2\| = \|f_1\| \|f_2\|$. For $A_1 \subseteq H(\mathcal{A}_1)$, $A_2 \subseteq H(\mathcal{A}_2)$, define

$$A_1 A_2 = \{f_1 f_2 : f_1 \in A_1, f_2 \in A_2\} \subseteq H(\mathcal{A}_1 \times \mathcal{A}_2).$$

Lemma 4.12. *If $A_1 \in s(H(\mathcal{A}_1))$, $A_2 \in s(H(\mathcal{A}_2))$, then*

$$A_1 A_2 \in s(H(\mathcal{A}_1 \times \mathcal{A}_2)).$$

Proof. If $f_1 f_2, g_1 g_2 \in A_1 A_2$ and $E = E_1 E_2$, $F = F_1 F_2 \in \mathcal{A}_1 \times \mathcal{A}_2$, then

$$\sum_{x_1 x_2 \in E} f_1 f_2(x_1 x_2) \bar{g}_1 \bar{g}_2(x_1 x_2) = \sum_{x_1 \in E_1} f_1(x_1) \bar{g}_1(x_1) \sum_{x_2 \in E_2} f_2(x_2) \bar{g}_2(x_2)$$
$$= \sum_{x_1 \in F_1} f_1(x_1) \bar{g}_1(x_1) \sum_{x_2 \in F_2} f_2(x_2) \bar{g}_2(x_2)$$
$$= \sum_{x_1 x_2 \in F} f_1 f_2(x_1 x_2) \bar{g}_1 \bar{g}_2(x_1 x_2). \quad \square$$

If $A_1 \in s(H(\mathcal{A}_1))$, $A_2 \in s(H(\mathcal{A}_2))$, we define $A_1 \hat{\otimes} A_2 = [A_1 A_2]$. As we shall see, $A_1 \hat{\otimes} A_2$ can be thought of as a tensor product of A_1 and A_2 in the ordinary sense. If H_1 and H_2 are Hilbert spaces, we denote their usual tensor product by $H_1 \otimes H_2$.

Theorem 4.13. *If $A_1 \in s(H(\mathcal{A}_1))$, $A_2 \in s(H(\mathcal{A}_2))$ are Hilbert spaces, then $A_1 \otimes A_2$ and $A_1 \hat{\otimes} A_2$ are isomorphic and the map $f_1 \otimes f_2 \mapsto f_1 f_2$, $f_1 \in A_1$, $f_2 \in A_2$, extends to a unique isomorphism.*

Proof. Let ψ_α, ϕ_β, $\alpha \in A$, $\beta \in B$ be orthonormal bases for A_1 and A_2, respectively. It is well known that $M = \{\psi_\alpha \otimes \phi_\beta : \alpha \in A, \beta \in B\}$ is an orthonormal basis for $A_1 \otimes A_2$. If $E_1 \in \mathcal{A}_1$, $E_2 \in \mathcal{A}_2$, then we have for any $\alpha, \alpha' \in A$, $\beta, \beta' \in B$

$$\langle \psi_\alpha \phi_\beta, \psi'_\alpha \phi'_\beta \rangle = \sum_{x_1 x_2 \in E_1 E_2} \psi_\alpha(x_1) \phi_\beta(x_2) \bar{\psi}'_\alpha(x_1) \bar{\phi}'_\beta(x_2)$$

$$= \sum_{x_1 \in E_1} \psi_\alpha(x_1) \bar{\psi}'_\alpha(x_1) \sum_{x_2 \in E_2} \phi_\beta(x_2) \bar{\phi}'_\beta(x_2)$$

$$= \langle \psi_\alpha, \psi'_\alpha \rangle \langle \phi_\beta, \phi'_\beta \rangle = \delta_{\alpha \alpha'} \delta_{\beta \beta'}.$$

It now easily follows that

$$\tilde{M} = \{\psi_\alpha \phi_\beta : \alpha \in A, \quad \beta \in B\}$$

is an orthonormal basis for span $A_1 A_2$ and hence for $A_1 \hat{\otimes} A_2$. Define the map $T: \operatorname{span} M \to \operatorname{span} \tilde{M}$ by

$$T(\sum c_{\alpha\beta} \psi_\alpha \otimes \phi_\beta) = \sum c_{\alpha\beta} \psi_\alpha \phi_\beta.$$

Then it is clear that T is a linear map and $\langle Tf, Tg \rangle = \langle f, g \rangle$ for all $f, g \in \operatorname{span} M$. It follows that T has a unique extension to an isomorphism $\bar{T}$ from $A_1 \otimes A_2$ to $A_1 \hat{\otimes} A_2$. Moreover, it is easy to show that $\bar{T}(f_1 \otimes f_2) = f_1 f_2$ for all $f_1 \in A_1$, $f_2 \in A_2$. $\square$

Corollary 4.14. *If $A_i \in m(H(\mathcal{A}_i))$, $i = 1, 2, \ldots$, then the map $f_1 \otimes f_2 \mapsto f_1 f_2$ extends to a unique isomorphism from $A_1 \otimes A_2$ to $A_1 \hat{\otimes} A_2$.*

The next result shows that if $A_i \in s[H(\mathcal{A}_i)]$, $i = 1, 2$, are unital, then $A_1 \hat{\otimes} A_2$ is defined strictly in terms of the algebraic s relation. In this way the "tensor product" $A_1 \hat{\otimes} A_2$ is intrinsically defined in terms of superpositions.

Theorem 4.15. *If $A_i \in s[H(\mathcal{A}_i)]$, $i = 1, 2$, are unital, then $A_1 \hat{\otimes} A_2$ is the unique maximal s-set containing $A_1 A_2$. Moreover,*

$$A_1 \hat{\otimes} A_2 = (A_1 A_2)^{ss} = (A_1 A_2)^s.$$

Proof. Applying Theorem 4.8, it suffices to show that $A_1 A_2$ is unital in $H(\mathcal{A}_1 \times \mathcal{A}_1)$. If $x_i \in X_i$, $i = 1, 2$, then there exist $f_i \in A_i$ such that $f_i(x_i) \neq 0$, $f_i(y_i) = 0$ for all $y_i \perp x_i$, $i = 1, 2$. Then $f_1 f_2 \in A_1 A_2$ and $f_1 f_2(x_1 x_2) \neq 0$, $f_1 f_2(y_1 y_2) = 0$, for all $y_1 y_2 \perp x_1 x_2$. $\square$

4.3 Frame Manuals

Let K be a complex Hilbert space with unit sphere $S(K)$ and let $\mathcal{F} = \mathcal{F}(K)$ be the collection of all maximal orthogonal sets in $S(K)$. Then $\mathcal{F}(K)$ is a *frame manual* with outcome set $S(K)$. For $v \in K$, define the function $f_v \colon S(K) \to \mathbb{C}$ by $f_v(x) = \langle v, x \rangle$. It is easy to show that $f_v \in H(\mathcal{F})$ and $\|f_v\| = \|v\|$. Moreover, $f_v s f_u$ for all $v, u \in K$ and $\langle f_v, f_u \rangle = \langle v, u \rangle$. We denote the set of pure elements in $H(\mathcal{F})$ by $H_p(\mathcal{F})$. The next result characterizes $H(\mathcal{F})$ and $H_p(\mathcal{F})$ where $\dim K \geq 3$.

Theorem 4.16. *Let $\mathcal{F} = \mathcal{F}(K)$ be a frame manual with $\dim K \geq 3$. Then $f \in H(\mathcal{F})$ if and only if there exists a unique, positive, trace class operator T_f on K such that $f(x) = \chi(x) \langle T_f x, x \rangle^{1/2}$ where $\chi \in G$ is unique on $\operatorname{supp}(f)$. Also, $f \in H_p(\mathcal{F})$ if and only if $f = \chi f_v$ where $v \in K$ is unique within a multiple of modulus 1 and $\chi \in G$ is unique on $\operatorname{supp}(f)$.*

Proof. If $f \in H(\mathcal{F})$, then it follows from the proof of Gleason's theorem [Gleason, 1957] that there exists a unique, positive, trace class operator T_f such that $|f(x)|^2 = \langle T_f x, x \rangle$ for all $x \in S(K)$. For $x \in \operatorname{supp}(f)$, let $\chi(x) = f(x)/|f(x)|$ and let $\chi(x) = 1$ for $x \in \ker f$. Then $\chi \in G$ and

$$f(x) = \chi(x)|f(x)| = \chi(x)\langle T_f x, x \rangle^{\frac{1}{2}}$$

for all $x \in S(K)$. For uniqueness, suppose $f(x) = \chi_1(x)\langle Tx, x \rangle^{1/2}$ for a positive, trace class operator T and a $\chi_1 \in G$. Then

$$\langle Tx, x \rangle = |f(x)|^2 = \langle T_f x, x \rangle$$

for all $x \in S(K)$. Hence, $T = T_f$ and $\chi_1 = \chi$ on $\operatorname{supp}(f)$.

Conversely, suppose $f \colon S(K) \to \mathbb{C}$ has the form $f(x) = \chi(x)\langle Tx, x \rangle^{1/2}$ for a positive, trace class operator T and a $\chi \in G$. Then for any $E \in \mathcal{F}$ we have

$$\sum_{x \in E} |f(x)|^2 = \sum_{x \in E} \langle Tx, x \rangle = \operatorname{tr}(T).$$

Hence, $f \in H(\mathcal{F})$.

Now let $f \in H_p(\mathcal{F})$. Then from the above, $f(x) = \chi_0(x)\langle Tx, x \rangle^{1/2}$ for a positive, trace class operator T and a $\chi_0 \in G$. If $T = 0$, then $f(x) = \langle 0, x \rangle$ for all $x \in S(K)$. Otherwise, by Lemma 4.5b, $\mu_f/\|f\|^2$ is an extremal state of the form $x \mapsto \langle Tx, x \rangle / \operatorname{tr}(T)$. It follows that $T = \lambda P_u$ where P_u

is a one-dimensional projection onto some $u \in S(K)$ and $\lambda > 0$. Letting $v = \lambda^{1/2} u$ we have

$$f(x) = \chi_0(x) \langle \lambda P_u x, x \rangle^{\frac{1}{2}}$$
$$= \chi_0(x) |\langle v, x \rangle|$$

for all $x \in S(K)$. Define $\chi(x) = 1$ if $\langle v, x \rangle = 0$ and

$$\chi(x) = \frac{\chi_0(x) |\langle v, x \rangle|}{\langle v, x \rangle}$$

if $\langle v, x \rangle \neq 0$. Then $\chi \in G$. and $f(x) = \chi(x) \langle v, x \rangle$ for all $x \in S(K)$. For uniqueness, suppose $f(x) = \chi_1(x) \langle v_1, x \rangle$ for some $v_1 \in K$, $\chi_1 \in G$. Then $|\langle v_1, x \rangle| = |\langle v, x \rangle|$ for all $x \in S(K)$. Then $v_1 \perp x$ if and only if $v \perp x$. Hence, there exists an $\alpha \in \mathbb{C}$ such that $v_1 = \alpha v$. Moreover, $|\alpha| |\langle v, x \rangle| = |\langle v, x \rangle|$. Thus, if $v \neq 0$, $|\alpha| = 1$.

Conversely, it is clear that any function of the form $f = \chi f_v$ is in $H_p(\mathcal{F})$. □

The next result gives a large class of maximal s-sets in $H(\mathcal{F})$. For a fixed $\chi \in G$ define $H_\chi(\mathcal{F}) = \{\chi f_v : v \in K\}$.

Theorem 4.17. (a) $H_\chi(\mathcal{F})$ is a unital s-set. (b) $H_\chi(\mathcal{F})$ is a Hilbert space and the map $\phi: H_\chi(\mathcal{F}) \to K$ given by $\phi(\chi f_v) = v$ is an isomorphism. (c) $H_\chi(\mathcal{F}) \in m(H(\mathcal{F}))$.

Proof. (a) If $x \in S(K)$, then $|\chi f_x(x)| = \|f_x\|$. Hence, $H_\chi(\mathcal{F})$ is unital. Moreover, it is clear that $H_\chi(\mathcal{F})$ is an s-set. (b) The map ϕ is well-defined since $\chi f_x = \chi f_u$ implies $v = u$. It is clear that $H_\chi(\mathcal{F})$ is a linear space and that $\alpha \chi f_v = \chi f_{\alpha v}$, $\chi f_v + \chi f_u = \chi f_{v+u}$ for all $\alpha \in \mathbb{C}$, $v, u \in K$. Moreover, ϕ is surjective. To show that ϕ is injective, suppose $\chi f_v \neq \chi f_u$. Then there is an $x \in S(K)$ with

$$\chi(x) \langle v, x \rangle \neq \chi(x) \langle u, x \rangle$$

so $v \neq u$. To show that ϕ is an isomorphism, we have for any $E \in \mathcal{F}$, $v, u \in K$,

$$\langle v, u \rangle = \sum_{x \in E} \langle v, x \rangle \langle x, u \rangle$$
$$= \sum_{x \in E} \chi(x) \langle v, x \rangle \bar{\chi}(x) \overline{\langle u, x \rangle}$$
$$= \langle \chi f_v, \chi f_u \rangle.$$

The result now follows. (c) This follows from (a), (b) and Corollary 4.9b. □

Recall that a quasimanual $\mathcal{A}'$ is a *subquasimanual* of a quasimanual $\mathcal{A}$ if $\mathcal{A}' \subseteq \mathcal{A}$. A subquasimanual of a frame manual is called a *frame subquasimanual*. For a quasimanual $\mathcal{A}$, a set $A \subseteq H(\mathcal{A})$ is *strongly separating* if for $x \neq y \in X(\mathcal{A})$ there exists an $f \in A$ such that $f(x) = \|f\| \neq f(y)$. Notice that a strongly separating set is unital. Since frame subquasimanuals form the structural basis for traditional quantum mechanics, it is important to characterize such quasimanuals. It turns out that they are precisely those quasimanuals which admit sufficiently rich superpositions. This richness is given in terms of the strongly separating condition.

Theorem 4.18. *A quasimanual $\mathcal{A}$ is isomorphic to a frame subquasimanual if and only if $H(\mathcal{A})$ contains a strongly separating s-set.*

Proof. Suppose $\mathcal{A}$ is isomorphic to a frame subquasimanual $\mathcal{A}' \subseteq \mathcal{F}(K)$ under an isomorphism $\phi: X(\mathcal{A}) \to X(\mathcal{A}')$. Let $X = X(\mathcal{A})$ and $X' = X(\mathcal{A}')$. For each $v \in X'$, define $g_v: X \to \mathbb{C}$ by $g_v(x) = \langle v, \phi(x) \rangle$, and let $A = \{g_v : v \in X'\}$. It is clear that $A \subseteq H(\mathcal{A})$. To show that $A \in s(H(\mathcal{A}))$, suppose that $g_v, g_u \in A$, and $E, F \in \mathcal{A}$. Then since $\phi(E), \phi(F) \in \mathcal{A}'$, we have

$$\sum_{x \in E} g_v(x) \bar{g}_u(x) = \sum_{x \in E} \langle v, \phi(x) \rangle \langle \phi(x), u \rangle$$

$$= \langle v, u \rangle = \sum_{x \in F} \langle v, \phi(x) \rangle \langle \phi(x), u \rangle$$

$$= \sum_{x \in F} g_v(x) \bar{g}_u(x).$$

To show that A is strongly separating, suppose that $x \neq y \in X$. Then $\phi(x) \neq \phi(y)$ and

$$g_{\phi(x)}(x) = \|\psi(x)\|^2 - 1 - \|g_{\phi(x)}\|.$$

Now suppose that $g_{\phi(x)}(y) = 1$. Then

$$\|\phi(x)\| \, \|\phi(y)\| = g_{\phi(x)}(y)$$
$$= |\langle \phi(x), \phi(y) \rangle|.$$

Since we have equality in Schwarz's inequality, there exists an $\alpha \in \mathbb{C}$ such that $\phi(x) = \alpha \phi(y)$. But then

$$1 = g_{\phi(x)}(y) = \langle \phi(x), \phi(y) \rangle$$
$$= \alpha \langle \phi(y), \phi(y) \rangle = \alpha.$$

Hence, $\phi(x) = \phi(y)$ which is a contradiction.

Conversely, suppose that $H(\mathcal{A})$ contains a strongly separating s-set A. Since A is unital, it follows from Theorem 4.8 that $K = A^s$ is a Hilbert space. For $x \in X$, we show that there exists a unique $g \in K$ such that $g(x) = \|g\| = 1$. Since A is strongly separating, there exists a $0 \neq f \in A$ such that $f(x) = \|f\|$. Then if $g = f/\|f\|$, we have $g \in K$ and $g(x) = \|g\| = 1$. Now suppose $g, h \in K$ with

$$g(x) = h(x) = \|g\| = \|h\| = 1.$$

If $y \perp x$, then it follows that $g(y) = h(y) = 0$. Thus, if $x \in E \in \mathcal{A}$, we have

$$\langle g, h \rangle = \sum_{y \in E} g(y)\bar{h}(y)$$
$$= g(x)\bar{h}(x)$$
$$= \|g\| \|h\|.$$

We conclude that there exists an $\alpha \in \mathbb{C}$ such that $g = \alpha h$. Hence, $1 = \alpha\|h\|^2 = \alpha$ so $g = h$. Define $f: X \to S(K)$ by $\phi(x) = g_x$ where $g_x(x) = \|g_x\| = 1$. To show that ϕ is injective, suppose that $x, y \in X$ with $x \neq y$. Since A is strongly separating, there exists a $0 \neq f \in A$ such that $f(x) = \|f\| \neq f(y)$. Then $g_x = f/\|f\|$ and $g_x(y) \neq 1$. Hence,

$$\phi(x) = g_x \neq g_y = \phi(y).$$

Now consider the quasimanual $\phi(\mathcal{A})$. To complete the proof, it suffices to show that $\phi(E)$ is an orthonormal basis in K for each $E \in \mathcal{A}$. For $x \neq y \in E$, we have

$$\langle g_x, g_y \rangle = \sum_{z \in E} g_x(z)\bar{g}_y(z)$$
$$= \sum_{z \in E} \delta_{xz}\delta_{yz} = 0$$

so $\phi(x) \perp \phi(y)$. Now let $f \in K$ and suppose $f \perp \phi(E)$. If $x \in E$, then $f \perp g_x$. Hence,

$$0 = \langle f, g_x \rangle = \sum_{y \in E} f(y)\bar{g}_x(y) = f(x).$$

It follows that $f(x) = 0$ for all $x \in E$, so

$$\|f\|^2 = \sum_{x \in E} |f(x)|^2 = 0.$$

Hence, $f = 0$ and $\phi(E)$ is an orthonormal basis. □

We have actually proved a stronger result than that given in Theorem 4.18.

Corollary 4.19. *If $\mathcal{A}$ admits a strongly separating s-set $A \subseteq H(\mathcal{A})$, then there exists an isomorphism ϕ from $\mathcal{A}$ to a subquasimanual of the frame manual $\mathcal{F}(A^s)$. Moreover, for each $f \in A^s$ we have $f(x) = \langle f, \phi(x) \rangle$ for all $x \in X(\mathcal{A})$.*

4.4 Transition Amplitudes

Section 3.7 considered amplitude functions of the form f_x where $f_x(y)$ was the amplitude of a transition of the system from outcome x to outcome y. These amplitudes were then used to compute quantum probabilities. In this context, the transition amplitudes $g(x,y) = f_x(y)$ play a fundamental role in quantum probability. The amplitude functions then arise when one keeps the initial outcome fixed. We thus see that transition amplitude is a stronger concept than amplitude function. For this reason we can obtain stronger results using the former than we have previously obtained using the latter. The situation is similar to the relationship between conditional probability and probability in the classical theory. The present section motivates the basic properties of transition amplitudes and the following section incorporates these properties to obtain a definition of a transition amplitude space.

In previous sections, transition amplitudes (and amplitude functions) still played a secondary role since they were defined as complex-valued functions on a quasimanual with specified properties. If we are serious about the fact that the essence of quantum mechanics is given by transition amplitudes (or amplitude functions), then they should be the basic elements for an axiomatic development. The properties of these elements should be delineated and the axiomatic structure should be built upon these properties. We now proceed to this end.

Let X be the set of outcomes for the experiments that can be performed on a quantum system. For example, X might consist of the outcomes of compound Stern-Gerlach experiments on a spin-$\frac{1}{2}$ particle. If A is to be a transition amplitude on X, then A should be a map from $X \times X$ into $\mathbb{C}$, where we interpret $A(x, y)$ as the transition amplitude from outcome x to outcome y. To understand what this means, it is important to distinguish between two types of transition amplitudes, a static transition amplitude A_0, and a dynamic transition amplitude A_t, $t > 0$. Dynamic transition amplitudes are physically more transparent so we begin with these. It is clear that $A_t(x, y)$ should be interpreted as follows. If outcome x results initially, then the amplitude that outcome y results at time $t > 0$ is $A_t(x, y)$. We would then define the probability $P_t(x, y)$ of a transition from x to y in time t to be $|A_t(x, y)|^2$.

There are strong physical reasons that our system acts like a Markov process at the amplitude level. If outcome x results at time t_0, then future outcomes for $t > t_0$ depend only on x and not on outcomes prior to t_0. In other words, quantum evolutions do not appear to possess memory. An important property of Markov processes is that they satisfy the Chapman-Kolmogorov equation (see Section 1.6). In our framework this may be stated as follows. There exists a set E of intermediate outcomes such that

$$A_{t_1+t_2}(x, y) = \sum_{z \in E} A_{t_1}(x, z) A_{t_2}(z, y) \tag{4.4}$$

for all $t_1, t_2 > 0$ and for all $x, y \in X$. As far as we are concerned, (4.4) is the most important basic property of A_t.

Now we may view the static transition amplitude $A_0(x, y)$ as the limit $A_0(x, y) = \lim_{t \to 0} A_t(x, y)$. Taking the limit of (4.4) gives

$$A_0(x, y) = \sum_{z \in E} A_0(x, z) A_0(z, y) \tag{4.5}$$

for all $x, y \in X$. It is convenient and useful to extend the process A_t, $t > 0$, to negative t. This is usually done by defining $A_{-t}(x, y) = \bar{A}_t(y, x)$, $t > 0$. Thus, if the initial outcome is x and $t > 0$, we interpret $\bar{A}_t(y, x)$ to be the amplitude that the outcome was y at t units of time previously. If we let $t \to 0$, we obtain $A_0(x, y) = \bar{A}_0(y, x)$ and we can write (4.5) in the form

$$A_0(x, y) = \sum_{z \in E} A_0(x, z) \bar{A}_0(y, z) \tag{4.6}$$

TRANSITION AMPLITUDES   139

for all $x, y \in X$. Equation (4.6) will be used as the main defining relation for A_0 in Section 4.5. The other defining relation will be

$$A_0(x, x) = 1. \tag{4.7}$$

Equation (4.7) has a clear physical interpretation. We could also derive (4.7) using the plausible assumption that from an initial outcome x an intermediate outcome in E results with certainty. Then (4.6) gives

$$1 = \sum_{z \in E} P_0(x, z) = \sum_{z \in E} A_0(x, z) \bar{A}_0(x, z) = A_0(x, x).$$

Other interpretations of A_0 are possible, and these give alternative ways of justifying (4.5) (and subsequently (4.6)). We can view A_0 as a transmission amplitude. In this viewpoint [Mielnik, 1968, 1969], we represent the physical system as a particle. In fact, since we are making a statistical analysis, it is better to represent the system as an ensemble or beam of noninteracting particles. An outcome x can be thought of as a filter which transmits only particles with certain specified properties. We call this an x-filter. If an x-filter is placed in the path of the particle beam, then only particles which give an x result are transmitted. Suppose we now place an x-filter preceding a y-filter in the beam path. Then $A_0(x, y)$ is interpreted as the amplitude of transmission for the beam through the y-filter given that it is transmitted through the x-filter. The transmission probability $P_0(x, y) = |A_0(x, y)|^2$ is the ratio of the number of particles which pass through the y-filter to the number of particles which pass through the x-filter in the two filter experiment.

Now suppose we place a z-filter between the x and y-filters. Let $A_0(x, z, y)$ be the transmission amplitude for this set of three filters and let $P_0(x, z, y) = |A_0(x, z, y)|^2$ be the corresponding transmission probability. It is then clear that

$$P_0(x, z, y) = P_0(x, z) P_0(z, y). \tag{4.8}$$

However, at a deeper level we propose that

$$A_0(x, z, y) = A_0(x, z) A_0(z, y). \tag{4.9}$$

Property 4.9 has been emphasized by Feynman and others [Feynman and Hibbs, 1965; Landé, 1965] (cf. Section 3.7) and is called the product rule for transmission amplitudes. Of course, (4.8) easily follows from (4.9). We next propose the existence of a complete set E of z-filters. Such a set E has the property that an individual particle of the beam is transmitted by precisely one of the filters in E. That is, E is a selection process that classifies particles into distinct categories. There may be many such selection processes, but the important point is that there exists at least one. It is now natural to assume that $A_0(x,y)$ is the sum of the transmission amplitudes over the various particle categories in E. That is,

$$A_0(x,y) = \sum_{z \in E} A_0(x,z,y). \tag{4.10}$$

Then (4.5) follows from (4.9) and (4.10).

It should be mentioned that (4.5) does not hold for transition probabilities. For these, only the much weaker relation $\sum_{z \in E} P_0(x,z) = 1$, holds. Because of this less restrictive condition, there are pathological examples of transition probabilities which do not come from transition amplitudes. Although some of these examples may have physical significance [Mielnik, 1968, 1969], the situation is not entirely clear. In any case (4.5) is not only physically plausible for a large class of situations, it has strong mathematical consequences.

Still another interpretation is that $A_0(x,y)$ gives an amplitude for the information that y has in common with x [Cantoni, 1975]. An analysis similar to that given previously can be used to justify (4.5) in this situation.

In Hilbert space quantum mechanics, (4.4), (4.5), (4.6), and (4.7) follow from the Hilbert space structure. In this case, an outcome (or equivalently a pure state) is represented by a unit vector x in a complex Hilbert space $\mathcal{H}$. We then define $A_0(x,y)$ to be the inner product $\langle x,y \rangle$ (see Section 2.3). Then (4.7) is just the relation $\|x\| = 1$. If E is an orthonormal basis for $\mathcal{H}$, then (4.5) (or (4.6)) is Parseval's equality. The time evolution for outcomes (or pure states) is given by a one-parameter unitary group U_t and the dynamic transition amplitude is defined as $A_t(x,y) = \langle U_t x, y \rangle$. Then (4.4) follows from

$$\begin{aligned} A_{t_1+t_2}(x,y) &= \langle U_{t_1+t_2} x, y \rangle = \langle U_{t_1} x, U_{t_2}^* y \rangle \\ &= \sum_{z \in E} \langle U_{t_1} x, z \rangle \langle z, U_{t_2}^* y \rangle \\ &= \sum_{z \in E} \langle U_{t_1} x, z \rangle \langle U_{t_2} z, y \rangle \\ &= \sum_{z \in E} A_{t_1}(x,z) A_{t_2}(z,y). \end{aligned}$$

4.5 Transition Amplitude Spaces

Let X be a nonempty set and let $A: X \times X \to \mathbb{C}$. We call a set $E \subseteq X$ an *A-set* if for every $x, y \in X$ we have $\sum_{z \in E} |A(x,z)A(y,z)| < \infty$ and

$$A(x,y) = \sum_{z \in E} A(x,z)\bar{A}(y,z) \tag{4.11}$$

Denote the collection of A-sets by $\mathcal{F}(A)$. We call $A: X \times X \to \mathbb{C}$ a *transition amplitude* if (i) $\mathcal{F}(A) \neq \emptyset$, and (ii) $A(x,x) = 1$ for all $x \in X$. These conditions were motivated in Section 4.4. From a mathematical standpoint, property (ii) is a mild normalization condition since it already follows from (i) and (4.11) that $A(x,x) \geq 0$. Also, notice that from (4.11) we have $\bar{A}(x,y) = A(y,x)$ for all $x,y \in X$ so we can rewrite (4.11) as

$$A(x,y) = \sum_{z \in E} A(x,z)A(z,y) \tag{4.12}$$

for all $x, y \in X$. If A is a transition amplitude on X we call (X, A) a *transition amplitude space* (*tas*). In the sequel, (X, A) will denote a tas. We say that $x, y \in X$ are *orthogonal* and write $x \perp y$ if $A(x,y) = 0$. It follows from Zorn's lemma that the collection of maximal orthogonal sets $\mathcal{M}(A)$ covers X; that is, $X = \cup \mathcal{M}(A)$. We call (X, A) *total* if $\mathcal{M}(A) = \mathcal{F}(A)$. The following lemma shows that (X, A) is total if $\mathcal{M}(A) \subseteq \mathcal{F}(A)$.

Lemma 4.20. *If (X, A) is a tas, then $\mathcal{F}(A) \subseteq \mathcal{M}(A)$.*

Proof. Let $E \in \mathcal{F}(A)$ and suppose $x, y \in E$ with $x \neq y$. Denoting $E \setminus \{x\}$ by E' we have

$$1 = A(x,x) = \sum_{z \in E} |A(x,z)|^2 = 1 + \sum_{x \in E'} |A(x,z)|^2.$$

Hence, $A(x,y) = 0$ so $x \perp y$. To show that $E \in \mathcal{M}(A)$ suppose $x \perp E$. Since $A(x,z) = 0$ for all $x \in E$ we have

$$1 = A(x,x) = \sum_{z \in E} |A(x,z)|^2 = 0.$$

This is a contradiction. □

A tas (X, A) is *strong* if $A(x,y) = 1$ implies $x = y$.

Example 4.4. The unit sphere $X(\mathcal{H})$ of a complex Hilbert space $\mathcal{H}$ with $A(x,y) = \langle x, y \rangle$ is a total, strong tas. In this case we use the notation $\mathcal{F}(\mathcal{H}) = \mathcal{F}(A) = \mathcal{M}(A)$. Of course, $\mathcal{F}(\mathcal{H})$ is the set of orthonormal bases for $\mathcal{H}$. According to our previous terminology, $\mathcal{F}(\mathcal{H})$ is the frame manual on $\mathcal{H}$. □

Example 4.5. Let X be a nonempty set and let $A \colon X \times X \to \mathbb{C}$ be defined by $A(x, y) = 1$ if $x = y$ and $A(x, y) = 0$ if $x \neq y$. Then X itself is the only A-set. Indeed, for every $x, y \in X$ we have

$$\sum_{z \in X} A(x, z) A(z, y) = A(x, y)$$

so X is an A-set. Moreover, if B is a proper subset of X and $x \notin B$, then

$$A(x, x) = 1 \neq 0 = \sum_{z \in B} A(x, z) A(z, x)$$

so $B \notin \mathcal{F}(A)$. It follows that (X, A) is a total, strong tas. □

Example 4.6. Let $\mathcal{H}$ be a complex Hilbert space and let $X = \{x \in \mathcal{H} : x \neq 0\}$. Define $A \colon X \times X \to \mathbb{C}$ by

$$A(x, y) = \frac{\langle x, y \rangle}{\|x\| \, \|y\|}.$$

Then $A(x, x) = 1$ and for any $E \in \mathcal{M}(A)$ we have

$$\sum_{z \in E} A(x, z) A(z, y) = \sum_{z \in E} \frac{\langle x, z \rangle}{\|x\| \, \|z\|} \frac{\langle z, y \rangle}{\|z\| \, \|y\|}$$

$$= \frac{\langle x, y \rangle}{\|x\| \, \|y\|} = A(x, y)$$

$x, y \in X$. Hence, (X, A) is a total tas. However, (X, A) is not strong since $A(x, y) = 1$ if and only if $x/\|x\| = y/\|y\|$ so, for example, $A(x, 2x) = 1$. □

Example 4.7. Let $A(x, y) = \langle x, y \rangle$ be the usual inner product on $\mathbb{C}^2$. Let $x = (1, 0)$, $y = (0, 1)$, $z = (1, 1)/\sqrt{2}$ be elements of $\mathbb{C}^2$ and let $X = \{x, y, z\}$. Then (X, A) is a strong tas with $\mathcal{F}(A) = \{\{x, y\}\}$. Moreover, $\{z\} \in \mathcal{M}(A)$ but $\{z\} \notin \mathcal{F}(A)$ so (X, A) is not total. □

TRANSITION AMPLITUDE SPACES 143

If (X, A) is a tas, it is easy to show that $\mathcal{F}(A)$ is a premanual. Thus, all the motivation and results that have been presented earlier for premanuals can be applied. However, the important point now is that the transition amplitude A is taken as basic and the manual structure has been derived from A. In this way the physics of the system is determined by the set of possible outcomes X and the transition amplitude A.

We now show that there is a close relationship between a tas and a corresponding natural frame manual. Mathematically, this result shows that the generalized Parseval equality (4.12) essentially characterizes the inner product and orthonormal bases of a Hilbert space. This happens even though there is no linear structure postulated for a tas.

A *representation* of a tas (X, A) into a Hilbert space $\mathcal{H}$ is a map $\phi \colon X \to \mathcal{H}$ such that $A(x,y) = \langle \phi(x), \phi(y) \rangle$ for all $x, y \in X$ and $\phi(E) \in \mathcal{F}(\mathcal{H})$ for some $E \in \mathcal{F}(A)$. Notice that if ϕ is a representation, then $\|\phi(x)\| = 1$ for all $x \in X$ and $x \perp y$ if and only if $\phi(x) \perp \phi(y)$. It follows that ϕ is injective on sets of mutually orthogonal elements. However, ϕ need not be injective on X. For instance, in Example 4.6 if we define $\phi \colon X \to \mathcal{H}$ by $\phi(x) = x/\|x\|$, then ϕ is a representation which is not injective. It follows from the next lemma that a representation of (X, A) into $\mathcal{H}$ is an outcome-preserving morphism from $\mathcal{F}(A)$ into $\mathcal{F}(\mathcal{H})$.

Lemma 4.21. *Let $\phi \colon X \to \mathcal{H}$ be a representation. Then $\phi(E) \in \mathcal{F}(\mathcal{H})$ for all $E \in \mathcal{F}(A)$. Conversely, if $E \subseteq X$ satisfies $\phi \mid E$ is injective and $\phi(E) \in \mathcal{F}(\mathcal{H})$, then $E \in \mathcal{F}(A)$.*

Proof. Since ϕ is a representation, there is an $F \in \mathcal{F}(A)$ such that $\phi(F) \in \mathcal{F}(\mathcal{H})$. Now let $E \in \mathcal{F}(A)$. By Lemma 4.20, $\phi(E)$ is an orthonormal set in $\mathcal{H}$. Moreover, for any $x \in F$ we have

$$\|\phi(x)\|^2 = A(x,x) = \sum_{z \in E} |A(x,z)|^2 = \sum_{z \in E} |\langle \phi(x), \phi(z) \rangle|^2.$$

It follows that

$$\phi(x) = \sum_{z \in E} \langle \phi(x), \phi(z) \rangle \phi(z)$$

for all $x \in F$. For $f \in \mathcal{H}$, if $f \perp \phi(z)$ for all $z \in E$, then $f \perp \phi(x)$ for all $x \in F$. Since $\phi(F) \in \mathcal{F}(\mathcal{H})$, $f = 0$. Thus, $\phi(E)$ is a maximal orthonormal set so $\phi(E) \in \mathcal{F}(\mathcal{H})$. Conversely, suppose $\phi \mid E$ is injective and $\phi(E) \in \mathcal{F}(\mathcal{H})$. Then for any $x, y \in X$ we have

$$A(x,y) = \langle \phi(x), \phi(z) \rangle = \sum_{z \in E} \langle \phi(x), \phi(z) \rangle \langle \phi(z), \phi(y) \rangle$$
$$= \sum_{z \in E} A(x,z) A(z,y).$$

Hence, $E \in \mathcal{F}(A)$. □

If (X, A) is a tas, then $\mathcal{F}(A)$ is a quasimanual (actually a premanual) with outcome set $X(\mathcal{F}) = \cup \mathcal{F}(A)$. For $x \in X$, define the function $f_x: X(\mathcal{F}) \to \mathbb{C}$ by $f_x(y) = A(x,y)$. Then $f_x \in H(\mathcal{F})$; that is, f_x is an amplitude function. Indeed, for $E \in \mathcal{F}(A)$ we have

$$\sum_{y \in E} |f_x(y)|^2 = \sum_{y \in E} |A(x,y)|^2 = A(x,x) = 1.$$

In this way, A generates a large collection $S = \{f_x : x \in X\}$ of amplitude functions. Since $f_x(x) = 1 = \|f_x\|$, S is unital. Moreover, S is an s-set in $H(\mathcal{F})$ since for any $E \in \mathcal{F}(A)$ we have

$$\sum_{z \in E} f_x(z) \bar{f}_y(z) = \sum_{z \in E} A(x,z) \bar{A}(y,z) = A(x,y).$$

In general, S need not be strongly separating unless (X, A) is strong. Because of this we can not apply Theorem 4.18 to prove that $\mathcal{F}(A)$ is isomorphic to a frame subquasimanual. However, we can prove the following.

Theorem 4.22. *A tas (X, A) admits a representation.*

Proof. Let $E \in \mathcal{F}(A)$ and let $\mathcal{H}$ be the Hilbert space

$$\mathcal{H} = \ell^2(E) = \{f: E \to \mathbb{C} : \sum_{x \in E} |f(x)|^2 < \infty\}$$

with inner product $\langle f, g \rangle = \sum_{x \in E} f(x) \bar{g}(x)$. For $x \in X$, define $f_x \in \mathcal{H}$ by $f_x(z) = A(x,z)$. Define $\phi: X \to \mathcal{H}$ by $\phi(x) = f_x$. Then for $x, y \in X$ we have

$$\langle \phi(x), \phi(y) \rangle = \sum_{z \in E} A(x,z) \bar{A}(y,z) = A(x,y).$$

If $z_1 \neq z_2 \in E$, by Lemma 4.20

$$\langle \phi(z_1), \phi(z_2) \rangle = A(z_1, z_2) = 0$$

so $\phi(E)$ is an orthonormal set. Also, for $z, z' \in E$,
$$f_z(z') = A(z, z') = \delta_{zz'}. \tag{4.13}$$
If $f \in \mathcal{H}$, then applying (4.13) gives
$$\langle f, f_z \rangle = \sum_{z' \in E} f(z') f_z(z') = f(z). \tag{4.14}$$
It follows from (4.14) that
$$\|f\|^2 = \sum_{z \in E} |f(z)|^2$$
$$= \sum_{z \in E} |\langle f, f_z \rangle|^2$$
$$= \sum_{z \in E} |\langle f, \phi(z) \rangle|^2.$$
We now conclude that $\phi(E) \in \mathcal{F}(\mathcal{H})$. □

The following corollary characterizes injective representations.

Corollary 4.23. *If (X, A) is a tas, the following statements are equivalent*

(i) *There exists an injective representation $\phi\colon X \to \mathcal{H}$.*

(ii) *(X, A) is strong.*

(iii) *$A(x, z) = A(y, z)$ for every $z \in X$ implies $x = y$.*

(iv) *Every representation of (X, A) is injective.*

Proof. (i)⇒(ii) Suppose (i) holds and $A(x, y) = 1$. Since $\langle \phi(x), \phi(y) \rangle = \|\phi(x)\| \|\phi(y)\|$, it follows from Schwarz's inequality that $\phi(x) = \alpha \phi(y)$ for some $\alpha \in \mathbb{C}$. But then
$$\alpha = \alpha \langle \phi(y), \phi(y) \rangle = \langle \phi(x), \phi(y) \rangle = 1$$
so $\phi(x) = \phi(y)$. Since ϕ is injective, $x = y$. (ii)⇒(iii) Suppose (ii) holds and $A(x, z) = A(y, z)$ for all $z \in X$. Then $A(x, y) = A(y, y) = 1$ so $x = y$. (iii)⇒(iv) Suppose (iii) holds and $\phi\colon X \to \mathcal{H}$ is a representation. If $\phi(x) = \phi(y)$, then for every $z \in X$ we have
$$A(x, z) = \langle \phi(x), \phi(z) \rangle = \langle \phi(y), \phi(z) \rangle = A(y, z).$$
Hence, $x = y$ so ϕ is injective. (iv)⇒(i) follows from Theorem 4.22. □

Since every tas admits a representation, we can immediately obtain additional properties of transition amplitudes.

Corollary 4.24. *If (X, A) is a tas, then the following conditions hold.*

(i) $|A(x, y)| \leq 1$ *for every* $x, y \in X$ *and* $|A(x, y)| = 1$ *if and only if* $A(x, z) = \overline{A(x, y)} A(y, z)$ *for all* $z \in X$.

(ii) *A is positive definite. That is, for any* $\alpha_1, \ldots, \alpha_n \in \mathbb{C}, x_1, \ldots, x_n \in X$, *we have*

$$\sum_{i,j} \alpha_i \bar{\alpha}_j A(x_i, x_j) \geq 0.$$

(iii) *Every $E \in \mathcal{F}(A)$ has the same cardinality.*

Proof. Let $\phi: X \to \mathcal{H}$ be a representation. (i) We have

$$|A(x, y)| = |\langle \phi(x), \phi(y) \rangle| \leq \|\phi(x)\| \|\phi(y)\| = 1.$$

If $|A(x, y)| = 1$, then there is an $\alpha \in \mathbb{C}$ such that $\phi(y) = \alpha \phi(x)$. Now

$$1 = \langle \phi(y), \phi(y) \rangle = \alpha \langle \phi(x), \phi(y) \rangle = \alpha A(x, y).$$

Hence, for any $z \in X$

$$A(x, z) = \alpha^{-1} A(y, z) = \overline{A(x, y)} A(y, z).$$

(ii) We have

$$\sum_{i,j} \alpha_i \bar{\alpha}_j A(x_i, x_j) = \sum_{i,j} \alpha_i \bar{\alpha}_j \langle \phi(x_i), \phi(x_j) \rangle$$

$$= \left\langle \sum \alpha_i \phi(x_i), \sum \alpha_j \phi(x_j) \right\rangle \geq 0.$$

(iii) If $E \in \mathcal{F}(A)$, then $\phi(E)$ is an orthonormal basis for $\mathcal{H}$ and all orthonormal bases have the same cardinality. □

The *dimension* $\dim(X, A)$ of a tas (X, A) is the cardinality of any $E \in \mathcal{F}(A)$.

Theorem 4.25. *Let (X, A) and (Y, B) be tas's. Then $\dim(X, A) = \dim(Y, B)$ if and only if (X, A) and (Y, B) admit representations in the same Hilbert space.*

Proof. If (X, A) and (Y, B) admit representations in $\mathcal{H}$, then

$$\dim(X, A) = \dim \mathcal{H} = \dim(Y, B).$$

Conversely, suppose $\dim(X,A) = \dim(Y,B)$ and let $\phi\colon X \to \mathcal{H}$ and $\psi\colon Y \to \mathcal{K}$ be representations. Since $\dim \mathcal{H} = \dim \mathcal{K}$, there exists a unitary transformation $U\colon \mathcal{K} \to \mathcal{H}$. Define $\psi'\colon Y \to \mathcal{H}$ by $\psi' = U\psi$. Then ψ' is a representation of Y since $\psi'(E) \in \mathcal{F}(\mathcal{H})$ for all $E \in \mathcal{F}(B)$ and for every $x,y \in Y$ we have
$$B(x,y) = \langle \psi(x), \psi(y) \rangle$$
$$= \langle U\psi(x), U\psi(y) \rangle$$
$$= \langle \psi'(x), \psi'(y) \rangle. \quad \square$$

4.6 Isomorphisms and A-Forms

Two tas's (X,A) and (Y,B) are *isomorphic* if there exists a bijection $J\colon X \to Y$ such that $A(x,y) = B(Jx,Jy)$ for all $x,y \in X$. We then call J an *isomorphism*. An isomorphism from X to X is called an *automorphism*. Notice that if J is an isomorphism from (X,A) to (Y,B), then $J\mathcal{F}(A) = \mathcal{F}(B)$.

Theorem 4.26. *Let (X,A) and (Y,B) be tas's. (a) If $J\colon X \to Y$ is an isomorphism, then for any representations $\phi\colon X \to \mathcal{H}$, $\psi\colon Y \to \mathcal{K}$ there exists a unique unitary transformation $U\colon \mathcal{H} \to \mathcal{K}$ such that $\psi J = U\phi$. (b) Conversely, if (X,A), (Y,B) are strong, $\phi\colon X \to \mathcal{H}$, $\psi\colon Y \to \mathcal{K}$ are representations and there exists a unitary transformation $U\colon \mathcal{H} \to \mathcal{K}$ such that $U\phi(X) = \psi(Y)$, then (X,A) and (Y,B) are isomorphic.*

Proof. (a) If $E \in \mathcal{F}(A)$, then $\phi(E) \in \mathcal{F}(\mathcal{H})$. Now $J(E) \in \mathcal{F}(B)$ and $\psi[J(E)] \in \mathcal{F}(\mathcal{K})$. Define the unitary transformation $U\colon \mathcal{H} \to \mathcal{K}$ by letting $U\phi(z) = \psi(Jz)$ for all $z \in E$ and extending by linearity and closure. Now for every $x \in X$, we have
$$\phi(x) = \sum_{z \in E} A(x,z)\phi(z)$$
and hence,
$$U\phi(x) = \sum_{z \in E} A(x,z)\psi(Jz).$$
Moreover,
$$\psi(Jz) = \sum_{z \in E} B(Jx, Jz)\psi(Jz)$$
$$= \sum_{z \in E} A(x,z)\psi(Jz).$$

Hence, $U\phi = \psi J$. To show that U is unique, let $V: \mathcal{H} \to \mathcal{K}$ be unitary and suppose $V\phi = \psi J$. Then $U \mid \phi(X) = V \mid \phi(X)$ and in particular U and V agree on $\phi(E) \in \mathcal{F}(\mathcal{H})$. It follows that $U = V$. (b) Define $J: X \to Y$ by $Jx = \psi^{-1} U\phi(x)$. Then J is a bijection and for all $x, y \in X$ we have

$$A(x, y) = \langle \phi(x), \phi(y) \rangle = \langle U\phi(x), U\phi(y) \rangle$$
$$= B(\psi^{-1} U\phi(x), \psi^{-1} U\phi(y))$$
$$= B(Jx, Jy). \quad \square$$

We say that two representations $\phi: X \to \mathcal{H}$ and $\psi: X \to \mathcal{K}$ are *unitarily equivalent* if there exists a unitary transformation $U: \mathcal{H} \to \mathcal{K}$ such that $\psi = U\phi$.

Corollary 4.27. *Any two representations of a tas are unitarily equivalent.*

Proof. Let J be the identity map I in Theorem 4.26a. $\quad \square$

Corollary 4.28. *Let (X, A) be a tas and let $\phi: X \to \mathcal{H}$ be a representation. If $J: X \to X$ is an automorphism, then there exists a unique unitary operator U on $\mathcal{H}$ such that $\phi J = U\phi$. Conversely, if (X, A) is strong, $U: \mathcal{H} \to \mathcal{H}$ is a unitary operator and $U\phi(X) = \phi(X)$, then $J = \phi^{-1} U\phi$ is an automorphism on X.*

We next consider the mathematical description of the time evolution of a system. A *one-parameter group of automorphisms* on a tas (X, A) is a map J from $\mathbb{R}$ to the automorphisms of X such that $J(0) = I$ and $J(s + t) = J(s)J(t)$ for all $s, t \in \mathbb{R}$. A one-parameter group of automorphisms J is *continuous* if $t \mapsto A(J(t)x, y)$ is continuous for all $x, y \in X$. The next result shows that there is a Hamiltonian operator associated with such a time evolution.

Theorem 4.29. *Let $\phi: X \to \mathcal{H}$ be a representation of the tas (X, A). If J is a continuous one-parameter group of automorphisms on (X, A), then there exists a unique self-adjoint operator T on $\mathcal{H}$ such that $\phi J(t) = e^{-itT} \phi$ for all $t \in \mathbb{R}$.*

Proof. By Corollary 4.28, there exist unique unitary operators $U(t)$ on $\mathcal{H}$ such that $\phi J(t) = U(t)\phi$ for all $t \in \mathbb{R}$. Since $U(0)\phi = \phi$ and

$$U(s + t)\phi = \phi J(s + t) = \phi J(s)J(t) = U(s)\phi J(t) = U(s)U(t)\phi$$

ISOMORPHISMS AND A-FORMS

we see that $U(0) = I$ and $U(s+t) = U(s)U(t)$. Let $E \in \mathcal{F}(A)$ and hence $\phi(E) \in \mathcal{F}(\mathcal{H})$. If $f, g \in \mathcal{H}$, we have

$$\langle U(t)f, g\rangle = \left\langle U(t) \sum_{z \in E} \langle f, \phi(z)\rangle \phi(z), \sum_{z \in E} \langle g, \phi(z)\rangle \phi(z)\right\rangle$$

$$= \sum_{z,z'} \langle f, \phi(z)\rangle\langle \phi(z'), g\rangle\langle U(t)\phi(z), \phi(z')\rangle$$

$$= \sum_{z,z'} \langle f, \phi(z)\rangle\langle \phi(z'), g\rangle\langle \phi[J(t)z], \phi(z')\rangle$$

$$= \sum_{z,z'} \langle f, \phi(z)\rangle\langle \phi(z'), g\rangle A(J(t)z, z').$$

It follows that $t \mapsto \langle U(t)f, g\rangle$ is continuous, so $U(t)$ is a weakly continuous one-parameter group of unitary operators on $\mathcal{H}$. By Stone's theorem there exists a unique self-adjoint operator T on $\mathcal{H}$ such that $U(t) = e^{-itT}\phi$ for all $t \in \mathbb{R}$. □

If (X, A) is a tas, a map $B: X \times X \to \mathbb{C}$ is called a *form*. A sequence $(\alpha_i, x_i) \in \mathbb{C} \times X$ is called a *null sequence* if for every $y \in X$ we have

$$\sum_i |\alpha_i A(x_i, y)| < \infty \text{ and } \sum_i \alpha_i A(x_i, y) = 0. \quad (4.15)$$

We denote the set of null sequences by $n(A)$. We call a form B an *A-form* if for all $y \in X$ and $(\alpha_i, x_i) \in n(A)$ we have

$$\sum_i \alpha_i B(x_i, y) = \sum_i \bar{\alpha}_i B(y, x_i) = 0. \quad (4.16)$$

Notice that (4.16) is a type of absolute continuity condition. We now prove a simple variant of the Radon-Nikodym theorem.

Lemma 4.30. *A form B is an A-form if and only if for all $E \in \mathcal{F}(A)$ and $x, y \in X$ we have*

$$B(x, y) = \sum_{z \in E} B(x, z)A(z, y) = \sum_{z \in E} A(x, z)B(z, y). \quad (4.17)$$

Proof. Suppose B is an A-form and let $E \in \mathcal{F}(A)$, $x, y \in X$. Since $\sum_{z \in E} |A(x, z)|^2 < \infty$, $A(x, z) = 0$ except for a countable set $z_2, z_3, \ldots$. Let

$x_1 = x$, $x_i = z_i$, $i = 2, 3, \ldots$, and let $\alpha_1 = 1$, $\alpha_i = -A(x, z_i)$, $i = 2, 3, \ldots$. We now show that $(\alpha_i, x_i) \in n(A)$. Indeed,

$$\sum_i |\alpha_i A(x_i, y)| = |A(x, y)| + \sum_i |A(x, z_i) A(z_i, y)|$$

$$\leq |A(x, y)| + \left[\sum |A(x, z_i)|^2\right]^{\frac{1}{2}} \left[\sum |A(z_i, y)|^2\right]^{\frac{1}{2}}$$

$$\leq |A(x, y)| + 1$$

and

$$\sum \alpha_i A(x_i, y) = A(x, y) - \sum A(x, z_i) A(z_i, y)$$

$$= A(x, y) - \sum_{z \in E} A(x, z) A(z, y) = 0.$$

It follows from (4.16) that

$$B(x, y) - \sum A(x, z_i) B(z_i, y) = 0.$$

A similar method gives the other equality in (4.17). Conversely, suppose (4.17) holds and $(\alpha_i, x_i) \in n(A)$. Then for $E \in \mathcal{F}(A)$ we have

$$\sum_i \alpha_i B(x_i, y) = \sum_i \alpha_i \sum_{z \in E} A(x_i, z) B(z, y)$$

$$= \sum_{z \in E} B(z, y) \sum_i \alpha_i A(x_i, z) = 0.$$

Again, the other equality in (4.16) is similar. □

A form B is *bounded*, if there exists a $b \geq 0$ such that for any $E \in \mathcal{F}(A)$ and $(\alpha_i, z_i) \in \mathbb{C} \times E$, $i = 1, \ldots, n$, with $z_i \neq z_j$, $i \neq j$, we have

$$\sum_{z \in E} \left|\sum_i \alpha_i B(z_i, z)\right|^2 \leq b^2 \sum_i |\alpha_i|^2. \tag{4.18}$$

Notice that A itself is bounded since for a representation $\phi \colon X \to \mathcal{H}$ we have

$$\sum_{z \in E} \left|\sum_i \alpha_i A(z_i, z)\right|^2 = \sum_{z \in E} \left|\left\langle \sum_i \alpha_i \phi(z_i), \phi(z)\right\rangle\right|^2$$

$$= \left\|\sum \alpha_i \phi(z_i)\right\|^2 = \sum |\alpha_i|^2.$$

ISOMORPHISMS AND A-FORMS   151

Theorem 4.31. *Let (X, A) be a tas and let $\phi: X \to \mathcal{H}$ be a representation. There exists a bijection $B \mapsto \hat{B}$ from the set of bounded A-forms onto the set of bounded linear operators on $\mathcal{H}$ such that $B(x, y) = \langle \hat{B}\phi(x), \phi(y) \rangle$ for every $x, y \in X$.*

Proof. Let B be a bounded A-form and let $E \in \mathcal{F}(A)$. Let $\mathcal{H}_0 \subseteq \mathcal{H}$ be the linear hull of $\{\phi(z): z \in E\}$. On $\mathcal{H}_0$ define

$$\hat{B}\sum_{i=1}^{n} \alpha_i \phi(z_i) = \sum_{z \in E} \sum_{i} \alpha_i B(z_i, z)\phi(z).$$

Then $\hat{B}$ is a bounded linear operator on $\mathcal{H}_0$ since

$$\left\| \hat{B}\sum \alpha_i \phi(z_i) \right\|^2 = \sum_{z \in E} \left| \sum \alpha_i B(z_i, z) \right|^2$$

$$\leq b^2 \sum |\alpha_i|^2 = b^2 \left\| \sum_i \alpha_i \phi(z_i) \right\|^2.$$

Since $\mathcal{H}_0$ is dense in $\mathcal{H}$, $\hat{B}$ has a unique bounded linear extension to $\mathcal{H}$, that we also denote by $\hat{B}$. If $x, y \in X$, $z' \in E$, since

$$\hat{B}\phi(z') = \sum_{z \in E} B(z', z)\phi(z)$$

we have by Lemma 4.30

$$\langle \hat{B}\phi(x), \phi(y) \rangle = \left\langle \hat{B} \sum_{z' \in E} A(x, z')\phi(z'), \sum_{z \in E} A(y, z)\phi(z) \right\rangle$$

$$= \left\langle \sum_{z, z' \in E} A(x, z')B(z', z)\phi(z), \sum_{z \in E} A(y, z)\phi(z) \right\rangle$$

$$= \left\langle \sum_{z \in E} B(x, z)\phi(z), \sum_{z \in E} A(y, z)\phi(z) \right\rangle$$

$$= \sum_{z \in E} B(x, z)A(z, y) = B(x, y).$$

It is clear that $B \mapsto \hat{B}$ is injective. To show surjectivity, suppose L is a bounded linear operator on $\mathcal{H}$ and define a form by $B(x, y) = \langle L\phi(x), \phi(y) \rangle$.

To show that B is bounded, let $E \in \mathcal{F}(A)$ and $(\alpha_i, z_i) \in \mathbb{C} \times E$, $i = 1, \ldots, n$, with $z_i \neq z_j$, $i \neq j$. Then

$$\sum_{z \in E} \left| \sum \alpha_i B(z_i, z) \right|^2 = \sum_{z \in E} \left| \sum \alpha_i \langle L\phi(z_i), \phi(z) \rangle \right|^2$$

$$= \sum_{z \in E} \left| \langle L \sum \alpha_i \phi(z_i), \phi(z) \rangle \right|^2$$

$$= \left\| L \sum \alpha_i \phi(z_i) \right\|^2 \leq \|L\|^2 \left\| \sum \alpha_i \phi(z_i) \right\|^2$$

$$= \|L\|^2 \sum |\alpha_i|^2.$$

To show that B is an A-form, we have for $E \in \mathcal{F}(A)$

$$\sum_{z \in E} B(x, z) A(z, y) = \sum_{z \in E} \langle L\phi(x), \phi(z) \rangle \langle \phi(z), \phi(y) \rangle$$

$$= \langle L\phi(x), \phi(y) \rangle = B(x, y)$$

and

$$\sum_{z \in E} A(x, z) B(z, y) = \sum_{z \in E} \langle \phi(x), \phi(z) \rangle \langle L\phi(z), \phi(y) \rangle$$

$$= \sum_{z \in E} \langle \phi(x), \phi(z) \rangle \langle \phi(z), L^* \phi(y) \rangle$$

$$= \langle \phi(x), L^* \phi(y) \rangle$$

$$= \langle L\phi(x), \phi(y) \rangle = B(x, y).$$

We conclude from Lemma 4.30 that B is an A-form. □

Let $B(X, A)$ be the set of bounded A-forms on the tas (X, A). It is easy to check that $B(X, A)$ is a complex linear space under pointwise addition and scalar multiplication. For $B \in B(X, A)$, define $B^*(x, y) = \bar{B}(y, x)$. Then $B^* \in B(X, A)$. Moreover, for $B \in B(X, A)$ define $\|B\|$ to be the infimum of the b's in (4.18). Finally, for $B, C \in B(X, A)$, $E \in \mathcal{F}(A)$, define the form BC by

$$BC(x, y) = \sum_{z \in E} C(x, z) B(z, y).$$

It is not hard to show that $BC \in B(X, A)$ and is independent of the $E \in \mathcal{F}(A)$. The following result is a straightforward corollary of Theorem 4.31.

ISOMORPHISMS AND A-FORMS 153

Corollary 4.32. $B(X, A)$ is a C^*-algebra under the above operations with identity A, and $B \mapsto \hat{B}$ is an isometric $*$-isomorphism.

A C^*-state on $B(X, A)$ is a positive linear functional $\mu: B(X, A) \to \mathbb{R}$ such that $\mu(A) = 1$. An important class of C^*-states are given as follows. For $x \in X$, define the C^*-state μ_x by $\mu_x(B) = B(x, x)$. If $\phi: X \to \mathcal{H}$ is a representation, we see that $\mu_x(B) = \langle \hat{B}\phi(x), \phi(x) \rangle$. This observation together with Corollary 4.32 gives a connection between the present framework and the algebraic (or C^*-algebra) approach to quantum mechanics (see preface for references).

Although not every element of $B(X, A)$ has physical significance, we now consider an important class that does. If B is an event in the premanual $\mathcal{F}(A)$, we call the form A_B defined by

$$A_B(x, y) = \sum_{z \in B} A(x, z) A(z, y)$$

the A-transition amplitude conditioned by B. Let $\phi: X \to \mathcal{H}$ be a representation, let $f \in \mathcal{H}$ be a unit vector and let P_f be the one-dimensional projection onto P_f. If B is an event, form the projection operator $P_B = \sum_{z \in B} P_{\phi(z)}$. We then have

$$A_B(x, y) = \sum_{z \in B} \langle \phi(x), \phi(z) \rangle \langle \phi(z), \phi(y) \rangle$$
$$= \langle P_B \phi(x), \phi(y) \rangle.$$

It follows from Theorem 4.31 that $A_B \in B(X, A)$ and $P_B = \hat{A}_B$. Let

$$\mathcal{E} = \{A_B : B \text{ is an event }\}.$$

For $A_B, A_C \in \mathcal{E}$, define $A'_B = A - A_B$ and $A_B \leq A_C$ if $A_B A_C = A_B$.

Corollary 4.33. If (X, A) is a total tas, then $(\mathcal{E}, \leq, ')$ is an atomistic, σ-orthocomplete, orthomodular poset.

4.7 Sums and Tensor Products

Direct sums and tensor products are two important constructions in Hilbert space quantum mechanics. As we have seen, direct sums are necessary in describing systems that have superselection rules, while tensor products are used for describing combined systems. As we shall show, both of these constructions proceed quite naturally in the present framework.

Let (X_1, A_1) and (X_2, A_2) be tas's where $X_1 \cap X_2 = \emptyset$. Let $X = X_1 \cup X_2$ and define $A: X \times X \to \mathbb{C}$ by $A(x,y) = A_i(x,y)$ if $x, y \in X_i$, $i = 1, 2$ and $A(x, y) = 0$, otherwise. We use the notation $X = X_1 \oplus X_2$, $A = A_1 \oplus A_2$, and call $(X_1 \oplus X_2, A_1 \oplus A_2)$ the *sum* of (X_1, A_1) and (X_2, A_2).

Lemma 4.34. $(X_1 \oplus X_2, A_1 \oplus A_2)$ *is a tas and* $E \in \mathcal{F}(A_1 \oplus A_2)$ *if and only if* $E = E_1 \cup E_2$ *where* $E_i \in \mathcal{F}(A_i)$, $i = 1, 2$.

Proof. Let $X = X_1 \oplus X_2$ and $A = A_1 \oplus A_2$. Clearly $A(x,x) = 1$ for all $x \in X$. Let $E = E_1 \cup E_2$ where $E_i \in \mathcal{F}(A_i)$, $i = 1, 2$. Then

$$\sum_{z \in E} A(x,z)\bar{A}(y,z) = \sum_{z \in E_1} A(x,z)\bar{A}(y,z) + \sum_{z \in E_2} A(x,z)\bar{A}(y,z). \quad (4.19)$$

If x and y are not in the same X_i, then both sums on the right side of (4.19) vanish so we obtain

$$\sum_{z \in E} A(x,z)\bar{A}(y,z) = 0 = A(x,y).$$

If x and y are in the same X_i, then one of the sums on the right side of (4.19) vanishes and the other equals $A_i(x,y) = A(x,y)$. We conclude that A is a transition amplitude on X and that $E_1 \cup E_2 \in \mathcal{F}(A)$. Finally, suppose $E \in \mathcal{F}(A)$ and let $E_i = E \cap X_i$, $i = 1, 2$. Then for all $x, y \in X$ we have

$$A(x,y) = \sum_{z \in E_1} A(x,z)\bar{A}(y,z) + \sum_{z \in E_2} A(x,z)\bar{A}(y,z). \quad (4.20)$$

If $x, y \in X_i$, $i = 1$ or 2, then we obtain from (4.20)

$$A_i(x,y) = \sum_{z \in E_i} A_i(x,z)\bar{A}_i(y,z).$$

Hence, $E_i \in \mathcal{F}(A_i)$, $i = 1, 2$, and $E = E_1 \cup E_2$. □

The following result characterizes those tas's that are the sum of two tas's.

SUMS AND TENSOR PRODUCTS   155

Lemma 4.35. *A tas (X, A) is isomorphic to a sum of two tas's if and only if there is a proper, nonempty subset X_1 of X such that $A(x,y) = 0$ whenever $x \in X_1$, $y \notin X_1$.*

Proof. Let X_1 be a proper, nonempty subset of X such that $A(x,y) = 0$ whenever $x \in X_1$, $y \notin X_1$. Define $A_1: X_1 \times X_1 \to \mathbb{C}$ by $A_1 = A \mid X_1 \times X_1$, let $X_2 = X \setminus X_1$ and define $A_2: X_2 \times X_2 \to \mathbb{C}$ by $A_2 = A \mid X_2 \times X_2$. We now show that (X_1, A_1) is a tas (the proof for (X_2, A_2) being similar). Clearly, $A_1(x, x) = 1$ for all $x \in X_1$. Let $E \in \mathcal{F}(A)$ and let $E_1 = E \cap X_1$. For $x, y \in X_1$ we have

$$A_1(x,y) = A(x,y) = \sum_{z \in E} A(x,z) \bar{A}(y,z)$$

$$= \sum_{z \in E_1} A(x,z) \bar{A}(y,z)$$

$$= \sum_{z \in E_1} A_1(x,z) \bar{A}_1(y,z).$$

Hence, $E_1 \in \mathcal{F}(A_1)$ and (X_1, A_1) is a tas. We now show that the identity map $I: X \to X_1 \oplus X_2$ is an isomorphism from (X, A) to $(X_1 \oplus X_2, A_1 \oplus A_2)$. Indeed, if $x, y \in X_i$, $i = 1$ or 2, then

$$A(x,y) = A_i(x,y) = (A_1 \oplus A_2)(x,y)$$

and otherwise

$$A(x,y) = 0 = (A_1 \oplus A_2)(x,y).$$

Conversely, suppose there is an isomorphism $J: X \to X_1 \oplus X_2$ where (X, A), (X_i, A_i), $i = 1, 2$, are tas's. Then $J^{-1}(X_1)$ is a proper, nonempty subset of X. Moreover, if $x \in J^{-1}(X_1)$, $y \notin J^{-1}(X_1)$, we have

$$A(x,y) = (A_1 \oplus A_2)(Jx, Jy) = 0. \quad \square$$

If X_1 is a proper, nonempty subset of X satisfying the condition of Lemma 4.35, we call $(X_1, A \mid X_1)$ a *subtas* of (X, A).

Corollary 4.36. (a) $\dim(X_1 \oplus X_2, A_1 \oplus A_2) = \dim(X_1, A_1) + \dim(X_2, A_2)$.
(b) *If $(X_1, A \mid X_1)$ is a subtas of (X, A), then*

$$\dim(X, A) = \dim(X_1, A \mid X_1) + \dim(X \setminus X_1, A \mid (X \setminus X_1)).$$

If $\mathcal{H}_1$ and $\mathcal{H}_2$ are Hilbert spaces, we denote the usual direct sum by $\mathcal{H}_1 \oplus \mathcal{H}_2$.

Theorem 4.37. *A map $\phi: X_1 \oplus X_2 \to \mathcal{H}$ is a representation of $(X_1 \oplus X_2, A_1 \oplus A_2)$ if and only if there are Hilbert spaces, $\mathcal{H}_1, \mathcal{H}_2$ such that $\mathcal{H} = \mathcal{H}_1 \oplus \mathcal{H}_2$ and representations $\phi_i: X_i \to \mathcal{H}_i$, $i = 1, 2$, such that $\phi(x) = \phi_i(x)$ for all $x \in X_i$, $i = 1, 2$.*

Proof. Let $\phi: X_1 \oplus X_2 \to \mathcal{H}$ be a representation. If $\mathcal{H}_i$ is the closed span of $\phi(X_i)$, $i = 1, 2$, then $\mathcal{H}_i$ is a Hilbert space and is a closed subspace of $\mathcal{H}$. If $x \in X_1$, $y \in X_2$, then

$$\langle \phi(x), \phi(y) \rangle = A(x, y) = 0$$

so $\mathcal{H}_1 \perp \mathcal{H}_2$. If $E \in \mathcal{F}(A)$, then by Lemma 4.34, $E = E_1 \cup E_2$ where $E_i \in \mathcal{F}(A_i)$, $i = 1, 2$. Since $\phi(E) \in \mathcal{F}(\mathcal{H})$ we have for any $f \in \mathcal{H}$

$$f = \sum_{z \in E} \langle f, \phi(z) \rangle \phi(z)$$
$$= \sum_{z \in E_1} \langle f, \phi(z) \rangle \phi(z) + \sum_{z \in E_2} \langle f, \phi(z) \rangle \phi(z).$$

It follows that $\mathcal{H} = \mathcal{H}_1 \oplus \mathcal{H}_2$. Define $\phi_i: X_i \to \mathcal{H}_i$ as $\phi_i = \phi \mid X_i$, $i = 1, 2$. We now show that ϕ_i is a representation of (X_i, A_i), $i = 1, 2$. Indeed, if $x, y \in X_i$ then for $i = 1$ or 2 we have

$$A_i(x, y) = A(x, y) = \langle \phi(x), \phi(y) \rangle$$
$$= \langle \phi_i(x), \phi_i(y) \rangle.$$

If $E_i \in \mathcal{F}(A_i)$, $i = 1, 2$, then by Lemma 4.34, $E_1 \cup E_2 \in \mathcal{F}(A)$. Since $\phi(E_1 \cup E_2) \in \mathcal{F}(\mathcal{H})$, it follows that $\phi(E_i) \in \mathcal{F}(\mathcal{H}_i)$, $i = 1, 2$. It is clear that $\phi(x) = \phi_i(x)$ for all $x \in X_i$, $i = 1, 2$.

Conversely, suppose $\phi_i: X_i \to \mathcal{H}_i$, $i = 1, 2$ are representations and $\phi: X_1 \oplus X_2 \to \mathcal{H}_1 \oplus \mathcal{H}_2$ satisfies $\phi(x) = \phi_i(x)$ for all $x \in X_i$, $i = 1, 2$. If $x, y \in X_i$, $i = 1$ or 2, then

$$A(x, y) = A_i(x, y) = \langle \phi_i(x), \phi_i(y) \rangle$$
$$= \langle \phi(x), \phi(y) \rangle$$

and otherwise $A(x, y) = 0 = \langle x, y \rangle$. If $E \in \mathcal{F}(A)$, then by Lemma 4.34, $E = E_1 \cup E_2$ where $E_i \in \mathcal{F}(A_i)$, $i = 1, 2$. Since $\phi(E_i) \in \mathcal{F}(\mathcal{H})$, $i = 1, 2$, we have $\phi(E) \in \mathcal{F}(\mathcal{H})$. □

SUMS AND TENSOR PRODUCTS

If we define the sum of an arbitrary collection of tas's in the natural way, then it is straightforward to generalize the above results to this situation. Moreover, it is easy to show that $(X_1 \oplus X_2, A_1 \oplus A_2)$ is strong if and only if (X_1, A_1) and (X_2, A_2) are strong.

If (X_1, A_1) and (X_2, A_2). are tas's, define $X_1 \otimes X_2 = X_1 \times X_2$ and define $A_1 \otimes A_2 \colon X_1 \otimes X_2 \to \mathbb{C}$ by

$$A_1 \otimes A_2((x_1, x_2)(y_1, y_2)) = A_1(x_1, y_1) A_2(x_2, y_2).$$

We call $(X_1 \otimes X_2, A_1 \otimes A_2)$ the *tensor product* of (X_1, A_1) and (X_2, A_2).

Lemma 4.38. *If $(X_1, A_1), (X_2, A_2)$ are tas's, then $(X_1 \otimes X_2, A_1 \otimes A_2)$ is a tas and $E \in \mathcal{F}(A_1 \otimes A_2)$ if $E = E_1 \times E_2$, where $E_i \in \mathcal{F}(A_i)$, $i = 1, 2$.*

Proof. Let $X = X_1 \otimes X_2$, $A = A_1 \otimes A_2$. It is clear that $A(x, x) = 1$ for all $x \in X$. Now let $E_i \in \mathcal{F}(A_i)$, $i = 1, 2$, and let $E = E_1 \times E_2$. Then

$$\sum_{(z_1, z_2) \in E} A((x_1, x_2), (z_1, z_2)) \bar{A}((y_1, y_2), (z_1, z_2))$$

$$= \sum_{(z_1, z_2) \in E} A_1(x_1, z_1) A_2(x_2, z_2) \bar{A}_1(y_1, z_1) \bar{A}_2(y_2, z_2)$$

$$= \sum_{z_1 \in E_1} A_1(x_1, z_1) \bar{A}_1(y_1, z_1) \sum_{z_2 \in E_2} A_2(x_2, z_2) \bar{A}_2(y_2, z_2)$$

$$= A_1(x_1, y_1) A_2(x_2, y_2)$$

$$= A((x_1, x_2), (y_1, y_2)).$$

We conclude that (X, A) is a tas and $E_1 \times E_2 \in \mathcal{F}(A)$. □

Corollary 4.39. $\dim(X_1 \otimes X_2, A_1 \otimes A_2) = \dim(X_1, A_1) \dim(X_2, A_2)$

The converse of Lemma 4.38 does not hold. For example, let $\mathcal{H}_1, \mathcal{H}_2$ be two-dimensional Hilbert spaces with unit spheres X_1, X_2 respectively. Define $A_i(x, y) = \langle x, y \rangle$ for $x, y \in X_i$, $i = 1, 2$. Let $\{x_1, y_1\}$ be an orthonormal basis for $\mathcal{H}_1$ and let $\{x_2, y_2\}$, $\{x_2', y_2'\}$ be distinct orthonormal bases for $\mathcal{H}_2$. Define

$$E = \{(x_1, x_2), (x_1, y_2), (y_1, x_2'), (y_1, y_2')\}.$$

Then $E \in \mathcal{F}(A_1 \otimes A_2)$ but $E \neq E_1 \times E_2$, with $E_i \in \mathcal{F}(A_i)$, $i = 1, 2$.

Theorem 4.40. *Let (X_1, A_1), (X_2, A_2) be tas's. A map $\phi: X_1 \otimes X_2 \to \mathcal{H}$ is a representation of $(X_1 \otimes X_2, A_1 \otimes A_2)$ if and only if there are Hilbert spaces $\mathcal{H}_1, \mathcal{H}_2$ such that $\mathcal{H} = \mathcal{H}_1 \otimes \mathcal{H}_2$ and representations $\phi_i: X_i \to \mathcal{H}_i$, $i = 1, 2$, such that $\phi(x_1, x_2) = \phi_1(x_1) \otimes \phi_2(x_2)$.*

Proof. Let $\phi: X_1 \otimes X_2$ be a representation. Let $E_i \in \mathcal{F}(A_i)$, $i = 1, 2$, $u \in E_1$, $v \in E_2$ be fixed, let $\mathcal{H}_1 \subseteq \mathcal{H}$ be the closed span of $\{\phi(x_1, v): x_1 \in X_1\}$, and let $\mathcal{H}_2 \subseteq \mathcal{H}$ be the closed span of $\{\phi(u, x_2): x_2 \in X_2\}$. Define the maps $\phi_i: X_i \to \mathcal{H}_i$, $i = 1, 2$, by $\phi_1(x_1) = \phi(x_1, v)$, $\phi_2(x_2) = \phi(u, x_2)$. We now show that ϕ_1 is a representation of (X_1, A_1) (the proof for ϕ_2 being similar). For $x_1, y_1 \in X_1$ we have

$$\langle \phi_1(x_1), \phi_1(y_1) \rangle = \langle \phi(x_1, v), \phi(y_1, v) \rangle$$
$$= A_1 \otimes A_2((x_1, v), (y_1, v)) = A_1(x_1, y_1).$$

If $z, z' \in E_1$, $z \neq z'$, then $\phi_1(z) \perp \phi_1(z')$ so $\{\phi(z): z \in E_1\}$ is an orthonormal set in $\mathcal{H}_1$. Moreover, for any $x_1 \in X_1$, since $E_1 \times E_2 \in \mathcal{F}(A_1 \otimes A_2)$, we have

$$\phi(x_1, v) = \sum_{(z,z') \in E_1 \times E_2} \langle \phi(x_1, v), \phi(z, z') \rangle \phi(z, z')$$
$$= \sum_{(z,z') \in E_1 \times E_2} A_1(x_1, z) A_2(v, z') \phi(z, z')$$
$$= \sum_{z \in E_1} A_1(x_1, z) \phi(z, v)$$
$$= \sum_{z \in E_1} A_1(x_1, z) \phi_1(z).$$

Hence, if $f \in \mathcal{H}_1$ satisfies $f \perp \phi_1(z)$ for all $z \in E_1$, we have $\langle f, \phi(x_1, v) \rangle = 0$ for all $x_1 \in X_1$. Since $\{\phi(x_1, v): x_1 \in X_1\}$ is dense in $\mathcal{H}_1$, we conclude that $f = 0$ hence $\phi_1(E_1) \in \mathcal{F}(\mathcal{H}_1)$. Define the unitary transformation $U: \mathcal{H} \to \mathcal{H}_1 \otimes \mathcal{H}_2$ by

$$U\phi(z_1, z_2) = \phi_1(z_1) \otimes \phi_2(z_2), \qquad z_i \in E_i, \quad i = 1, 2$$

and extend by linearity and closure. We can thus identify $\mathcal{H}$ with $\mathcal{H}_1 \otimes \mathcal{H}_2$. Moreover, for any $(x_1, x_2) \in X_1 \otimes X_2$ we have

SUMS AND TENSOR PRODUCTS   159

$$U\phi(x_1,x_2) = \sum_{(z,z')\in E_1\times E_2} \langle\phi(x_1,x_2),\phi(z,z')\rangle\phi_1(z)\otimes\phi_2(z')$$

$$= \sum_{(z,z')\in E_1\times E_2} \langle\phi_1(x_1),\phi_1(z)\rangle\langle\phi_2(x_2),\phi_2(z')\rangle\phi_1(z)\otimes\phi_2(z')$$

$$= \sum_{z\in E_1} \langle\phi_1(x_1),\phi_1(z)\rangle\phi_1(z)\otimes \sum_{z\in E_2} \langle\phi_2(x_2),\phi_2(z')\rangle\phi_2(z')$$

$$= \phi_1(x_1)\otimes\phi_2(x_2).$$

Conversely, suppose $\phi_i\colon X_i \to \mathcal{H}_i$, $i = 1, 2$, are representations and $\phi\colon X_1\otimes X_2 \to \mathcal{H} = \mathcal{H}_1\otimes\mathcal{H}_2$ satisfies $\phi(x_1,x_2) = \phi_1(x_1)\otimes\phi_2(x_2)$. We now show that ϕ is a representation. For $(x_1,x_2),(y_1,y_2)\in X_1\otimes X_2$ we have

$$\langle\phi(x_1,x_2),\phi(y_1,y_2)\rangle = \langle\phi_1(x_1)\otimes\phi_2(x_2),\phi_1(y_1)\otimes\phi_2(y_2)\rangle$$

$$= \langle\phi_1(x_1),\phi_1(y_1)\rangle\langle\phi_2(x_2),\phi_2(y_2)\rangle$$

$$= A_1(x_1,y_1)A_2(x_2,y_2)$$

$$= A_1\otimes A_2((x_1,x_2),(y_1,y_2)).$$

Finally, if $E_i \in \mathcal{F}(A_i)$, $i = 1, 2$, then by Lemma 4.38, $E_1\times E_2 \in \mathcal{F}(A_1\otimes A_2)$. For $(x_1,x_2)\in X_1\otimes X_2$ we have

$$\sum_{(z,z')\in E_1\times E_2}|\langle\phi(x_1,x_2),\phi(z,z')\rangle|^2 = \sum|A_1(x_1,z)|^2\sum|A_2(x_2,z')|^2$$

$$= 1 = \|\phi(x_1,x_2)\|^2.$$

Hence,
$$\phi(x_1,x_2) = \sum_{(z,z')\in E_1\times E_2}\langle\phi(x_1,x_2),\phi(z,z')\rangle\phi(z,z').$$

If $f\in\mathcal{H}$ satisfies $f\perp\phi(z,z')$ for all $(z,z')\in E_1\times E_2$, then $\langle f,\phi(x_1,x_2)\rangle = 0$ for all $(x_1,x_2)\in X_1\otimes X_2$. Since the elements $\phi(x_1,x_2) = \phi_1(x_1)\otimes\phi_2(x_2)$, $(x_1,x_2)\in X_1\otimes X_2$, are total in $\mathcal{H}$, we conclude that $f = 0$. Hence, $\phi(E_1\times E_2)\in\mathcal{F}(\mathcal{H})$ and ϕ is a representation. □

If we define the tensor product of a finite number of tas's in the natural way, then it is straightforward to generalize the above results to this situation.

4.8 Quantum Probability Spaces

Classical probability theory is based upon a probability space (Ω, Σ, μ) where Σ is a σ-algebra of subsets of Ω and $\mu(A)$ is interpreted as the probability that an event $A \in \Sigma$ occurs. In quantum probability theory, the situation is quite different. We still have a triple (Ω, Λ, μ'), but in general, Λ is not a σ-algebra and μ' is not a measure on Λ. As we have discussed earlier, Λ is a σ-algebra and μ' is a measure on Λ only when attention is restricted to a single operation (or measurement or experiment). When multiple operations are considered, interference effects remove us from the realm of classical probability theory. The interference effects occur because, unlike μ which can be quite arbitrary, the quantum probability function μ' is obtained in a specific way. Quantum mechanics begins with an amplitude function $f: \Omega \to \mathbb{C}$ (or in some cases, a transition amplitude). If $A \in \Lambda$, we define the amplitude of A by $\hat{f}(A) = \sum_{\omega \in A} f(\omega)$ and we define the probability that A occurs by

$$\mu'(A) = |\hat{f}(A)|^2 = \left|\sum_{\omega \in A} f(\omega)\right|^2.$$

Thus, a quantum probability function is induced from an amplitude function. If $A, B \in \Lambda$ and $A \cap B = \emptyset$, we need not have $\mu'(A \cup B) = \mu'(A) + \mu'(B)$. In fact, we need not even have $A \cup B \in \Lambda$. It is because of such differences between classical and quantum probability theory that we are confronted with certain so-called "paradoxes" of quantum mechanics such as the two-hole experiment discussed in Section 2.5. These paradoxes vanish when it is realized that probabilities are computed differently in quantum mechanics than in classical probability theory.

We now give precise definitions corresponding to the above motivating remarks. These definitions provide an axiomatic framework for quantum mechanics based on amplitude functions. Let Ω be a nonempty set and let $f: \Omega \to \mathbb{C}$. We call the elements $\omega \in \Omega$ *sample points*, the function f a *probability amplitude* and (Ω, f) a *quantum probability space*. We say that (Ω, f) is a *point closed* if $\sum_{\omega \in \Omega} |f(\omega)|^2 = 1$. A subset $A \subseteq \Omega$ is *summable* if $\sum_{\omega \in A} |f(\omega)| < \infty$ and we denote the collection of summable sets by Σ_0. We define the set function $\hat{f}: \Sigma_0 \to \mathbb{C}$ by

$$\hat{f}(A) = \sum_{\omega \in A} f(\omega) \quad [\hat{f}(\emptyset) = 0]$$

and call $\hat{f}(A)$ the *amplitude* of A.

QUANTUM PROBABILITY SPACES

Lemma 4.41. *Σ_0 is a ring of sets with the property that $A \in \Sigma_0$ and $B \subseteq A$ implies $B \in \Sigma_0$. Moreover, $\hat{f}$ is a σ-finite complex measure on Σ_0.*

Proof. The proof that Σ_0 is a ring with the above property is straightforward. To show that $\hat{f}$ is a complex measure on Σ_0, suppose $A_i \in \Sigma_0$, $A_i \cap A_j = \emptyset$, $i \neq j = 1, 2, \ldots$, and $\cup A_i = A \in \Sigma_0$. Denumerate the elements of A_i on which f is nonvanishing by ω_{ij}, $j = 1, 2, \ldots$. Then since

$$\sum_{ij} |f(\omega_{ij})| = \sum_{\omega \in A} |f(\omega)| < \infty$$

it follows from Fubini's theorem that

$$\hat{f}(A) = \sum_{ij} f(\omega_{ij})$$
$$= \sum_i [\sum_j f(\omega_{ij})] = \sum_i \hat{f}(A_i).$$

Hence, $\hat{f}$ is countably additive on Σ_0. To show that $\hat{f}$ is σ-finite, let $A \in \Sigma_0$ and let ω_i, $i = 1, 2, \ldots$, be the elements of A on which f is nonvanishing. Let $B = A \setminus \{\omega_i : i = 1, 2, \ldots\}$. Then $A = \cup\{\omega_i\} \cup B$ and $\hat{f}(B) = 0$, $\hat{f}(\{\omega_i\}) \in \mathbb{C}$ so $\hat{f}$ is σ-finite. □

This section will only introduce the concept of a quantum probability space and the theory will be developed in later sections. A quantum probability space (Ω, f) is an extremely general concept and without additional physically motivated structure it would be devoid of interest. We now discuss such additional structure.

An *operation* on (Ω, f) is a collection of mutually disjoint sets $A_i \in \Sigma_0$ such that $\sum |\hat{f}(A_i)|^2 = 1$. Notice that (Ω, f) is point closed if and only if the set $\{\{w\} : \omega \in \Omega\}$ is an operation. Denote the set of operations on (Ω, f) by $\mathcal{A}(\Omega, f)$. An *outcome* is a set $A \in \Sigma_0$ such that $A \in E$ for some $E \in \mathcal{A}(\Omega, f)$. Of course, $\mathcal{A}(\Omega, f)$ is a quasimanual and an outcome corresponds to the usual operational statistics definition. If we denote the set of outcomes by $X(\Omega, f)$, then $\hat{f} : X \to \mathbb{C}$ is an amplitude function in our previously defined sense. Of course, an outcome can belong to more than one operation. If a distinction is needed, we say that A is an *E-outcome* for $E \in \mathcal{A}(\Omega, f)$. If $\hat{f}(A) = 0$, A is called a *trivial* outcome and an operation is *proper* if it contians no trivial outcomes.

If $E \in \mathcal{A}(\Omega, f)$ and $B \subseteq E$ we call B an *event* of E and denote the set of events of E by $\mathcal{E}(E)$. The set of all events is denoted by $\mathcal{E}(\Omega, f)$.

As usual, the amplitude function $\hat{f}$ determines an $\mathcal{A}$-state μ_f, and for any $B \in \mathcal{E}(\Omega, f)$ we have

$$\mu_f(B) = \sum_{A_i \in B} \mu_f(A_i) = \sum_{A_i \in B} |\hat{f}(A_i)|^2.$$

If $B = \{A_i\}$ is an event, we frequently identify B with $\cup A_i \subseteq \Omega$ and this identification is unique as long as the operation E is specified. We then define $\mu_f^E(\cup A_i) = \mu_f(B)$ and interpret this as the probability that $\cup A_i$ occurs upon execution of the operation E.

Notice that in this formalism the roles of quasimanual and amplitude function are reversed from those in Section 4.2. In Section 4.2, the quasimanual is given and the amplitude function is a derived concept. In the present section, we begin with the amplitude function and then derive the quasimanual. In this sense, the amplitude function becomes the basic axiomatic element.

Suppose a physical system is described by a quantum probability space (Ω, f) and we perform a measurement on this system. Corresponding to this measurement there should be an operation $E \in \mathcal{A}(\Omega, f)$ whose outcomes specify the values of the measurement. In this way a measurement corresponds to a certain kind of "measurable" function. Motivated by this, we call a function $h: D(h) \to \mathbf{R}$ with domain $D(h) \subseteq \Omega$ *observable* if

$$\{h^{-1}(\lambda): \lambda \in \mathbf{R}\} \in \mathcal{A}(\Omega, f).$$

Thus, h is observable if and only if

$$\sum_{\lambda \in \mathbf{R}} |\hat{f}[h^{-1}(\lambda)]|^2 = 1. \tag{4.21}$$

Notice that (4.21) is just the probabilistically reasonable requirement

$$\sum_{\lambda \in \mathbf{R}} \mu_f[h^{-1}(\lambda)] = 1.$$

Of course, we are now considering measurements with a countable number of values. A more general definition could be given for measurements with a continuum of values but the above definition will suffice for our present purposes. More generally we say that a finite set of functions $h_i: D(h_i) \to \mathbf{R}$, $i = 1, \ldots, n$, is *jointly observable* if

$$\{h_1^{-1}(\lambda_1) \cap \cdots \cap h_n^{-1}(\lambda_n): \lambda_1, \ldots, \lambda_n \in \mathbf{R}\} \in \mathcal{A}(\Omega, f).$$

Example 4.8. Let (Ω, μ) be a finite probability space with $\Omega = \{\omega_1, \ldots, \omega_n\}$. Define $f: \Omega \to \mathbb{C}$ by $f(\omega) = \mu(\omega)^{\frac{1}{2}}$. Then (Ω, f) becomes a point closed quantum probability space. The collection

$$E = \{\{\omega_1\}, \ldots, \{\omega_n\}\}$$

is an operation and $\mu_f^E = \mu$ on the power set $P(\Omega) = \mathcal{E}(E)$. In general, E is the only operation in (Ω, f) and the only outcomes are the singleton sets $\{\omega_i\}$, $i = 1, \ldots, n$. Thus, there is a bijection between the sample points ω_i and the outcomes $\{\omega_i\}$. In general, the only observable functions are the real-valued injections defined on all of Ω. □

Example 4.9. Let (X, Σ, μ) be a probability space. Define $\Omega = \Sigma$ and $f: \Omega \to \mathbb{C}$ by $f(\omega) = \mu(\omega)^{\frac{1}{2}}$. Then (Ω, f) is a quantum probability space which is not point closed in general. If E is a measurable partition of X, then $E \in \mathcal{A}(\Omega, f)$ and $\mu_f^E = \mu$. The sample points, outcomes, and events correspond to elements of Σ. Any random variable with countably many values is observable. □

Example 4.10. Let $\mathcal{H}$ be a separable complex Hilbert space with unit sphere Ω and let $\phi \in \Omega$. Define $f: \Omega \to \mathbb{C}$ by $f(\omega) = \langle \phi, \omega \rangle$. Then (Ω, f) is a quantum probability space which is not point closed. If e_i, $i = 1, 2, \ldots$, is an orthonormal basis for $\mathcal{H}$, then $E = \{\{e_1\}, \{e_2\}, \ldots\}$ is an operation. Indeed,

$$\sum |f(e_i)|^2 = \sum |\langle \phi, e_i \rangle|^2 = \|f\|^2 = 1.$$

Thus, any sample point ω corresponds naturally to the outcome $\{\omega\}$. If $h: \{e_i\} \to \mathbb{R}$ is injective, then h is an observable function. If P_i is the projection onto e_i, we can identify h with the self-adjoint operator $\sum h(c_i) P_i$. □

Example 4.11. Let $\mathcal{H}$ be a complex Hilbert space and let Ω be the lattice of orthogonal projections on $\mathcal{H}$. Let $\phi \in \mathcal{H}$ with $\|\phi\| = 1$. Define $f: \Omega \to \mathbb{C}$ by $f(P) = \langle P\phi, \phi \rangle^{\frac{1}{2}} = \|P\phi\|$. Again (Ω, f) is a quantum probability space which is not point closed. If $P_i \in \Omega$, with $P_i P_j = 0$, $i \neq j = 1, 2, \ldots$, and $\sum P_i = I$, then $\{\{P_1\}, \{P_2\}, \ldots\}$ is an operation. Let $D(h) = \{P_i : i = 1, 2, \ldots\}$ and suppose $h: D(h) \to \mathbb{R}$ is injective. Then h is observable and can be identified with the self-adjoint operator $\sum h(P_i) P_i$. □

Example 4.12. Let $\Omega = \{\omega_1, \omega_2, \omega_3\}$ and define $f: \Omega \to \mathbb{C}$ by $f(\omega_1) = (2-i)/3$, $f(\omega_2) = (1+i)/3$, $f(\omega_3) = \sqrt{2}/3$. Then (Ω, f) is a point closed quantum probability space and the only operations are

$$E = \{\{\omega_1\}, \{\omega_2\}, \{\omega_3\}\} \text{ and } F = \{\{\omega_1, \omega_2\}\}.$$

Notice that $\{\omega_1, \omega_2\}$ is an outcome in F and also can be identified with the event $\{\{\omega_1\}, \{\omega_2\}\}$ in E. However, it has a different probability of occurring depending on which operation is executed. Indeed,

$$\mu_f^E[\{\omega_1, \omega_2\}] = \frac{7}{9} \neq 1 = \mu_f^F[\{\omega_1, \omega_2\}].$$

The only observable functions defined on all of Ω are the injective real-valued functions. If g and h are real-valued functions defined on all of Ω and at least one of them is injective, then they are jointly observable. If g and h are constant on Ω, they are not jointly observable. $\square$

We now consider an example that is closely related to traditional quantum dynamics. Let $\mathcal{H}$ be a separable, complex Hilbert space and let e_i, $i = 1, 2, \ldots$, be an orthonormal basis for $\mathcal{H}$. Let ϕ_0 be a unit vector in $\mathcal{H}$ which represents the initial state of a quantum system and let H be a self-adjoint operator on $\mathcal{H}$ representing the Hamiltonian of the system. If we establish a standard unit of time, then according to traditional quantum mechanics, the probability amplitude that the system which is initially in the state ϕ_0 is in the state e_j after n time steps is given by

$$\langle U^n \phi_0, e_j \rangle = \langle e^{-inH} \phi_0, e_j \rangle. \tag{4.22}$$

The probability that the above outcome occurs is then given by

$$\left| \langle U^n \phi_0, e_j \rangle \right|^2.$$

Expanding (4.22) in terms of U gives

$$\langle U^n \phi_0, e_j \rangle = \sum_{i_1, \ldots, i_{n-1}} \langle U \phi_0, e_{i_1} \rangle \langle U e_{i_1}, e_{i_2} \rangle \cdots \langle U e_{i_{n-1}}, e_j \rangle. \tag{4.23}$$

We now interpret (4.23) in terms of quantum probability theory. Let $\omega(i_1, \ldots, i_n)$ represent the sample point that a quantum system, which is initially in the state ϕ_0, moves to state e_{i_1} in one time step, then to state e_{i_2} in one more time step and continues in this way until it arrives at state

QUANTUM PROBABILITY SPACES   165

e_{i_n} after a total of n time steps. We denote the set of these sample points by
$$\Omega = \{\omega(i_1, \ldots, i_n): i_1, \ldots, i_n = 1, 2, \ldots\}.$$

The amplitude of a sample point is given by

$$f[\omega(i_1, \ldots, i_n)] = \langle U\phi_0, e_{i_1}\rangle\langle Ue_{i_1}, e_{i_2}\rangle \cdots \langle Ue_{i_{n-1}}, e_{i_n}\rangle \quad (4.24)$$

and (Ω, f) becomes a quantum probability space. Let $A(j)$ be the outcome that the quantum system in the initial state ϕ_0 moves through various combinations of the states e_i and arrives at the state e_j after n time steps. The outcome $A(j)$ consists of the sample points $\omega(i_1, \ldots, i_{n-1}, j)$, $i_1, \ldots, i_{n-1} = 1, 2, \ldots$. Applying (4.23) and (4.24) gives

$$\hat{f}[A(j)] = \sum_{i_1, \ldots, i_{n-1}} f[\omega(i_1, \ldots, i_{n-1}, j)]$$
$$= \langle U^n \phi_0, e_j\rangle$$

which agrees with the usual formula (4.22). Moreover, $A(j)$ is indeed an outcome, $j = 1, 2, \ldots$, and $\{A(j): j = 1, 2, \ldots\}$ is an operation since

$$\sum_j |\hat{f}[A(j)]|^2 = \|U^n \phi_0\|^2 = \|\phi_0\|^2 = 1.$$

We also have that the quantum probability space (Ω, f) is point closed since

$$\sum |f[\omega(i_1, \ldots, i_n)]|^2 = \|U\phi_0\|^2 = \|\phi_0\|^2 = 1.$$

Notice that a quantum probability space (Ω, f) frequently induces three probabilistic levels. At the lowest level we have the set of sample points Ω. The sample points are the carriers of the probability amplitude f and in a sense they can be thought of as hidden variables since they may not be subject to direct observation. For example, they may correspond to specific trajectories of a quantum particle. The next level is the set of outcomes X. In general, an outcome A is a set of sample points and the amplitude of A is defined as $\hat{f}(A) = \sum_{\omega \in A} f(\omega)$. The outcomes are the observable results of an experiment and it is here that quantum interference effects are felt. For example, an outcome might be the set of trajectories from point a to point b. We may know that a particle has moved from a to b, but we can not tell which of the various possible paths the particle actually traversed. The third level is the set of events $\mathcal{E}$. An event B is a set of outcomes $B = \{A_i\}$. There is no quantum interference at this level since the probability of B is the sum of the probabilities of its consituent A_i's. Roughly speaking, an operation is then a maximal event and we obtain a quasimanual $\mathcal{A}(\Omega, f)$ with outcome set X. The map $\hat{f}: X \to \mathbb{C}$ is what we earlier called an amplitude function so our present terminology is consistent.

We conclude this section with a discussion of conditional amplitudes. Let (Ω, f) be a quantum probability space and let $A \in \Sigma_0$. We define the map $B \mapsto \hat{f}(B \mid A)$ from the power set $P(\Omega)$ to $\mathbb{C}$ by $\hat{f}(B \mid A) = \hat{f}(B \cap A)/\hat{f}(A)$ if $\hat{f}(A) \neq 0$ and $\hat{f}(B \mid A) = 0$, otherwise. Notice that if $\hat{f}(A) \neq 0$, then $\hat{f}(\cdot \mid A)$ is a complex measure on $P(\Omega)$ with $\hat{f}(\Omega \mid A) = 1$. We call $\hat{f}(B \mid A)$ the *conditional amplitude of B given A* and A is the *conditioning set*. Care must be taken with conditional amplitudes since $\hat{f}(A) = 0$ need not imply that $\hat{f}(B \cap A) = 0$ even if the sets are outcomes. Because of this possibility, formulas such as $\hat{f}(B \cap A) = \hat{f}(A)\hat{f}(B \mid A)$ need not hold when $\hat{f}(A) = 0$. However, when the conditioning sets have nonzero amplitude, the following more general formula holds:

$$\hat{f}(A_1 \cap \cdots \cap A_n)$$
$$= \hat{f}(A_1)\hat{f}(A_2 \mid A_1)\hat{f}(A_3 \mid A_1 \cap A_2) \cdots \hat{f}(A_n \mid A_1 \cap \cdots \cap A_{n-1}). \quad (4.25)$$

Moreover, if $\{A_i: i = 1, \ldots, n\}$ is a Σ_0-partition of Ω, we have

$$\hat{f}(B) = \sum_{i=1}^{n} \hat{f}(A_i)\hat{f}(B \mid A_i). \quad (4.26)$$

In later sections we shall need some additional terminology. If $g, h: A \subseteq \Omega \to \mathbb{R}$, then h is g *observable* if

$$\sum_\lambda \left|\hat{f}[h^{-1}(\lambda) \mid g^{-1}(\alpha)]\right|^2 = 1$$

for every $\alpha \in \mathbb{R}$ such that $\hat{f}[g^{-1}(\alpha)] \neq 0$. If g is observable and h is g observable, then h and g are jointly observable. Indeed, when $\hat{f}[g^{-1}(\alpha)] \neq 0$ we have

$$\hat{f}[h^{-1}(\lambda) \cap g^{-1}(\alpha)] = \hat{f}[g^{-1}(\alpha)]\hat{f}[h^{-1}(\lambda) \mid g^{-1}(\alpha)]$$

and the result easily follows. We say that h is g *orthogonal* if

$$\sum_\lambda \hat{f}[h^{-1}(\lambda) \mid g^{-1}(\alpha)]\bar{\hat{f}}[h^{-1}(\lambda) \mid g^{-1}(\alpha')] = 0 \quad (4.27)$$

for every $\alpha \neq \alpha'$.

4.9 Notes and References

The material in Section 4.1 can be found in [Gudder, 1986b], and we refer the reader to that paper for more details. Section 4.2 introduces the idea of amplitude functions on a quasimanual and this material is also contained in [Gudder, 1986b]. This section shows that a quasimanual is associated with a loosely connected collection of Hilbert spaces. These Hilbert spaces exhibit Dirac's superposition principle and they "explain" why Hilbert space structure is so important in quantum mechanics. Section 4.3 shows that if there is a rich enough supply of amplitude functions, then the collection of Hilbert spaces can be replaced by a single Hilbert space. This work (also found in [Gudder, 1986b]) generalizes a result in [Gudder, 1982].

In certain situations we have a stronger structure than that given in terms of amplitude functions. This is the case when we have a transition amplitude. Section 4.5 shows that when a transition amplitude is present, Hilbert space is not far behind. The concept of a transition amplitude space strengthens that of a transition probability space. Some references for the latter are [Cantoni, 1975; Mielnik, 1968, 1969; Pulmannová, 1986; Uhlmann, 1976]. For more details on the former and a comparison of the two we refer the reader to the work of S. Pulmannová and the author [Gudder and Pulmannová, 1987]. Sections 4.4–4.7 present an axiomatic structure based upon transition amplitudes. Moreover, connections are made to other axiomatic approaches to quantum mechanics.

We think that the essence of quantum mechanics is directly captured in Section 4.8. This section introduces the concept of a quantum probability space (Ω, f) whose main constituent is the probability amplitude f. Although the material presented here is mainly discrete, a more general theory can be developed at the cost of measure theoretic complications. In fact, this will be done in a later section when we introduce probability amplitude densities. The work in Section 4.8 can be found in [Gudder, 1984b].

One of the first investigators to develop an axiomatic framework for quantum mechanics based on amplitudes and transition amplitudes was A. Landé. In fact, he emphasized the importance of the fundamental equation (4.5) (or (4.12)) many years ago. For insights to Landé's motivation and reasoning, we refer the reader to his book [Landé, 1965].

5

Generalized Probability Spaces

In Section 3.4 (Corollary 3.12) we showed that a quantum logic L admits hidden variables if and only if L is isomorphic to a σ-additive class Λ of subsets of a set X. In a certain sense we may think of X as a phase space representation of our physical system and the sets in Λ as measurable sets. However, one should not get the impression that we now obtain a classical description of the physical system. Indeed, a σ-additive class is much more general than a σ-algebra so we do not have a classical measurable space. In fact, as we shall show in this chapter and the next one, much of quantum mechanics can be described in the framework of σ-additive classes. In particular, σ-additive classes will provide us with the measure theoretic background necessary for the introduction of probability amplitude densities mentioned in Section 4.9. This will ultimately result in a framework for a general theory of quantum probability spaces.

Motivated by the above, we define a *generalized probability space* to be a triple (X, Λ, μ) where X is a nonempty set, Λ is a σ-additive class of subsets of X and μ is a probability measure on Λ. Of course, the probability measure μ is what we have previously called a regular state and its definition is the same as that for a probability measure on a σ-algebra. There are philosophical reasons why a generalized probability space (X, Λ, μ) has a more natural place than a classical probability space (X, Σ, μ). The main reason for introducing the σ-algebra Σ on X is to provide sets on which to define the measure μ. But since the countable additivity of μ requires only closure under countable *disjoint unions*, there is no reason for requiring the stronger condition of closure under arbitrary countable unions. In this sense, the defining axioms of a σ-additive class are the most reasonable minimal set of axioms needed for defining a measure and the σ-algebra axioms are unnecessarily restrictive. From a probabilistic viewpoint, there seems to be no convincing reason why the union of two arbitrary events should be an event. However, the rule of additivity of probabilities makes it reasonable that the union of two disjoint events should be an event.

5.1 Embedding Generalized Measure Spaces

It is sometimes convenient to consider a generalized measure space (gms) (X, Λ, μ) rather than a slightly more restrictive generalized probability space (gps). Of course, in this case Λ is again a σ-additive class on X and a (positive) measure μ is defined in the usual way. We begin with some examples which motivate the study of gms's.

Example 5.1. Suppose a statistician is interested in two events A, B and their probabilities, where $A \cap B \neq \emptyset$, but has no interest in $A \cap B$. He would like to work only with the events $\emptyset, X, A, B, A^c, B^c$ which do not form an algebra (in general, they do form an additive class). But to do classical probability theory, he is forced to work with a much more complicated algebra consisting of the sets $\emptyset, X, A, B, A^c, B^c, A \cap B, A \cup B, A \cap B^c, A^c \cap B, A^c \cap B^c, A \cup B^c, A^c \cup B, A^c \cup B^c, (A \cap B^c) \cup (A^c \cap B), (A \cap B) \cup (A^c \cap B^c)$ and assign arbitrary probabilities to many unwanted sets. Actually the situation is even worse than this. We shall see that in slightly more complicated situations, there may not exist an extension of the probability from the wanted to the unwanted sets. In short, the σ-algebra framework may require the introduction of artificial events and probabilities that have no relevance to the problem at hand. For more details about such problems see [Fine, 1973]. □

Example 5.2. Suppose we are making length measurements with a micrometer that is accurate to within λcm (e.g., $\lambda = 0.001$) and can measure lengths up to $N\lambda$ where N is a positive integer (e.g., $N = 1,000$). Since we must round off measurements to the nearest λcm, the micrometer in effect is able to give lengths only of the form $n\lambda$, where n is an integer between 0 and N. In order to do measure theory, we form the triple $([0, N\lambda], \Lambda, \mu)$

where μ is Lebesgue measure and Λ is the collection of Borel subsets of $[0, N\lambda]$ with μ measure $n\lambda$, $n = 0, 1, \ldots, N$. But Λ is not an algebra (it is an additive class) so classical measure theory is stuck. Why can't we remedy this situation by adjoining more subsets to Λ? For example, why can't we assume that any subinterval of $[0, N\lambda]$ is in Λ and that the length of a subinterval is that measured by the micrometer rounded off to the nearest λcm? The reason we can't is that we would lose additivity of length. For example, the length of the subintervals $[0, \lambda/3)$, $[\lambda/3, 2\lambda/3)$, $[2\lambda/3, \lambda]$ would be zero and if length is additive, we are faced with the contradictory conclusion that the length of $[0, \lambda]$ is zero. As another example of what can go wrong, suppose we assume that the length of $[0, \lambda/2]$ is λ. Using additivity we would conclude that the length of $(\lambda/2, \lambda]$ is zero. But then by symmetry, the length of $[0, \lambda/2)$ is also zero. Hence, the length of the point $\{\lambda/2\}$ is one! □

One might argue that the situation in Example 5.2 is unrealistic in that we have unduly restricted ourselves. We do not have to use micrometers with an accuracy of λ. We can, in principle, construct length-measuring apparata with increasingly fine accuracy. We can then attribute to any interval an arbitrarily fine length accuracy. Our answer to this argument is twofold. First, if we are to describe a particular measuring apparatus, we must be content with its inherent accuracy. Second, it is possible that nature has forced upon us an intrinsic limit to accurate length measurements. There is experimental as well as theoretical evidence [Atkison and Halpern, 1967; Blumethal, et al, 1966; Gamow, 1966; Gudder, 1968b; Heisenberg, 1930] suggesting the existence of an elementary length. This would be a length λ such that no smaller length measurement is attainable and all length measurements must be an integer multiple of λ.

There are, in fact, instances in nature in which an ultimate accuracy is known to obtain. It is accepted that the charge on the electron e is the smallest charge obtainable and all charge measurements must be integer multiples of e (quarks have charge $\pm e/3$, $\pm 2e/3$ but they probably can never be isolated as individual particles). There are other important situations in which the apparatus under consideration has inherent accuracy limitations. For example, high-speed digital computers have such limitations. The computer will accept only a certain quantity of significant digits and all numbers must be rounded off to this quantity. Numbers are rounded off with each internal operation performed by the computer, and this sometimes results in considerable round-off error.

Example 5.3. Another example of the above phenomenon is motivated by quantum mechanics. Suppose we are considering a particle that is constrained to move along the x-axis. According to classical mechanics, if we form a two-dimensional phase space X with coordinate axes x, p, where p is the x-momentum of the particle, then the mechanics of the particle is completely described by a point in X together with the laws of motion. The evolution of the particle in time is then given by a trajectory in X. If we want to describe quantum mechanical effects however, we would have to contend with the Heisenberg uncertainty principle. In fact, if $\Delta x, \Delta p$ are the errors made in an x and p measurement respectively, then $\Delta x \Delta p \geq \hbar/2$. For this reason, a point in X is physically unobservable since one can never determine whether the phase space coordinates (x, p) of a particle are at a particular point. The best we can do is determine whether (x, p) is contained in a rectangle with sides of length $\Delta x, \Delta p$ where $\Delta x \Delta p = \hbar/2$. Physically, these elementary rectangles in X become the basic elements of the theory instead of the points in X. Generalizing this slightly, we say that a set $A \subseteq X$ is *admissible* if the area of A (strictly speaking, we mean the Lebesgue measure of A) is an integer multiple of $\hbar/2$. An admissible set A is one for which it is physically meaningful to say that $(x, p) \in A$. For simplicity, let us assume that X is a large rectangle of area $N\hbar/2$ where N is a large integer. Now it is clear that the collection of admissible sets Λ is an additive class but it is not an algebra. This is the basis of "microcanonical analysis" in the theory of partial differential equations. □

A situation similar to that in Example 5.3 occurs in pattern recognition problems. In such problems one frequently measures the intensity of light reflected from certain cells on the subject. One would like to apply measure theory to analyze the problem but the collection of cells does not form an algebra although it does form an additive class [Watanabe, 1972].

Let (X, Λ, μ) be a gms and let $\Sigma(\Lambda)$ be the σ-algebra generated by Λ. If μ can be extended to a measure μ' on $\Sigma(\Lambda)$, then we say that (X, Λ, μ) can be *embedded* in the measure space $(X, \Sigma(\Lambda), \mu')$. It is also of interest to know whether μ has a unique extension to $\Sigma(\Lambda)$ and we shall consider this question also. If a gms can be embedded in a measure space, life is much simpler. To prove a result in (X, Λ, μ), go up to $(X, \Sigma(\Lambda), \mu')$, use classical methods, and then come back down to (X, Λ, μ). Unfortunately, as we shall now show, an embedding may not exist.

A measure μ on a σ-algebra Σ is always subadditive; that is $\mu(\cup A_i) \leq$

EMBEDDING GENERALIZED MEASURE SPACES 173

$\Sigma \mu(A_i)$ for any $A_i \in \Sigma$, $i = 1, 2, \ldots$. The following example shows that a measure on a σ-additive class need not have this property when the left-hand side is defined.

Example 5.4. Consider the following subsets of $\mathbb{R}$:
$$A = [0,4], \quad B = [2,5], \quad C = [0,1] \cup [2,3] \cup [5,6].$$
Let
$$\Lambda = \{\emptyset, [0,6], A, B, C, A^c, B^c, C^c\}.$$
Then Λ is an additive class of subsets of [0,6]. Define the measure μ on Λ as follows:
$$\mu(\emptyset) = 0, \quad \mu(A) = \mu(B) = \mu(C) = 1$$
$$\mu([0,6]) = 4, \quad \mu(A^c) = \mu(B^c) = \mu(C^c) = 3.$$
Now $A \cup B \cup C = [0,6] \in \Lambda$ and yet
$$\mu([0,6]) = 4 > 3 = \mu(A) + \mu(B) + \mu(C). \quad \square$$

The previous example shows that a measure μ on a σ-additive class Λ cannot, in general, be extended to a measure on $\Sigma(\Lambda)$. Indeed, if such an extension were possible, then μ would be subadditive on Λ when the left-hand side is in Λ. Since a signed measure need not be subadditive, another question is suggested. Can μ be extended to a signed measure on $\Sigma(\Lambda)$? The next example shows that, in general, the answer to this question is also negative.

Example 5.5. [R. Prather, 1980] Let $X = \{0, 1, 2, 3, 4, 5\}$ and let Λ be the additive class
$$\Lambda = \{A \subseteq X : |A| = 0, 3 \text{ or } 6\}.$$
Define the measure μ on Λ by $\mu(\emptyset) = 0$, $\mu(X) = 4$, $\mu(\{0,2,4\}) = 1$, $\mu(\{1,3,5\}) = 3$, and $\mu(A) = 2$ for all other $A \in \Lambda$. Then μ has no extension to a signed measure on $\Sigma(\Lambda)$. Indeed, suppose μ' is such an extension of μ to $\Sigma(\Lambda)$. It is clear that $\Sigma(\Lambda) = P(X)$. We then have
$$4 = \mu'(\{0,1,2\}) + \mu'(\{2,3,4\})$$
$$= \mu'(\{0,2,4\}) + \mu'(\{1,2,3\}) = 1 + 2 = 3$$
which is a contradiction. $\square$

There are examples of physical interest in which an extension to a signed measure does exist. These situations stem from a generalized measure theory in which there is an "elementary length" L.

Example 5.6. Let $X = \{0, 1, 2, 3, 4, 5\}$ and let Λ be the additive class

$$\Lambda = \{A \subseteq X : |A| = 0, 2, 4 \text{ or } 6\}.$$

This corresponds to a discrete situation in which the "elementary length" L is two. The following theorem due to R. Prather shows that any measure on Λ has a unique extension to a signed measure on $\Sigma(\Lambda) = P(X)$. A proof of this theorem is given in [Prather, 1980]. □

Theorem 5.1. *Let* $X = \{0, 1, \ldots, N - 1\}$, $N = nL$, $L \geq 2$, $n > 2$ *integers, let* Λ *be the additive class*

$$\Lambda = \{A \subseteq X : |A| = n_1 L, \ 0 \leq n_1 \leq n\}$$

and let μ be a measure on Λ. Then μ has a unique extension to a signed measure on $\Sigma(\Lambda) = P(X)$.

Example 5.7. In Example 5.5 we considered a finite set X with an "elementary length" three. In this case the additive class was generated by the collection of subsets with cardinality three. J.P. Marchand and the author [Gudder and Marchand, 1980] have pointed out that it is more physical to only consider the cyclic subsets of X. That is, the additive class is now generated by the "translates" of the set $\{0, 1, 2\}$. In this case the additive class becomes

$$\Lambda = \{\emptyset, X, \{0,1,2\}, \{1,2,3\}, \{2,3,4\}, \{3,4,5\}, \{4,5,1\}, \{5,1,2\}\}$$

which is much smaller that the additive class in Example 5.5. We now show that also in this case a measure on Λ has an extension to a signed measure on $\Sigma(\Lambda) = P(X)$, although the extension may not be unique. □

Let $X = \{0, 1, \ldots, N - 1\}$, $N = nL$, $n, L \geq 2$ integers. For $j \in X$ define the translation map $\tau_j : X \to X$ by $\tau_j(i) = i + j \pmod{N}$. Let $I_0 = \{0, 1, \ldots, L - 1\}$ and for $j \in X$ define

$$I_j = \tau_j I_0 = \{\tau_j(0), \tau_j(1), \ldots, \tau_j(L-1)\}.$$

We define Λ to be the smallest additive class in $P(X)$ containing I_j, $j = 0, \ldots, N - 1$. It is easy to show that $\Sigma(\Lambda) = P(X)$.

EMBEDDING GENERALIZED MEASURE SPACES 175

Theorem 5.2. *If μ is a measure on Λ, then μ can be extended to a signed measure μ' on $P(X)$.*

Proof. Let V be the real inner product space $\mathbf{R}^N$ with the usual inner product

$$\langle x, y \rangle = \sum_{i=0}^{N-1} x_i y_i.$$

We assume that addition and subtraction in the indices are computed modulo N. Define the linear operator $T: V \to V$ by

$$(Tx)_i = \sum_{j=0}^{L-1} x_{i+j}.$$

The adjoint of T is

$$(T^*x)_i = \sum_{j=0}^{L-1} x_{i-j}$$

since

$$\langle T^*x, y \rangle = \sum_i \sum_{j=0}^{L-1} x_{i-j} y_i$$

$$= \sum_i \sum_{j=0}^{L-1} x_i y_{i+j} = \langle x, Ty \rangle.$$

The elements of the nullspace $N(T^*)$ of T^* are periodic in L; that is, if $x \in N(T^*)$ then for any $i \in X$ we have

$$x_i - x_{i-L} = \sum_{j=0}^{L-1} x_{i-j} - \sum_{j=0}^{L-1} x_{i-1-j} \qquad (5.1)$$

$$= (T^*x)_i - (T^*x)_{i-1} = 0.$$

Let $z \in V$ be given by $z_i = \mu(I_i)$. We now show that $z \perp N(T^*)$. Indeed, let $x \in N(T^*)$. Then from (5.1) we have

$$\langle x, z \rangle = \sum_{i=0}^{N-1} x_i z_i$$

$$= \sum_{j=0}^{n-1} \sum_{i=0}^{L-1} x_{jL+i} z_{jL+i} \qquad (5.2)$$

$$= \sum_{i=0}^{L-1} x_i \sum_{j=0}^{n-1} z_{jL+i}.$$

Moreover, for all $i \in X$ we have

$$\sum_{j=0}^{n-1} z_{jL+i} = \sum_{j=0}^{n-1} \mu(I_{jL+i})$$
$$= \mu(\bigcup_{j=0}^{n-1} I_{jL+i}) = \mu(X). \qquad (5.3)$$

Combining (5.2) and (5.3) gives

$$\langle x, z \rangle = \mu(X) \sum_{i=0}^{L-1} x_i$$
$$= \mu(X) \sum_{i=0}^{L-1} x_{L-1-i}$$
$$= \mu(X)(T^*x)_{L-1} = 0.$$

Since $z \perp N(T^*)$, z is in the range of T. Let $y \in V$ be the vector satisfying $Ty = z$, and define a signed measure μ' on $P(X)$ by $\mu'(A) = \sum_{j \in A} y_j$. We then have

$$\mu'(I_i) = \sum_{j \in I_i} y_j$$
$$= \sum_{j=0}^{L-1} y_{i+j} = (Ty)_i = z_i = \mu(I_i).$$

Taking disjoint unions and complementations gives $\mu'(A) = \mu(A)$ for all $A \in \Lambda$. Hence, μ' is an extension of μ to $P(X)$. □

The next result shows that (X, Λ, μ) has an embedding for any measure μ if and only if the "elementary length" is two. For a proof of this result see [Gudder and Marchand, 1980].

Theorem 5.3. *If $L = 2$, then any measure μ on Λ can be extended to a measure on $P(X)$. If $L \geq 3$, then there exists a measure μ on Λ that can not be extended to a measure on $P(X)$.*

A result analogous to Theorem 5.3 in the continuum case $X = [0, 2\pi]$ is given in [Gudder and Marchand, 1980].

5.2 Probability Models

This section considers some simple probability models for quantum mechanics. We begin with a discussion of J.S. Bell's investigations on hidden variables [Bell, 1966]. One way of interpreting Bell's work is the following. Quantum mechanics can only give statistical predictions. If hidden variables exist, then these statistical predictions are averages over the precise values of the hidden variables. Thus, in a hidden variables theory (according to Bell) we can represent quantum events $A, B, \ldots$ as events $A', B', \ldots$ in a probability space (X, Σ, μ) and their probabilities are given by $\mu(A'), \mu(B'), \ldots$. We emphasize that hidden variables, according to Bell, is different than our previous formulation of hidden variables, and in fact, we shall compare the two notions.

In order to study Bell's work we need to answer the following interesting question. If p, q, and r are numbers between 0 and 1, what conditions must they satisfy in order to be the probabilities of the simultaneous occurrence for three events? The precise form of this question together with its answer is contained in the following theorem.

Theorem 5.4. *Let $0 \leq p, q, r \leq 1$. Then there exist events A, B, C in a probability space (X, Σ, μ) such that $\mu(A \cap B) = p$, $\mu(A \cap C) = q$, $\mu(B^c \cap C) = r$ if and only if $q \leq p + r \leq 1$.*

Proof. Suppose there exist events A, B, C such that $\mu(A \cap B) = p$, $\mu(A \cap C) = q$, $\mu(B^c \cap C) = r$. Since $A \cap B$ and $B^c \cap C$ are disjoint, we have

$$p + r = \mu(A \cap B) + \mu(B^c \cap C) = \mu[(A \cap B) \cup (B^c \cap C)] \leq 1.$$

Moreover, since

$$A \cap C \subseteq (A \cup B^c) \cap (A \cup C) \cap (B \cup C) = [(A \cap B) \cup B^c] \cap [(A \cap B) \cup C]$$
$$= (A \cap B) \cup (B^c \cap C)$$

we have
$$q = \mu(A \cap C) \le \mu(A \cap B) + \mu(B^c \cap C) = p + r.$$

Conversely, suppose $0 \le p, q, r \le 1$ satisfy $q \le p + r \le 1$. Let $X = \{x_1, x_2, x_3, x_4, x_5\}$, $\Sigma = P(X)$, $A = \{x_1, x_2, x_3\}$, $B = \{x_1, x_2, x_5\}$, $C = \{x_1, x_3, x_4\}$, $B^c = \{x_3, x_4\}$. Let $m = \min(p, q)$ and define the function $f: X \to \mathbf{R}$ by $f(x_1) = m$, $f(x_2) = p - m$, $f(x_3) = q - m$, $f(x_4) = r - q + m$, $f(x_5) = 1 - (p + r)$. Then $0 \le f(x) \le 1$ for all $x \in X$ and $\sum_{x \in X} f(x) = 1$. If we define $\mu: \Sigma \to \mathbf{R}$ by $\mu(D) = \sum_{x \in D} f(x)$, the μ becomes a probability measure on Σ so (X, Σ, μ) is a probability space and A, B, C are events in this space. Now

$$\mu(A \cap B) = \mu(\{\omega_1, \omega_2\}) = p$$
$$\mu(A \cap C) = \mu(\{\omega_1, \omega_3\}) = q$$
$$\mu(B^c \cap C) = \mu(\{\omega_3, \omega_4\}) = r$$

which completes the proof. $\square$

The condition $\mu(B^c \cap C) = r$ in Theorem 5.4 is used instead of the more natural condition $\mu(B \cap C) = r$ because it makes the inequalities simpler and is a mere technicality. The next result is a different version of Theorem 5.4 in which the probabilities of the events A, B, C are prescribed in advance. Since there are more constraints, a stronger necessary and sufficient condition is required.

Theorem 5.5. Let $0 \le p_1, p_2, p_3 \le 1$. Then there exist events A, B, C in a probability space (X, Σ, μ) such that $\mu(A) = \mu(B) = \mu(C) = \frac{1}{2}$ and $\mu(A \cap B) = p_1$, $\mu(A \cap C) = p_2$, $\mu(B^c \cap C) = p_3$ if and only if $p_i \le p_j + p_k$, $i \ne j \ne k = 1, 2, 3$, and $p_1 + p_2 + p_3 \le 1$.

Proof. Suppose the events A, B, C exist with the given properties. As in the proof of Theorem 5.4, we conclude that $p_2 \le p_1 + p_3$ and $p_1 \le p_2 + p_3$. Moreover, since

$$B^c \cap C \subseteq (B^c \cap A) \cup (A^c \cap C)$$

we have

$$p_3 = \mu(B^c \cap C) \le \mu(B^c \cap A) + \mu(A^c \cap C)$$
$$= \mu(A) - \mu(B \cap A) + \mu(C) - \mu(A \cap C)$$
$$= \tfrac{1}{2} - p_1 + \tfrac{1}{2} - p_2 = 1 - p_1 - p_2.$$

Hence, $p_1 + p_2 + p_3 \leq 1$. Finally, applying the inclusion-exclusion principle, we have

$$\tfrac{3}{2} - (p_1 + p_2 + \tfrac{1}{2} - p_3) = \mu(A) + \mu(B) + \mu(C) - \mu(A \cap B) - \mu(A \cap C)$$
$$- [\mu(C) - \mu(B^c \cap C)]$$
$$= \mu(A) + \mu(B) + \mu(C)$$
$$- \mu(A \cap B) - \mu(A \cap C) - \mu(B \cap C)$$
$$= \mu(A \cup B \cup C) - \mu(A \cap B \cap C) \leq 1.$$

Hence, $p_3 \leq p_1 + p_2$.

Conversely, suppose p_1, p_2, p_3 satisfy the given conditions. Let $\Sigma = P(X)$ where

$$X = \{(i, j, k): i, j, k = 1, 2\}$$

and let

$$A = \{(1, j, k): j, k = 1, 2\}$$
$$B = \{(i, 1, k): i, k = 1, 2\}$$
$$C = \{(i, j, 1): i, j = 1, 2\}.$$

Define $f: X \to \mathbb{R}$ by

$$f(1, 1, 1) = f(2, 2, 2) = \tfrac{1}{2}(p_1 + p_2 - p_3)$$
$$f(1, 1, 2) = f(2, 2, 1) = \tfrac{1}{2}(p_1 + p_3 - p_2)$$
$$f(1, 2, 1) = f(2, 1, 2) = \tfrac{1}{2}(p_2 + p_3 - p_1)$$
$$f(2, 1, 1) = f(1, 2, 2) = \tfrac{1}{2} - \tfrac{1}{2}(p_1 + p_2 + p_3).$$

Then $0 \leq f(x) \leq 1$ for every $x \in X$ and $\sum_{x \in X} f(x) = 1$. If we define $\mu: \Sigma \to \mathbb{R}$ by $\mu(D) = \sum_{x \in D} f(x)$, then μ becomes a probability measure on Σ and (X, Σ, μ) is a probability space containing the events A, B, C. Now

$$\mu(A \cap B) = \mu\{(1,1,1),(1,1,2)\} = p_1$$
$$\mu(A \cap C) = \mu\{(1,1,1),(1,2,1)\} = p_2$$
$$\mu(B^c \cap C) = \mu\{(1,2,1),(2,2,1)\} = p_3$$
$$\mu(A) = f(1,1,1) + f(1,1,2) + f(1,2,1) + f(1,2,2) = \tfrac{1}{2}$$
$$\mu(B) = f(1,1,1) + f(1,1,2) + f(2,1,1) + f(2,1,2) = \tfrac{1}{2}$$
$$\mu(C) = f(1,1,1) + f(1,2,1) + f(2,1,1) + f(2,2,1) = \tfrac{1}{2}$$

which completes the proof. □

The inequalities $p_i \leq p_j + p_k$ are a version of the so-called *Bell's inequalities*. Theorems 4.4 and 4.5 are related to a result of Accardi-Fedulla [1982], although the approach and methods here are different. We now interpret Theorems 4.4 and 4.5 in terms of a spin-$\tfrac{1}{2}$ experiment.

Let x, y, z be three arbitrary directions in $\mathbb{R}^3$ and A_x, A_y, A_z be the quantum events for which the spin of an electron is $\frac{1}{2}$ in the x, y, z directions, respectively. Assuming the existence of hidden variables (according to Bell), we may consider A_x, A_y, A_z as being events in a probability space (X, Σ, μ). We now assume reality of the system. That is, we assume that the spin has a definite value in two directions simultaneously. Now this amounts to postulating that $A_x \cap A_y, A_y \cap A_z, \ldots$ are events. The probabilities of these events are denoted by

$$p(x,y) = \mu(A_x \cap A_y), p(y,z) = \mu(A_y \cap A_z), \ldots$$

where, for example, $p(x,y)$ is the probability that the spin is $\frac{1}{2}$ in the x and y directions simultaneously. If the spin is $-\frac{1}{2}$ in the x direction, the corresponding event is given by A_x^c and the simultaneous probabilities are denoted by $p(-x, y) = \mu(A_x^c \cap A_y), \ldots$.

We make one additional assumption, namely that space is isotropic. This amounts to postulating that $p(x, y)$ depends only on the angle between x and y. This assumption is eminently reasonable since otherwise there would be a preferred pair of directions in space. It follows from this assumption that $\mu(A_x) = \frac{1}{2}$ for any direction x. Indeed,

$$1 = \mu(A_x) + \mu(A_x^c) = p(x,x) + p(-x,-x) = 2p(x,x) = 2\mu(A_x).$$

Now from Theorem 4.4 or 4.5 we have

$$\mu(A_x \cap A_z) \leq \mu(A_x \cap A_y) + \mu(A_y^c \cap A_z)$$

or equivalently, $p(x, z) \leq p(x, y) + p(-y, z)$. Replacing x by $-x$ gives

$$p(-x, z) \leq p(-x, y) + p(-y, z). \tag{5.4}$$

We now rewrite (5.4) in terms of the function $p(\theta) = p(x, y)$ where $\theta = \text{angle}(x, y)$. Let $\theta_1, \theta_2, \theta_3$ be the angles between xy, yz, and xz, respectively. Then $\text{angle}(-y, z) = \pi - \theta_2$ and we have

$$p(\pi - \theta_2) + p(\theta_2) = p(-y, z) + p(y, z) = p(z, z) = \tfrac{1}{2}.$$

Hence, $p(\pi - \theta_2) = \frac{1}{2} - p(\theta_2)$. Inequality (5.4) becomes

$$\tfrac{1}{2} - p(\theta_3) \leq \tfrac{1}{2} - p(\theta_1) + \tfrac{1}{2} - p(\theta_2)$$

which gives

$$p(\theta_1) + p(\theta_2) - p(\theta_3) \leq \tfrac{1}{2} \quad \text{whenever} \quad \theta_3 \leq \theta_1 + \theta_2. \tag{5.5}$$

Following I. Pitowski [1983b] we can now prove the next lemma.

Lemma 5.6. $p(\pi/n) \leq (n-1)/2n$, $n = 1, 2, \ldots$.

Proof. Fix a positive integer n. We shall prove by induction that

$$kp\left(\frac{\pi}{n}\right) - p\left(\frac{k\pi}{n}\right) \leq \frac{1}{2}(k-1), \quad k = 1, 2, \ldots, n. \tag{5.6}$$

Equation (5.6) holds for $k = 1$. Suppose (5.6) holds for an integer $k < n$. Letting $\theta_1 = \pi/n$, $\theta_2 = k\pi/n$, $\theta_3 = (k+1)\pi/n$ in (5.5) gives

$$p\left(\frac{\pi}{n}\right) + p\left(\frac{k\pi}{n}\right) - p\left[\frac{(k+1)\pi}{n}\right] \leq \frac{1}{2}. \tag{5.7}$$

Adding (5.6) and (5.7) we obtain

$$(k+1)p\left(\frac{\pi}{n}\right) - p\left[\frac{(k+1)\pi}{n}\right] \leq \frac{k}{2}$$

which completes the proof by induction. Since $p(\pi) = p(-x, x) = 0$, letting $k = n$ in (5.6) gives our result. $\square$

Pitowski [1983b] now makes the following observation. Quantum mechanics predicts that $p(\theta) = \frac{1}{2}\cos^2(\theta/2)$ (see Section 2.4), and this has been verified experimentally. If we let $\theta = 2\pi/3$, then

$$p\left(\frac{2\pi}{3}\right) = \frac{1}{2}\cos^2\left(\frac{\pi}{3}\right) = \frac{1}{8}.$$

However, according to Lemma 5.6 we have

$$p\left(\frac{2\pi}{3}\right) = p\left(\pi - \frac{\pi}{3}\right) = \frac{1}{2} - p\left(\frac{\pi}{3}\right) \geq \frac{1}{2} - \frac{1}{2}\left(1 - \frac{1}{3}\right) = \frac{1}{6}.$$

We then conclude that reality and the existence of hidden variables (according to Bell) together are incompatible with the predictions of quantum mechanics.

The conclusion in the previous sentence was obtained within the framework of classical probability theory. However, as we have stressed earlier, there is no convincing reason to do this, and we could just as well use generalized probability theory. We would then represent quantum events by events in a gps and this is precisely our previous notion of hidden variables. As we shall show, in generalized probability theory, Bell's inequalities are no longer a logical consequence and reality, hidden variables and the predictions of quantum mechanics live in harmony together. This will be further elaborated in Chapter 6.

Let (X, Λ, μ) be a gps. If $A, B \in \Lambda$, then unfortunately $A \cap B$ need not be in Λ so we cannot yet discuss the probability $\mu(A \cap B)$ of a simultaneous occurrence. However, it is frequently possible to extend μ to the larger family $\Sigma(\Lambda)$ which certainly includes $A \cap B$. To be precise, we say that (X, Λ, μ) is *extendable* if there exists a signed measure μ' on $\Sigma(\Lambda)$ such that $\mu' \mid \Lambda = \mu$ (that is, μ' is an extension of μ). As we saw in Section 5.1, there are examples of gps's which are not extendable and even if (X, Λ, μ) is extendable, μ' need not be unique. If (X, Λ, μ) is extendable, we call $(X, \Lambda, \mu; \Sigma(\Lambda), \mu')$ an *extended* gps. The point is that although μ' may have negative values on some sets in $\Sigma(\Lambda)$ and cannot be interpreted as a probability, it may have values in $[0, 1]$ for the "important" sets $A \cap B$, $A, B \in \Lambda$, and can be interpreted as a probability for these sets. The next theorem should be compared to Theorem 5.5.

Theorem 5.7. *If $0 \leq p_1, p_2, p_3 \leq 1$, then there exists an extended gps $(X, \Lambda, \mu; \Sigma(\Lambda), \mu')$ and events $A, B, C \in \Lambda$ such that $\mu(A) = \mu(B) = \mu(C) = \frac{1}{2}$ and $\mu'(A \cap B) = p_1$, $\mu'(A \cap C) = p_2$, $\mu'(B^c \cap C) = p_3$.*

Proof. Let $X = \{x_1, \ldots, x_7\}$ and let $A = \{x_1, x_2, x_3\}$, $B = \{x_1, x_4, x_5\}$, $C = \{x_2, x_4, x_6\}$. Then

$$\Lambda = \{\emptyset, X, A, A^c, B, B^c, C, C^c\}$$

is an additive class. Define the function $f: X \to \mathbb{R}$ by

$f(x_1) = p_1,$ $f(x_2) = p_2,$ $f(x_3) = \frac{1}{2} - p_1 - p_2,$
$f(x_4) = \frac{1}{2} - p_3,$ $f(x_5) = p_3 - p_1,$ $f(x_6) = p_3 - p_2,$
$f(x_7) = p_1 + p_2 - p_3.$

Define the signed measure μ' on $\Sigma(\Lambda) = P(X)$ by $\mu'(D) = \sum_{x \in D} f(x)$, $D \in \Sigma(\Lambda)$, and let $\mu = \mu' \mid \Lambda$. Then $\mu(\emptyset) = 0$, $\mu(X) = 1$, and

$$\mu(A) = \mu(A^c) = \mu(B) = \mu(B^c) = \mu(C) = \mu(C^c) = \tfrac{1}{2}$$

so μ is a probability measure on X. Hence, $(X, \Lambda, \mu; \Sigma(\Lambda), \mu')$ is an extended gps. Moreover,

$\mu'(A \cap B) = \mu'(x_1) = p_1$
$\mu'(A \cap C) = \mu'(x_2) = p_2$

$\mu'(B^c \cap C) = \mu'(\{x_2, x_6\}) = p_3$

which completes the proof. □

If we consider conditional probabilities, then we can do better than Theorem 5.7. A *conditional* gps is an extended gps $(X, \Lambda, \mu; \Sigma(\Lambda), \mu')$ such that $0 \leq \mu'(A \cap B)/\mu(B) \leq 1$ for every $A, B \in \Lambda$ with $\mu(B) \neq 0$. We then call $\mu(A \mid B) \equiv \mu'(A \cap B)/\mu(B)$ the *conditional probability of A given B*. Notice that $\mu(\cdot \mid B)$ is then a probability measure on Λ for all $B \in \Lambda$ with $\mu(B) \neq 0$. In a conditional gps, $\mu(A \mid B)$ can be consistently interpreted as a conditional probability whenever, $\mu(B) \neq 0$.

Theorem 5.8. *If $0 < p_1, p_2, p_3 < 1$, then there exists a conditional gps $(X, \Lambda, \mu; \Sigma(\Lambda), \mu')$ and events $A, B, C \in \Lambda$ such that $\mu(A) = \mu(B) = \mu(C) = \frac{1}{2}$ and $\mu(A \mid B) = p_1$, $\mu(A \mid C) = p_2$, $\mu(B \mid C) = p_3$.*

Proof. Let X, A, B, C be defined as in Theorem 5.7 and define $f: X \to \mathbf{R}$ by

$f(x_1) = p_1/2, \quad f(x_2) = p_2/2, \qquad f(x_3) = (1 - p_1 - p_2)/2,$
$f(x_4) = p_3/2, \quad f(x_5) = (1 - p_1 - p_3)/2, \quad f(x_6) = (1 - p_2 - p_3)/2,$
$f(x_7) = (p_1 + p_2 + p_3 - 1)/2.$

Define the signed measure μ' on $\Sigma(\Lambda) = P(X)$ in terms of f as in Theorem 5.7. The proof now proceeds as in Theorem 5.7. □

Notice in Theorem 5.8 that $\mu(\cdot \mid \cdot)$ is symmetric in the two variables and that $\mu(A^c \mid B) = 1 - p_1$, $\mu(A^c \mid C) = 1 - p_2$, $\mu(B^c \mid C) = 1 - p_3$.

In the previous theorems we have constructed probability and generalized probability models for three simultaneous probabilities or conditional probabilities. In particular, we saw that for probability models, the three numbers p, q, r must satisfy certain constraints. We now show that even in traditional quantum mechanics p, q, r must satisfy constraints although they are less stringent than for the classical probability models. In traditional quantum mechanics, conditional probabilities are replaced by transition probabilities $|\langle \phi, \psi \rangle|^2$. We have then motivated the following definition. If $0 < p, q, r < 1$, then p, q, r *admit a complex (real) Hilbert space model* if there exists a complex (real) two-dimensional Hilbert space $\mathcal{H}$ with three unit vectors ϕ, ψ, χ satisfying

$$|\langle \phi, \psi \rangle|^2 = p, \quad |\langle \phi, \chi \rangle|^2 = q, \quad |\langle \psi, \chi \rangle|^2 = r \tag{5.8}$$

Accardi and Fedullo [1982] have shown that p, q, r admit a complex (real) Hilbert space model if and only if they satisfy certain constraints. We now give a simpler proof of their result [Zanghi and Gudder, 1984].

Theorem 5.9. (a) p, q, r admit a complex Hilbert space model if and only if $|p + q + r - 1| \leq 2(pqr)^{\frac{1}{2}}$. (b) p, q, r admit a real Hilbert space model if and only if $p + q + r - 1 = \pm 2(pqr)^{1/2}$.

Proof. (a) We can assume without loss of generality that $\mathcal{H} = \mathbb{C}^2$. If $U: \mathcal{H} \to \mathcal{H}$ is unitary, then ϕ, ψ, χ give a solution to (5.8) if and only if $U\phi, U\psi, U\chi$ give a solution to (5.8). Hence, we can assume that $\phi = (1, 0) \in \mathbb{C}^2$. Moreover, we can multiply any of the above vectors by a scalar $c \in \mathbb{C}$ with $|c| = 1$, without changing the results of (5.8). Hence, (5.8) has a solution if and only if there exist vectors $\phi, \psi, \chi \in \mathbb{C}^2$ of the form

$$\phi = (1, 0)$$
$$\psi = (a_1, a_2 e^{i\alpha}) \quad a_1, a_2, \alpha \geq 0, \quad a_1^2 + a_2^2 = 1$$
$$\chi = (b_1, b_2 e^{i\beta}) \quad b_1, b_2, \beta \geq 0, \quad b_1^2 + b_2^2 = 1$$

which satisfy (5.8). Then (5.8) holds if and only if

$$a_1 = p^{\frac{1}{2}}, \quad a_2 = (1-p)^{\frac{1}{2}}, \quad b_1 = q^{\frac{1}{2}}, \quad b_2 = (1-q)^{\frac{1}{2}} \tag{5.9}$$

and

$$|a_1 b_1 + a_2 b_2 e^{i(\alpha - \beta)}|^2 = r. \tag{5.10}$$

Solving (5.10) we obtain

$$\cos(\beta - \alpha) = \frac{r - a_1^2 b_1^2 - a_2^2 b_2^2}{2 a_1 b_1 a_2 b_2}. \tag{5.11}$$

It follows from (5.11) that (5.8) has a solution if and only if

$$|r - a_1^2 b_1^2 - a_2^2 b_2^2| \leq 2 a_1 b_1 a_2 b_2. \tag{5.12}$$

Letting $a_1^2 b_1^2 = x$, $a_2^2 b_2^2 = y$ in (5.12) gives

$$[r - (x + y)]^2 \leq 4xy. \tag{5.13}$$

After some algebraic manipulation, (5.13) becomes $|r + x - y| \leq 2(rx)^{\frac{1}{2}}$. Substituting for x, y and applying (5.9) gives $|p + q + r - 1| \leq 2(pqr)^{\frac{1}{2}}$. (b) There exists a real solution to (5.8) if and only if α or β equal 0 or π. It follows from (5.11) that (5.8) has a real solution if and only if

$$|r - a_1^2 b_1^2 - a_2^2 b_2^2| = 2 a_1 b_1 a_2 b_2.$$

Using the same manipulations as in (a) we have

$$p + q + r - 1 = \pm 2(pqr)^{\frac{1}{2}}. \quad \square$$

The previous results can be generalized to four or more events. Of course, we then have more complicated constraints [Zanghi and Gudder, 1984].

5.3 Measurable Functions and Compatibility

We have already noted that additive classes are precisely those operational logics that admit hidden variables. We now study the general algebraic properties of an additive class Λ of subsets of a set X. First Λ is a poset under the order $\subseteq$ and Λ has first and last elements $\emptyset$ and X, respectively. Moreover, it is clear that the usual complementation $A \mapsto A^c$ makes Λ an orthocomplemented poset. As usual, if $A, B \in \Lambda$ satisfy $A \subseteq B^c$ we write $A \perp B$ and say that A and B are *orthogonal*. Of course, $A \perp B$ if and only if $A \cap B = \emptyset$. It is clear that Λ is finitely orthocomplete and if Λ is a σ-additive class, then Λ is σ-orthocomplete. Finally, Λ is an orthomodular poset. Indeed, if $A \subseteq B$, then $A \perp B^c$ so $A \cup B^c \in \Lambda$ and hence, $A^c \cap B \in \Lambda$. It follows that $B = A \vee (B \wedge A^c)$. The following examples show that Λ need have no other standard algebraic properties.

Example 5.8. Let $X = \{1, 2, 3, 4\}$ and let Λ be the additive class of subsets of X with an even number of elements. In this example $(X, \subseteq)$ is a lattice. Indeed, except for the trivial cases, $A \wedge B = 0$ and $A \vee B = X$. Notice if $A \perp B$, then $A \wedge B = \emptyset$. However, $A \wedge B = \emptyset$ does not imply $A \perp B$. For example, $\{1, 2\} \wedge \{2, 3\} = \emptyset$ but $\{1, 2\} \not\perp \{2, 3\}$. This also shows that $A \wedge B$ need not equal $A \cap B$ (and $A \vee B$ need not equal $A \cup B$). Even though $(\Lambda, \subseteq)$ is a lattice, it is not distributive. Indeed,

$$(\{1,2\} \vee \{1,3\}) \wedge (\{1,2\} \vee \{3,4\}) = X \neq \{1,2\}$$

$$= \{1,2\} \vee (\{1,3\} \wedge \{3,4\}). \quad \square$$

Let Σ_1 and Σ_2 be nontrivial algebras of subsets of X_1 and X_2, respectively. Let $X = X_1 \times X_2$ and let

$$L = \{A_1 \times X_2, X_1 \times A_2 : A_1 \in \Sigma_1, A_2 \in \Sigma_2\}.$$

Then Λ is an additive class in X which is not an algebra. Moreover, if Σ_1 and Σ_2 are σ-additive classes, then so is Λ. We call Λ the *additive product* of Σ_1 and Σ_2.

Example 5.9. Let Σ_1, Σ_2 be the algebras of all subsets of $\{1,2,3\}$ and $\{1,2\}$, respectively, and let Λ be the additive product of Σ_1 and Σ_2. It is not hard to show that $(\Lambda, \subseteq)$ is a lattice. Let $A = \{(1,1),(1,2)\}$, $B = \{(1,1),(2,1),(3,1)\}$, $C = \{(1,1),(1,2),(2,2)\}$. Then $A \subseteq C$ but

$$A \vee (B \wedge C) = A \neq C = (A \vee B) \wedge C.$$

It follows that $(\Lambda, \subseteq)$ is not a modular lattice [Birkhoff, 1948]. □

Example 5.10. Let $X = \{1,2,3,4,5,6\}$ and let Λ be the additive class of all subsets of X with an even number of elements. If $A = \{1,2,3,4\}$ and $B = \{2,3,4,5\}$, then $A \wedge B$ does not exist. Indeed, $\{2,3\}$ and $\{3,4\}$ are lower bounds of A and B, but there is no lower bound for A and B in Λ that is greater than both $\{2,3\}$ and $\{3,4\}$. It follows that $\{\Lambda, \subseteq\}$ is not a lattice. □

We say that two sets A, B in an additive class Λ are *compatible* (written $A \leftrightarrow B$) if $A \cap B \in \Lambda$. It is easy to show that this definition of compatibility is equivalent to the definition of compatibility in an operational logic (Section 3.2) when specialized to an additive class. A collection of sets $\Lambda_1 \subseteq \Lambda$ is said to be compatible provided any finite intersection of sets in Λ_1 belongs to Λ. If we think of sets in Λ as quantum events, then $A \leftrightarrow B$ means that $A \cap B$ is an event so that A and B are simultaneously testable. Thus, compatible sets correspond to noninterfering events in quantum mechanics. These interference effects cannot be described in classical probability theory since any collection of sets in a σ-algebra is compatible. In fact, this property characterizes σ-algebras since a σ-additive class is a σ-algebra if and only if all its elements are pairwise compatible [Gudder, 1969, 1979].

Example 5.11. Let $X = \{1,2,\ldots,8\}$ and let Λ be the additive class of subsets of X with an even number of elements. In this example there are pairwise compatible collections which are not compatible; for instance $\{1,2,3,4\}, \{1,2,5,6\}, \{1,3,6,8\}$ is such a collection. □

Example 5.12. Let (X_1, Σ_1) and (X_2, Σ_2) be nontrivial measurable spaces and let Λ be the additive product of Σ_1 and Σ_2. Then $\Sigma_1 \times X_2$ and $X_1 \times \Sigma_2$ are σ-subalgebras of Λ which are completely noncompatible. That is $A_1 \not\leftrightarrow A_2$ for every nontrivial $A_1 \in \Sigma_1 \times X_2$, and $A_2 \in X_1 \times \Sigma_2$. □

The σ-subalgebras of a σ-additive class are extremely important since they correspond to classical probability spaces and, in quantum mechanics, they represent classical manuals and single observables. It is important to know when a collection of σ-subalgebras is contained in a single σ-subalgebra. The following theorem, whose proof can be found in [Gudder, 1979] answers this question.

Theorem 5.10. *A collection of sets belonging to a σ-additive class Λ is contained in a σ-subalgebra of Λ if and only if it is compatible.*

Corollary 5.11. *Let Σ_1, Σ_2 be σ-subalgebras of a σ-additive class Λ. If $A_1 \leftrightarrow A_2$ for every $A_1 \in \Sigma_1$, $A_2 \in \Sigma_2$, then $\Sigma_1 \cup \Sigma_2$ is contained in a σ-subalgebra of Λ.*

If Λ is a σ-additive class on X, a function $f : X \to \mathbb{R}$ is *measurable* if $f^{-1}(A) \in \Lambda$ for every open set $A \subseteq \mathbb{R}$. It follows from Theorem 5.10 that the σ-additive class generated by the open sets of $\mathbb{R}$ is the Borel σ-algebra $B(\mathbb{R})$. We conclude that if f is measurable, then $f^{-1}(A) \in \Lambda$ for every $A \in B(\mathbb{R})$. If f is measurable, we use the notation $\Lambda_f = \{f^{-1}(A) : A \in B(\mathbb{R})\}$. It is easily seen that Λ_f is a σ-subalgebra of Λ. We say that two measurable functions f and g are *compatible* (written $f \leftrightarrow g$) if $\Lambda_f \cup \Lambda_g$ is compatible. It follows from Corollary 5.11 that $f \leftrightarrow g$ if and only if $f^{-1}(A) \leftrightarrow g^{-1}(B)$ for every $A, B \in B(\mathbb{R})$. The following theorem gives an important characterization of compatible measurable functions and is due to V. Varadarajan [1968].

Theorem 5.12. *Let f, g be measurable functions on a σ-additive class. Then $f \leftrightarrow g$ if and only if there exist a measurable function h and two real Borel functions u, v such that $f = u \circ h$ and $g = v \circ h$.*

The above theorem shows that two measurable functions are compatible if and only if they are both functions of a single measurable function. This result shows that compatible measurable functions correspond to simultaneously observable (or testable) quantities since to measure these two quantities, one need measure only a single quantity. Theorem 5.12 holds in the more general case of two observables on a quantum logic [Varadarajan, 1968].

It is well known that in a measure space the sum of any two measurable functions is measurable. It follows that in a gms the sum of two compatible measurable functions is measurable. However, the sum of two noncompatible measurable functions need not be measurable. This follows from the general fact that a σ-additive class Λ is a σ-algebra if and only if the sum of any two measurable functions on Λ is measurable [Gudder, 1979]. One can show that the sum of two noncompatible measurable indicator functions is never measurable [Gudder, 1979]. Nevertheless, there are noncompatible measurable functions whose sum is measurable as the following example shows.

Example 5.13. Let X and Λ be as in Example 5.11. Define f and g on X as follows

$$f(1) = f(2) = f(7) = f(8) = 0, \quad f(3) = f(4) = 1, \quad f(5) = f(6) = 2,$$
$$g(3) = g(5) = g(7) = g(8) = 0, \quad g(1) = g(6) = 1, \quad g(2) = g(4) = 2.$$

Then $f \not\leftrightarrow g$ since $f^{-1}(\{2\}) \cap g^{-1}(\{0\}) = \{5\} \notin \Lambda$. But $f + g$ is measurable since

$$(f+g)^{-1}(\{0\}) = \{7,8\}, \quad (f+g)^{-1}(\{1\}) = \{1,3\},$$
$$(f+g)^{-1}(\{2\}) = \{2,5\}, \quad (f+g)^{-1}(\{3\}) = \{4,6\}. \quad \square$$

The fact that the sum of noncompatible measurable functions may be measurable could be important if this theory is to describe situations that arise in quantum mechanics. For example, in this case the total energy is a measurable quantity that is the sum of the kinetic and potential energies, the latter, in general, being noncompatible measurable quantities.

We now consider integration in a gms (X, Λ, μ). If $f: X \to \mathbb{R}$ is measurable, then the restriction $\mu \mid \Lambda_f$ is a measure on the σ-algebra Λ_f. Hence, $(X, \Lambda_f, \mu \mid \Lambda_f)$ becomes a measure space and we define the integral $\int f \, d\mu$ as the usual Lebesgue integral. We now consider monotone convergence. If $f \leq g$ are measurable functions, we need not have $f \leftrightarrow g$ as the following example shows.

Example 5.14. Let $X = \{1,2,3,4\}$ and let Λ be the additive class of subsets of X with an even number of elements. Define $f(1) = f(4) = 0$, $f(2) = f(3) = 1$, and $g(1) = g(2) = 1$, $g(3) = g(4) = 2$. Then f and g are measurable and $f \leq g$, but $f \not\leftrightarrow g$ since

$$f^{-1}(\{1\}) \cap g^{-1}(\{1\}) = \{2\} \notin \Lambda. \quad \square$$

Unlike classical measure theory, there are examples in which the limit of a sequence of measurable functions on a gms is not measurable [Gudder, 1979]. Also, there are examples of a sequence of measurable functions f_i that converge to a measurable function f for which there exists a measurable function g such that $g \leftrightarrow f_i$ for every i and yet $g \not\leftrightarrow f$ [Gudder, 1979]. However, for increasing sequences we have the following result due to J. Zerbe and the author [Gudder and Zerbe, 1981].

ADDITIVITY OF THE INTEGRAL

Theorem 5.13. *Let $f_1 \leq f_2 \leq \ldots$ be measurable functions on (X, Λ) and let $f = \lim f_i$. Then f is measurable and if g is measurable and $g \leftrightarrow f_i$, for all i then $g \leftrightarrow f$.*

The *center* $Z(\Lambda)$ of a σ-additive class Λ is defined as

$$Z(\Lambda) = \{A \in \Lambda : A \leftrightarrow B \text{ for all } B \in \Lambda\}.$$

A measure μ on Λ is σ-*finite* if there exists a sequence of mutually disjoint sets $A_i \in Z(\Lambda)$ such that $\cup A_i = X$ and $\mu(A_i) < \infty$, $i = 1, 2, \ldots$. We now generalize the monotone convergence theorem [Gudder and Zerbe, 1981].

Theorem 5.14. *Let (X, Λ, μ) be a gms with μ σ-finite and suppose $g \leq f_1 \leq f_2 \leq \ldots$ are measurable functions with $\int g \, d\mu > -\infty$. If $f_i \to f$, then $\int f_i \, d\mu \to \int f \, d\mu$.*

There are examples which show that generalizations of the dominated convergence theorem, Fatou's lemma, and Egoroff's theorem do not hold on a gms [Mullikin and Gudder, 1973]. However, one can prove a generalization of the Radon-Nikodym theorem [Gudder and Zerbe, 1981].

5.4 Additivity of the Integral

Let (X, Λ, μ) be a gms. We say that two measurable functions f and g are *summable* if $f + g$ is measurable. If $f \leftrightarrow g$, then f and g are summable. However, we have seen in Section 5.3 that there are noncompatible summable functions. A natural question to ask is the following. If f and g are summable, is

$$\int (f+g) d\mu = \int f \, d\mu + \int g \, d\mu \tag{5.14}$$

whenever the expressions exist? If the answer is yes, we say that the integral is *additive* on f and g. In classical measure theory, it is a standard result that the integral is additive. Analogously, in Hilbert space quantum mechanics, the integral is also additive as we now show. In this case a measurable function is replaced by a self-adjoint operator T and the measure is replaced by a normal state μ. If P^T is the spectral measure for T, then the integral (or expectation) of T becomes

$$E_\mu(T) = \int \lambda \mu[P^T(d\lambda)].$$

Now let S and T be bounded self-adjoint operators (we consider bounded operators to avoid technical domain problems). Then $S+T$ is self-adjoint and we ask whether $E_\mu(S+T) = E_\mu(S) + E_\mu(T)$. Thus, (5.14) is replaced by the following equation.

$$\int \lambda \mu[P^{S+T}(d\lambda)] = \int \lambda \mu[P^S(d\lambda)] + \int \lambda \mu[P^T(d\lambda)]. \tag{5.15}$$

It would be very difficult to prove (5.15) without further knowledge. However, applying Gleason's theorem (Section 2.3) the solution is easy. From Gleason's theorem we conclude that there is a density operator W such that $E_\mu(T) = \text{tr}(WT)$ for any bounded self-adjoint adjoint operator T. Hence,

$$E_\mu(S+T) = \text{tr}[W(S+T)] = \text{tr}(WS) + \text{tr}(WT) = E_\mu(S) + E_\mu(T)$$

The additivity problem is entirely different in generalized measure theory. First of all, the integral need not be additive on a gms as the following example shows.

Example 5.15. Let Z denote the integers and let $X = Z \times Z$. Let
$R_m = \{(i,j) \in X : i = m\}$
$C_n = \{(i,j) \in X : j = n\}$
$D_t = \{(i,j) \in X : i+j = t\}$.
Let $\Lambda_R, \Lambda_C, \Lambda_D$ be the σ-algebras generated by $\{R_m : m \in Z\}$, $\{C_n : n \in Z\}$, and $\{D_t : t \in Z\}$, respectively. Since

$$R_m \cap C_n = R_m \cap D_t = C_n \cap D_t \neq \emptyset$$

for all m, n, t, it follows that $\Lambda = \Lambda_R \cup \Lambda_C \cup \Lambda_D$ is a σ-additive class on X. Define the measure μ on Λ by
$\mu(R_0) = \mu(C_0) = \mu(D_1) = 1$
$\mu(R_m) = \mu(C_n) = \mu(D_t) = 0$
for $m, n \neq 0$, $t \neq 1$. Define the measurable functions f, g by $f(i,j) = i$, $g(i,j) = j$. Then $(f+g)(i,j) = i+j$ is measurable. Now $\int f\, d\mu = \int g\, d\mu = 0$ but $\int (f+g)\, d\mu = 1$. □

In Example 5.15, f and g have an infinite number of values. Suppose f and g are simple functions (have a finite number of values) and are summable. The example does not rule out the possibility that the integral is additive on f and g; and suprisingly enough this turns out to be true. This result is due to J. Zerbe and the author [1985]. In order to prove it, we first establish a special case using combinatorial arguments. Second, using a limiting procedure, the special case is shown to imply the general result.

Let f, g be simple summable functions on a gms (X, Λ, μ) each of which assumes the N values, $0, 1, \ldots, N-1$. It follows that $R_i = f^{-1}(i)$, $C_j = g^{-1}(j)$, $i, j = 0, 1, \ldots, N-1$ and $D_t = (f+g)^{-1}(t)$, $t = 0, 1, \ldots, 2N-2$ are members of Λ. Letting $r_i = \mu(R_i)$, $c_j = \mu(C_j)$, and $d_t = \mu(D_t)$, clearly

$$\sum r_i = \sum c_j = \sum d_t = \mu(X).$$

The additivity equation (5.14) now becomes

$$\sum t d_t = \sum i r_i + \sum j c_j. \tag{5.16}$$

The choice of notation for R_i, C_j, and D_t is deliberate as these constructs can be effectively modeled with the N rows, N columns, and $2N-1$ lower left to upper right diagonals of an $N \times N$ matrix. For later reference, let $B_{ij} = f^{-1}(i) \cap g^{-1}(j)$, which corresponds to the i, jth block of X. Of course, B_{ij} need not be in Λ. Without loss of generality, we henceforth assume that Λ is the additive class generated by the collection

$$\{R_i, C_j, D_t \colon i, j = 0, \ldots, N-1, \ t = 0, \ldots, 2N-2\}.$$

In the sequel, $0 < \beta < 1$ will be a fixed irrational number and λ will be a variable in the interval $[0, 1]$. Define the sequence

$$n_j(\lambda) = [j\lambda + \beta], \qquad j = 0, 1, \ldots$$

where $[\ ]$ is the greatest integer part function. Each $\lambda \in [0, 1]$ determines a subset $S(\lambda)$ of X which is defined as $S(\lambda) = \cup B_{i,j}$, where $0 \le i, j \le N-1$ satisfy

$$n_i(\lambda) + n_j(\lambda) > n_s(\lambda) + n_{i+j-s}(\lambda)$$

for some $0 \le s \le i + j$.

Lemma 5.15. *For fixed t and λ, the set*

$$\{n_i(\lambda) + n_{t-i}(\lambda) \colon i = 0, 1, \ldots, t\}$$

consists either of a single integer or two consecutive integers.

Proof.   There exist numbers $0 \leq \delta_1, \delta_2 < 1$ such that

$$n_i(\lambda) + n_{t-i}(\lambda) = [i\lambda + \beta] + [(t-i)\lambda + \beta]$$
$$= (i\lambda + \beta) - \delta_1 + (t-i)\lambda + \beta - \delta_2$$
$$= t\lambda + 2\beta - (\delta_1 + \delta_2).$$

Hence, for every $i = 0, \ldots, t$, $n_i(\lambda) + n_{t-i}(\lambda)$ is an integer in the interval $(t\lambda + 2\beta - 2, t\lambda + 2\beta)$ and this interval contains only two integers. □

Lemma 5.16.   (a) *For fixed $\lambda \in [0,1]$, at most one number in the sequence $j\lambda + \beta$, $j = 0, 1, \ldots$, is an integer.* (b) *As λ increases from 0 to 1, the sequence*

$$(n_j(0)) = (0, 0, 0, \ldots)$$

monotonically increases to the sequence

$$(n_j(1)) = (0, 1, 2, \ldots)$$

and precisely one of the integers in a sequence increases by 1 with each sequence change.

Proof.   (a) Suppose $j_1\lambda + \beta = n_1$ and $j_2\lambda + \beta = n_2$, where $j_1 \neq j_2$ and n_1, n_2 are integers. Then

$$\beta = \frac{j_1 n_2 - j_2 n_1}{j_1 - j_2}$$

which contradicts the fact that β is irrational. (b) As λ increases, the terms of the sequence $(n_j(\lambda))$ remain constant until one of the numbers $j\lambda + \beta$ becomes an integer. This term then increases by 1 and by part (a) is the only term that does so. □

Lemma 5.17.   *Let $0 \leq \lambda < 1$ and assume $(n_j(\lambda)) \neq (n_j(1))$. Increase λ until exactly one of the numbers, say $j_0\lambda + \beta$, $j_0 \leq N-1$, becomes an integer. Call the λ at which this occurs λ_0. Then $S(\lambda_0)$ can be constructed from $S(\lambda)$ by the following sequence of set operations.*

(1) *Disjoint union with C_{j_0}.*

(2) *Proper subtraction of D_{2j_0}.*

ADDITIVITY OF THE INTEGRAL

(3) *Disjoint union with R_{j_0}.*

(4) *Proper subtraction of all possible D_t's.*

Proof. By the choice of λ_0, we have

$$[j_0\lambda + \beta] = [j_0\lambda_0 + \beta] - 1 \tag{5.17}$$

$$[j\lambda + \beta] = [j\lambda_0 + \beta] \text{ for all } j \neq j_0. \tag{5.18}$$

Applying Lemma 5.15, for $s \neq j_0$ we have

$$2[j_0\lambda_0 + \beta] - 2 < [s\lambda_0 + \beta] + [(2j_0 - s)\lambda_0 + \beta].$$

Hence, from (5.17) and (5.18) we obtain

$$2[j_0\lambda + \beta] < [s\lambda + \beta] + [(2j_0 - s)\lambda + \beta]. \tag{5.19}$$

Similarly, from Lemma 5.15, for $j \neq j_0$ we have

$$[j\lambda_0 + \beta] + [j_0\lambda_0 + \beta] - 1 \leq [s\lambda_0 + \beta] + [(j + j_0 - s)\lambda_0 + \beta].$$

Hence, from (5.17) and (5.18) we obtain

$$[j\lambda + \beta] + [j\lambda_0 + \beta] \leq [s\lambda + \beta] + [(j + j_0 - s)\lambda + \beta]. \tag{5.20}$$

Now (5.19) and (5.20) show that $S(\lambda)$ has the proper configuration for the specified set operations to be performed. Specifically, (5.19) shows that

$$D_{2j_0} \setminus B_{j_0,j_0} \subseteq S(\lambda), \qquad B_{j_0,j_0} \cap S(\lambda) = \emptyset.$$

This and (5.20) show that

$$C_{j_0} \cap S(\lambda) = R_{j_0} \cap S(\lambda) = \emptyset.$$

An examination of the diagonal sets shows that $S(\lambda_0)$ is the set that results from applying the specified set operations to $S(\lambda)$. First consider D_{2j_0}. From Lemma 5.15 and 5.19, for $s \neq j_0$, we have

$$2[j_0\lambda + \beta] + 1 = [s\lambda + \beta] + [(2j_0 - s)\lambda + \beta].$$

Hence, from (5.17) we obtain

$$2[j_0\lambda_0 + \beta] - 1 = [s\lambda_0 + \beta] + [(2j_0 - s)\lambda_0 + \beta].$$

Therefore, $B_{j_0,j_0} \subseteq S(\lambda_0)$, while

$$(D_{2j_0} \setminus B_{j_0,j_0}) \cap S(\lambda_0) = \emptyset.$$

This corresponds to the configuration of D_{2j_0} after the set operations have been applied. Note that when D_{2j_0} is the degenerate diagonal consisting only of B_{j_0,j_0}, D_{2j_0} is removed in Step 4. However, in this case, $D_{2j_0} \cap S(\lambda_0) = \emptyset$ since there are no other values with which to compare $2[j_0\lambda_0 + \beta]$ (see definition of $S(\lambda)$) and the same conclusion applies.

Second, consider D_{j+j_0}, for $j \neq j_0$, and use (5.20) again:

(i) Suppose there is at least one s, $s \neq j_0$, such that equality holds in (5.20). After the increase to λ_0,

$$[s\lambda_0 + \beta] + [(j + j_0 - s)\lambda_0 + \beta]$$

does not change, while

$$[j\lambda_0 + \beta] + [j_0\lambda_0 + \beta] = [j\lambda + \beta] + [j_0\lambda + \beta] + 1.$$

Hence,

$$S(\lambda_0) \cap D_{j+j_0} = (S(\lambda) \cap D_{j+j_0}) \cup B_{j_0,j} \cup B_{j,j_0}.$$

This corresponds exactly to the configuration of a diagonal which would not be removed in Step 4.

(ii) Otherwise

$$[j\lambda + \beta] + [j_0\lambda + \beta] < [s\lambda + \beta] + [(j + j_0 - s)\lambda + \beta]$$

for all $s \neq j_0$. When λ is increased to λ_0, all of the values

$$[s\lambda_0 + \beta] + [(j + j_0 - s)\lambda_0 + \beta]$$

including $s = j_0$, are equal. Hence, $D_{j+j_0} \cap S(\lambda_0) = \emptyset$. This corresponds exactly to a diagonal which would be removed in Step 4. $\square$

Corollary 5.18. $S(\lambda) \in \Lambda$ if and only if $S(\lambda_0) \in \Lambda$.

We now prove the result for the special case under consideration.

ADDITIVITY OF THE INTEGRAL

Lemma 5.19. *For f and g defined above, we have*

$$\int (f+g)\,d\mu = \int f\,d\mu + \int g\,d\mu.$$

Proof. Since

$(n_j(0)) = (0,0,0,\ldots)$
$(n_j(1)) = (0,1,2,\ldots)$

it is easy to check that $S(0) = S(1) = \emptyset$. Lemma 5.17 shows that by letting λ increase in a finite sequence of prescribed steps from 0 to 1, $S(1)$ can be constructed from $S(0)$ by adjoining row and column sets, and deleting diagonal sets. Each time the jth element of $(n_0(\lambda), n_1(\lambda), \ldots)$ increases by 1, C_j is adjoined. Since $n_j(0) = 0$ and $n_j(1) = j$, C_j is adjoined j times. A similar count shows that R_i is adjoined i times. We next show that D_t is deleted t times. Note that $D_t = \cup B_{i,j}$, where $i+j = t$, $0 \le i, j \le N-1$. Now $B_{i,j}$ is adjoined i times as a subset of R_i and j times as a subset of C_j. Since $S(1) = \emptyset$, $B_{i,j}$ must be deleted $i+j = t$ times. Since the only way it can be deleted is as a subset of D_t, D_t must be deleted t times. Equation 5.16 now follows since

$$0 = \mu(S(1)) = \sum i r_i + \sum j c_j - \sum t d_t. \quad \square$$

The following example illustrates the proofs of the previous lemmas for the special case $N = 4$.

Example 5.16. Let $N = 4$ and let f and g be summable simple functions with values $0, 1, 2, 3$. The sets $R_i, C_j, B_{i,j}$, $i, j = 0, 1, 2, 3$ are illustrated in Figure 5.1. The sets $D_t = \cup\{B_{i,j} : i+j = t\}$ are given by the lower left to upper right diagonals.

	C_0	C_1	C_2	C_3
R_0	B_{00}	B_{01}	B_{02}	B_{03}
R_1	B_{10}	B_{11}	B_{12}	B_{13}
R_2	B_{20}	B_{21}	B_{22}	B_{23}
R_3	B_{30}	B_{31}	B_{32}	B_{33}

Figure 5.1 (Values of f and g)

For concreteness, let $\beta = 1/\sqrt{2}$ and call the sequence $(n_j(\lambda))$ the *magic sequence*. The jumps in the magic sequence occur at the values

$$\lambda_1 = \frac{1}{3} - \frac{1}{3\sqrt{2}}, \quad \lambda_2 = \frac{1}{2} - \frac{1}{2\sqrt{2}}, \quad \lambda_3 = 1 - \frac{1}{\sqrt{2}}$$

$$\lambda_4 = \frac{2}{3} - \frac{1}{3\sqrt{2}}, \quad \lambda_5 = 1 - \frac{1}{2\sqrt{2}}, \quad \lambda_6 = 1 - \frac{1}{3\sqrt{2}}.$$

At these values and at $\lambda = 0$, the magic sequence becomes

$(n_j(0)) = (0,0,0,0) \quad (n_j(\lambda_1)) = (0,0,0,1)$
$(n_j(\lambda_2)) = (0,0,1,1) \quad (n_j(\lambda_3)) = (0,1,1,1)$
$(n_j(\lambda_4)) = (0,1,1,2) \quad (n_j(\lambda_5)) = (0,1,2,2)$
$(n_j(\lambda_6)) = (0,1,2,3)$

Starting from the set $S(0) = \emptyset$ and ending with the set $S(\lambda_6) = \emptyset$, the sets $S(\lambda_1), \ldots, S(\lambda_5)$ are indicated by + marks in Figure 5.2.

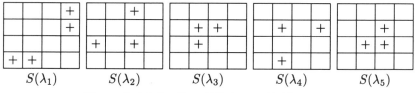

Figure 5.2 (The Sets $S(\lambda_1), \ldots, (\lambda_5)$)

Using + for disjoint union and − for proper subtraction, we see that the sets $S(\lambda_j)$ are related as follows:

$S(\lambda_1) = S(0) + C_3 - D_6 + R_3 - D_6 - D_5,$
$S(\lambda_2) = S(\lambda_1) + C_2 - D_4 + R_2 - D_5 - D_3,$
$S(\lambda_3) = S(\lambda_2) + C_1 - D_2 + R_1 - D_1 - D_4,$
$S(\lambda_4) = S(\lambda_3) + C_3 - D_6 + R_3 - D_6 - D_5 - D_3,$
$S(\lambda_5) = S(\lambda_4) + C_2 - D_4 + R_2 - D_5 - D_2,$
$S(\lambda_6) = S(\lambda_5) + C_3 - D_6 + R_3 - D_6 - D_5 - D_4 - D_3.$

It is thus seen that C_j is adjoined j times, R_j is adjoined j times, $j = 0, 1, 2, 3$, and D_t is deleted t times, $t = 0, 1, \ldots, 6$. □

We now use the special case proved in Lemma 5.19 to prove the result for arbitrary summable simple functions. But first we need

ADDITIVITY OF THE INTEGRAL

Lemma 5.20. *Let A be a $k \times k$ matrix with rational entries and let $y = (y_1, \ldots, y_k) \in \mathbb{R}^k$ satisfying $Ay = 0$. Then for any $\varepsilon > 0$, there exists a $z = (z_1, \ldots, z_k)$ such that $Az = 0$, z_i are rational, and $|z_i - y_i| < \varepsilon$, $i = 1, \ldots, k$.*

Proof. Let Q^k be the vector space of rational k-tuples. Then A defines a linear operator A' on Q^k. Since rank(A) equals the number of linearly independent columns of A, rank(A) = rank(A'). Denoting the null space of A by $N(A)$, we have $\dim N(A) = \dim N(A')$. Let $e_1, \ldots, e_n$ be a basis for $N(A')$. Then $e_1, \ldots, e_n$ is a basis for $N(A)$. Hence, $N(A')$ is dense in $N(A)$. Thus, for any $\varepsilon > 0$, if $(y_1, \ldots, y_k) \in N(A)$, there exists a $(z_1, \ldots, z_k) \in N(A')$ such that $|z_i - y_i| < \varepsilon$, $i = 1, \ldots, k$. $\square$

Corollary 5.21. *Suppose $y_1, \ldots, y_k$ are real numbers that satisfy a system of homogeneous linear equations having rational coefficients. Then for any $\varepsilon > 0$, there exist rational numbers $z_1, \ldots, z_k$ which satisfy the same system of equations and $|z_i - y_i| < \varepsilon$, $i = 1, \ldots, k$.*

Theorem 5.22. *If f and g are summable simple functions, then*

$$\int (f+g)\,d\mu = \int f\,d\mu + \int g\,d\mu.$$

Proof. In the proof of Lemma 5.19 it was not assumed that R_i, C_j, and D_t were nonempty. Thus the result holds if f and g have any nonnegative integer values. If a is constant, it is clear that $\int (a+f)\,d\mu = \int a\,d\mu + \int f\,d\mu$. Hence, the result holds if f and g have any integer values. Now suppose f and g have rational values. Then there exists a constant $b \neq 0$ such that bf and bg have integer values. Then

$$\int f\,d\mu + \int g\,d\mu = \left(\frac{1}{b}\right)\left[\int bf\,d\mu + \int bg\,d\mu\right]$$

$$= \left(\frac{1}{b}\right)\int b(f+g)\,d\mu = \int (f+g)\,d\mu.$$

Hence, the result holds if f and g have rational values. Finally, suppose that f and g have arbitrary distinct real values $\alpha_1, \ldots, \alpha_n$ and $\beta_1, \ldots, \beta_n$, respectively (there is no loss of generality in assuming that they have the same number of values.)

There may be indices $i \neq k$, $j \neq m$ such that $\alpha_i + \beta_j - \alpha_k - \beta_m = 0$. Let S be the system of all such equations. Choose an $\varepsilon > 0$ such that
$$\varepsilon < |\alpha_i - \alpha_j|, |\beta_i - \beta_j|, |\alpha_i + \beta_j - \alpha_k - \beta_m|$$
for all i, j, k, m for which the terms on the right of the inequality are nonzero. Applying Corollary 4.21, there exist rationals p_i, q_i, $i = 1, \ldots, n$ which satisfy S and $|\alpha_i - p_i|, |\beta_i - q_i| < \varepsilon/4$, $i = 1, \ldots, n$. If $\alpha_i + \beta_j - \alpha_k - \beta_m \neq 0$, then $p_i + q_j - p_k - q_m \neq 0$. Indeed, if $p_i + q_j - p_k - q_m = 0$, then
$$|\alpha_i + \beta_j - \alpha_k - \beta_m| = |(\alpha_i - p_i) + (\beta_j - q_j) - (\alpha_k - p_k) - (\beta_m - q_m)|$$
$$\leq |\alpha_i - p_i| + |\beta_j - q_j| + |\alpha_k - p_k| + |\beta_m - q_m| < \varepsilon.$$
This gives a contradiction.

Define f_1, g_1 by $f_1(\omega) = p_i$ if $f(\omega) = \alpha_i$ and $g_1(\omega) = q_i$ if $g(\omega) = \beta_i$. Because of the above inequalities, the p_i's are distinct and so are the q_i's. It follows that $f^{-1}(\alpha_i) = f_1^{-1}(p_i)$, $g^{-1}(\beta_i) = g_1^{-1}(q_i)$, $i = 1, \ldots, n$. Our work in the previous paragraph shows that
$$(f + g)^{-1}(\alpha_i + \beta_j) = (f_1 + g_1)^{-1}(p_i + q_j).$$
Since f_1 and g_1 are summable simple functions with rational values, we have
$$\sum p_i \mu[f_1^{-1}(p_i)] + \sum q_i \mu[g_1^{-1}(q_i)] = \sum (p_i + q_j) \mu[(f_1 + g_1)^{-1}(p_i + q_j)]$$
where the sum on the right-hand side is over the distinct $p_i + q_j$. We then obtain
$$\left| \sum \alpha_i \mu[f^{-1}(\alpha_i)] + \sum \beta_i \mu[g^{-1}(\beta_i)] \right.$$
$$\left. - \sum (\alpha_i + \beta_j) \mu[(f + g)^{-1}(\alpha_i + \beta_j)] \right|$$
$$\leq \sum |\alpha_i - p_i| \mu[f^{-1}(\alpha_i)] + \sum |\beta_i - q_i| \mu[g^{-1}(\beta_i)]$$
$$+ \sum |\alpha_i + \beta_j - p_i - q_j| \mu[(f + g)^{-1}(\alpha_i + \beta_j)]$$
$$\leq \varepsilon \mu(X).$$
Letting $\varepsilon \to 0$ completes the proof. □

We have shown that the generalized integral is additive on two summable simple functions. What about three summable simple functions? If f_1, f_2, and f_3 are simple functions and $f_1 + f_2 + f_3$ is measurable, then is
$$\int (f_1 + f_2 + f_3) \, d\mu = \int f_1 \, d\mu + \int f_2 \, d\mu + \int f_3 \, d\mu \, ?$$
The following example shows that in general the answer is negative.

Example 5.17. Let $X = \{1,2,3,4,5,6\}$ and let Λ be the additive class consisting of the subsets $\emptyset$, X, $\{1,2,3\}$, $\{4,5,6\}$, $\{1,2,4\}$, $\{3,5,6\}$, $\{1,3,6\}$, $\{2,4,5\}$, $\{1,4,6\}$, $\{2,3,5\}$. Define the measure μ on Λ by

$$\mu(X) = \mu(\{1,2,3\}) = \mu(\{1,2,4\}) = \mu(\{1,3,6\}) = \mu(\{2,3,5\}) = 1$$

and μ is zero on the other sets. Let f_1, f_2, f_3 be the indicator functions of $\{1,4,6\}$, $\{3,5,6\}$, $\{2,4,5\}$, respectively. Then f_1, f_2, f_3 are measurable functions. Also, $f = f_1 + f_2 + f_3$ is measurable since $f^{-1}(\{1\}) = \{1,2,3\}$ and $f^{-1}(\{2\}) = \{4,5,6\}$. Now

$$\int f_1 \, d\mu + \int f_2 \, d\mu + \int f_3 \, d\mu = 0 \neq 1 = \int (f_1 + f_2 + f_3) \, d\mu. \quad \square$$

5.5 Bayes' Rule and Hidden Variables

Let L be an orthomodular poset with state space $\Omega(L)$. Denote the set of dispersion-free states on L by $\Delta(L)$ and if L is σ-orthocomplete denote its set of normal (that is, countably additive) states by $\Omega_n(L)$. Recall from Section 3.4 that L *admits hidden variables* if $\Delta(L)$ is order-determining. In Theorem 3.11 we showed that L admits hidden variables if and only if L is isomorphic to an additive class. Moreover, in Corollary 3.12 we saw that if L is σ-orthocomplete, then L admits hidden variables if and only if L is σ-isomorphic to a σ-additive class. These results gave a connection between hidden variables and generalized probability spaces. Although the above characterization of hidden variables is useful, it is not directly related to the statistical properties of states. In this section we give another characterization which is related to the richness of the set of states that satisfy a Bayesian rule (see Section 1.2).

For $\omega \in \Omega(L)$, $p \in L$, we say that $\nu \in \Omega(L)$ is an (ω, p) *conditional state* if for every $q \in L$ with $q \leq p$ we have $\omega(q) = \omega(p)\nu(q)$. This notion of conditional state is essentially due to G. Cassinelli and N. Zanghi [Beltrametti and Cassinelli, 1981]. Denote the set of (ω, p) conditional states by $(\omega \mid p)$. One can give examples in which $(\omega \mid p) = \emptyset$ and in which $\omega(p) \neq 0$ but $|(\omega \mid p)| > 1$. (For the first, take an L with precisely one state [Greechie, 1966], for the second, take a horizontal sum of Boolean algebras.) Notice that if $\omega(p) = 0$, then $(\omega \mid p) = \Omega(L)$. For $p \in L$, we use the notation $\hat{p} = \{\omega \in \Omega(L) : \omega(p) = 1\}$. Notice that if $\omega(p) \neq 0$, then $\nu \in \hat{p}$ for all $\nu \in (\omega \mid p)$. Also, if $\omega \in \hat{p}$, then $\omega \in (\omega \mid p)$.

We say that a state ω is *order*-1 *Bayesian* if for any $p \in L$ we have

$$\omega = \omega(p)\omega_1 + \omega(p')\omega_2 \qquad (5.21)$$

where $\omega_1 \in (\omega \mid p)$, $\omega_2 \in (\omega \mid p')$. Denote the set of order-1 Bayesian states on L on $B_1(L)$. Continuing recursively, we say that a state ω is *order-n Bayesian*, $n > 1$, if (5.21) holds for any $p \in L$, where

$$\omega_1 \in (\omega \mid p) \cap B_{n-1}(L), \quad \omega_2 \in (\omega \mid p') \cap B_{n-1}(L). \qquad (5.22)$$

By convention we define $B_0(L) = \Omega(L)$. Moreover, we define $B_\infty(L) = \bigcap_{n=1}^\infty B_n(L)$ and call the elements of $B_\infty(L)$ *Bayesian states*.

Lemma 5.23. (a) $\Delta(L) \subseteq B_\infty(L)$. (b) $B_n(L) \subseteq B_{n-1}(L)$, $n = 1, 2, \ldots$. (c) $\omega \in B_n(L)$ *if and only if for every* $p \in L$ *there exists* $\omega_1, \omega_2 \in B_{n-1}(L)$, $\lambda \in [0,1]$ *such that* $\omega = \lambda\omega_1 + (1-\lambda)\omega_2$, *where* $\omega_1 \in \hat{p}$ *if* $\lambda \neq 0$ *and* $\omega_2 \in \hat{p}'$ *if* $\lambda \neq 1$.

Proof. (a) Let $\omega \in \Delta(L)$, and $p \in L$. If $\omega \in \hat{p}$, then $\omega \in (\omega \mid p)$ and $\omega = \omega(p)\omega + \omega(p')\omega$. If $\omega \in \hat{p}'$, then $\omega \in (\omega \mid p')$ and the same equation holds. Hence, $\omega \in B_1(L)$. By induction, $\omega \in B_n(L)$, $n = 1, 2, \ldots$.

(b) We proceed by induction on n. Clearly

$$B_2(L) \subseteq B_1(L) \subseteq B_0(L)$$

Suppose $B_n(L) \subseteq B_{n-1}(L)$, and let $\omega \in B_{n+1}(L)$. Then for any $p \in L$, $\omega = \omega(p)\omega_1 + \omega(p')\omega_2$, where ω_1, ω_2 satisfy (5.22) with $n-1$ replaced by n. By the induction hypothesis, ω_1, ω_2 satisfy (5.22). Hence, $\omega \in B_n(L)$.

(c) Suppose $\omega \in B_n(L)$ and $p \in L$, then $\omega = \omega(p)\omega_1 + \omega(p')\omega_2$, where ω_1, ω_2 satisfy (5.22). If $\lambda = \omega(p)$, then clearly $\lambda \in [0,1]$ and $\omega(p') = 1 - \lambda$. If $\lambda \neq 0$ then $\omega_1 \in \hat{p}$ and if $\lambda \neq 1$, then $\omega_2 \in \hat{p}'$. Conversely, suppose the given condition holds and $p \in L$. Then $\omega = \lambda\omega_1 + (1-\lambda)\omega_2$, where $\lambda, \omega_1, \omega_2$ satisfy the condition. If $\lambda = 0$ then $\omega_2 \in \hat{p}'$, so $\omega_2 \in (\omega \mid p')$. Since $\omega(p) = 0$, $\omega_1 \in (\omega \mid p)$. Hence, $\omega \in B_n(L)$. A similar result holds if $\lambda = 1$. Now suppose $0 < \lambda < 1$. Then $\omega_1 \in \hat{p}$, $\omega_2 \in \hat{p}'$. Hence, $\lambda = \omega(p)$ and $1 - \lambda = \omega(p')$. If $q \leq p$, we have $\omega_2(q) = 0$, so $\omega(q) = \omega(p)\omega_1(q)$. If $q \leq p'$, we have $\omega_1(q) = 0$, so $\omega(q) = \omega(p')\omega_2(q)$. Hence, $\omega_1 \in (\omega \mid p)$, $\omega_2 \in (\omega \mid p')$ and $\omega \in B_n(L)$. □

Corollary 5.24. *If $\Omega(L) = B_1(L)$, then $\Omega(L) = B_\infty(L)$.*

We now consider some examples. Let X be a nonempty set and let Σ be an algebra of subsets of X. Then $\Sigma = (\Sigma, \subseteq, {}^c)$ is an orthomodular poset and $\Omega(\Sigma)$ is the set of (finitely additive) probability measures on Σ. The next result shows that conditional states always exist and, except for trivial cases, are unique. Moreover, conditional states give the usual conditional probabilities, and every state is Bayesian.

Theorem 5.25. (a) *If $\mu \in \Omega(\Sigma)$ and $A \in \Sigma$ with $\mu(A) \neq 0$, then $|(\mu \mid A)| = 1$. Moreover, $\nu \in (\mu \mid A)$ if and only if*

$$\nu(B) = \frac{\mu(B \cap A)}{\mu(A)} = \mu(B \mid A).$$

(b) $B_\infty(\Sigma) = \Omega(\Sigma)$.

Proof. (a) Suppose $\mu(A) \neq 0$ and $\nu \in (\mu \mid A)$. Now for any $B \in \Sigma$ we have $\nu(B) = \nu(B \cap A) + \nu(B \cap A^c)$. Since $\nu(A) = 1$, we have $\nu(A^c) = 0$, so $\nu(B \cap A^c) = 0$. Since $B \cap A \subseteq A$, we have

$$\nu(B) = \nu(B \cap A) = \frac{\mu(B \cap A)}{\mu(A)}.$$

Conversely, if $\nu(B) = \mu(B \mid A)$, then clearly $\nu \in (\mu \mid A)$. The result now follows.

(b) Let $\mu \in \Omega(\Sigma)$, $A \in \Sigma$. If $\mu(A) = 0$ or 1, we have $\mu = \mu(A)\mu + \mu(A')\mu$. If $0 < \mu(A) \leq 1$, then for any $B \in \Sigma$ we have

$$\mu(B) = \mu(A)\mu(B \mid A) + \mu(A^c)\mu(B \mid A^c).$$

By part (a), the probability measures $\mu_1(B) = \mu(B \mid A)$ and $\mu_2(B) = \mu(B \mid A^c)$ satisfy $\mu_1 \in (\mu \mid A)$, $\mu_2 \in (\mu \mid A^c)$. Hence, $\Omega(\Sigma) \subseteq B_1(\Sigma)$. The result follows from Corollary 5.24. □

As another example, let $\mathcal{H}$ be a complex Hilbert space and let $P(\mathcal{H})$ be the set of all orthogonal projections on $\mathcal{H}$. We have already observed that $(P(\mathcal{H}, \leq,')$ is an orthomodular poset. If $3 \leq \dim \mathcal{H} < \infty$, it follows from Gleason's theorem (Section 2.3) that every $\mu \in \Omega[P(\mathcal{H})]$ has the form $\mu(P) = \operatorname{tr}(PW_\mu)$, where W_μ is a unique density operator. If $\dim \mathcal{H} = \infty$, then the same result holds for $\mu \in \Omega_n[P(\mathcal{H})]$. In this example, the next theorem shows that conditional states always exist and, except for trivial cases, are unique. However, Bayesian states do not exist if $\dim \mathcal{H} \geq 3$. In this theorem we assume that $\dim \mathcal{H} < \infty$. However, if one considers normal states, a similar proof applies to the infinite-dimensional case for parts (a) and (c). For another proof of part (a) see [Beltrametti and Cassinelli, 1981]. Also notice in part (a) that the conditional states are given by the usual von Neumann formula for conditional probabilities (see Section 2.3)

Theorem 5.26. *Let $\dim \mathcal{H} = N$, where $3 \leq N < \infty$. (a) If $\mu \in \Omega[P(\mathcal{H})]$, $P \in P(\mathcal{H})$ with $\mu(P) \neq 0$, then $|(\mu \mid P)| = 1$. Moreover, $\nu \in (\mu \mid P)$ if and only if*

$$\nu(Q) = \frac{\operatorname{tr}(QPW_\mu P)}{\mu(P)}$$

for all $Q \in P(\mathcal{H})$. (b) $|B_1[P(\mathcal{H})]| = 1$ and $\mu \in B_1[P(\mathcal{H})]$ if and only if $W_\mu = I/N$. (c) $B_2[P(\mathcal{H})] = \emptyset$.

Proof. (a) Let $\nu \in (\mu \mid P)$ where $\mu(P) \neq 0$. As noted above ν has the form $\nu(Q) = \operatorname{tr}(QW_\nu)$, where $W_\nu \geq 0$, $\operatorname{tr}(W_\nu) = 1$. Let $\psi_1, \ldots, \psi_N$ be an orthonormal basis for $\mathcal{H}$, where $P\psi_i = \psi_i$, $i = 1, \ldots, n$, and $P\psi_i = 0$, $i = n+1, \ldots, N$. Let $P_1, \ldots, P_N$ be the corresponding one-dimensional projections. Since $\nu(P) = 1$, we have

$$\operatorname{tr}(P_i W_\nu) = \nu(P_i) = 0, \qquad i = n+1, \ldots, N.$$

Also, for $i = 1, \ldots, n$ we have

$$\operatorname{tr}(P_i W_\nu) = \nu(P_i) = \frac{\operatorname{tr}(P_i W_\mu)}{\mu(P)} = \frac{\operatorname{tr}(P_i P W_\mu P)}{\mu(P)}.$$

We thus have

$$\langle W_\nu \psi_i, \psi_i \rangle = \frac{\langle P W_\mu P \psi_i, \psi_i \rangle}{\mu(P)}, \qquad i = 1, \ldots, N.$$

It follows that $W_\nu = PW_\mu P/\mu(P)$. Conversely, if ν has the given form, it is clear that $\nu \in (\mu \mid P)$.

(b) Suppose $W_\mu = I/N$, and $P \in P(\mathcal{H})$. As before we can dispense with the cases $\mu(P) = 0$ or 1 and assume that $0 < \mu(P) \leq 1$. Define $\mu_1, \mu_2 \in \Omega[P(\mathcal{H})]$ by $\mu_1(Q) = \operatorname{tr}(QP)/N\mu(P)$ and $\mu_2(Q) = \operatorname{tr}(QP')/N\mu(P')$. Then $\mu_1 \in (\mu \mid P)$, $\mu_2 \in (\mu \mid P')$, and for every $Q \in P(\mathcal{H})$

$$\mu(Q) = \frac{\operatorname{tr}(Q)}{N} = \frac{\operatorname{tr}(QP)}{N} + \frac{\operatorname{tr}(QP')}{N}$$
$$= \mu(P)\mu_1(Q) + \mu(P')\mu_2(Q).$$

Hence, $\mu \in B_1[P(\mathcal{H})]$. Conversely, suppose $\mu \in B_1[P(\mathcal{H})]$. If $\mu(P) = 0$ or 1, it follows that $PW_\mu = W_\mu P$. Assume that $0 < \mu(P) < 1$. Now there exist $\mu_1 \in (\mu \mid P)$, $\mu_2 \in (\mu \mid P')$ with $\mu = \mu(P)\mu_1 + \mu(P')\mu_2$. Hence, by part (a), for any $Q \in P(\mathcal{H})$ we have

$$\mathrm{tr}(QW_\mu) = \mu(Q) = \mathrm{tr}\big[Q(PW_\mu P + P'W_\mu P')\big].$$

It follows that $W_\mu = PW_\mu P + P'W_\mu P'$. Hence,

$$PW_\mu = PW_\mu P = W_\mu P.$$

Since W_μ commutes with every $P \in P(\mathcal{H})$, it follows that $W_\mu = cI$ for some $c \in \mathbb{C}$. Since $\mathrm{tr}(W_\mu) = 1$, $c = N$.

(c) Let $\mu \in B_2[P(\mathcal{H})]$ and suppose $\mu(P) \neq 0$. Now there exist $\mu_1 \in B_1[P(\mathcal{H})] \cap (\mu \mid P)$, $\mu_2 \in B_1[P(\mathcal{H})] \cap (\mu \mid P')$ with $\mu = \mu(P)\mu_1 + \mu(P')\mu_2$. By parts (a) and (b) we have

$$W_{\mu_1} = \frac{1}{N} = \frac{PW_\mu P}{\mu(P)} = \frac{P}{N\mu(P)} = \frac{P}{\mathrm{tr}(P)}.$$

It follows that $P = I$. But this is impossible since there exists a $Q \in P(\mathcal{H})$ with $Q \neq I$ and $\mu(Q) = \mathrm{tr}(Q)/N \neq 0$. □

It follows from Lemma 5.23a and Theorem 5.26c that $\Delta[P(\mathcal{H})] = \emptyset$ whenever $3 \leq \dim \mathcal{H} < \infty$. But, more generally, applying Gleason's theorem we conclude that $\Delta[P(\mathcal{H})] = \emptyset$ whenever $\dim \mathcal{H} \geq 3$. It will follow from our next result that $B_\infty[P(\mathcal{H})] = \emptyset$ if $\dim \mathcal{H} \geq 3$. Moreover, since our next result shows that $B_n[P(\mathcal{H})]$ is compact for all n, we conclude that $B_n[P(\mathcal{H})] = \emptyset$ for some n if $\dim \mathcal{H} \geq 3$.

The orthomodular posets in these last two examples correspond to classical probability theory and traditional quantum mechanics, respectively. In the first case every state is Bayesian, and in the second case no state is Bayesian. Our next theorem shows that there is a close relationship between Bayesian and dispersion-free states. This result is due to T. Armstrong and the author [Gudder and Armstrong, 1985].

Let L be an orthomodular poset and let $\mathbb{R}^L$ be the vector space of real-valued functions on L. We endow $\mathbb{R}^L$ with the product topology. In this topology a net $f_\alpha \in \mathbb{R}^L$ converges to $f \in \mathbb{R}^L$ if and only if $f_\alpha(p) \to f(p)$ for all $p \in L$. Since $[0,1]$ is compact in $\mathbb{R}$, it follows from Tychonoff's theorem [Dunford and Schwartz, 1958] that $[0,1]^L$ is compact in $\mathbb{R}^L$. Now $\Omega(L)$ is a convex subset of $[0,1]^L$ and it is easy to show that $\Omega(L)$ is closed in

$[0,1]^L$. Hence, $\Omega(L)$ is a compact convex subset of $\mathbf{R}^L$. It follows from the Krein-Milman theorem [Dunford and Schwartz, 1958] that $\Omega(L)$ is the closed convex hull of its extreme points. Denoting the extreme points of a convex set S by $\text{ext}(S)$ and the closed convex hull of S by $\overline{\text{conv}}(S)$, this result becomes $\Omega(L) = \overline{\text{conv}}\{\text{ext}[\Omega(L)]\}$.

Theorem 5.27. $B_\infty(L) = \overline{\text{conv}}[\Delta(L)]$.

Proof. We first show by induction that $B_n(L)$ is convex. We already know that $B_0(L) = \Omega(L)$ is convex. Suppose $B_n(L)$ is convex and $\omega_1, \omega_2 \in B_{n+1}(L)$, $\lambda \in [0,1]$, $\omega = \lambda\omega_1 + (1-\lambda)\omega_2$. If $p \in L$, there exist $\omega_{ij} \in B_n(L)$, $i, j = 1, 2$, where $\omega_{i1} \in (\omega_i \mid p)$, $\omega_{i2} \in (\omega_i \mid p')$, i=1,2, and $\omega_i = \omega_i(p)\omega_{i1} + \omega_i(p')\omega_{i2}$. We then have

$$\omega = \lambda\omega_1(p)\omega_{11} + (1-\lambda)\omega_2(p)\omega_{21} + \lambda\omega_1(p')\omega_{12} + (1-\lambda)\omega_2(p')\omega_{22}.$$

If $\omega(p) = 0$, we obtain $\omega = \lambda\omega_{12} + (1-\lambda)\omega_{22}$ and it follows from Lemma 5.23c that $\omega \in B_{n+1}(L)$. Similarly, if $\omega(p) = 1$, we have $\omega = \lambda\omega_{11} + (1-\lambda)\omega_{21}$ and again $\omega \in B_{n+1}(L)$. If $0 < \omega(p) < 1$, we can write

$$\omega = \alpha[\beta_1\omega_{11} + (1-\beta_1)\omega_{21}] + (1-\alpha)[\beta_2\omega_{12} + (1-\beta_2)\omega_{22}]$$
$$= \alpha\nu_1 + (1-\alpha)\nu_2$$

where $\alpha = \omega(p)$, $\beta_1 = \lambda\omega_1(p)\alpha^{-1}$, $\beta_2 = \lambda\omega_1(p')(1-\alpha)^{-1}$, and $\nu_i = \beta_i\omega_{1i} + (1-\beta_i)\omega_{2i}$, $i = 1, 2$. Since $B_n(L)$ is assumed convex, $\nu_1, \nu_2 \in B_n(L)$. Applying Lemma 5.23c gives $\omega \in B_{n+1}(L)$.

We next show by induction that $B_n(L)$ is compact in $\Omega(L)$. Since $\Omega(L)$ is compact, we need only show that $B_n(L)$ is closed. We already know that $B_0(L) = \Omega(L)$ is closed. Suppose that $B_n(L)$ is closed and let ω_α be a net in $B_{n+1}(L)$ with $\omega_\alpha \to \omega \in \Omega(L)$. Since $B_n(L)$ is closed, by Lemma 5.23b we have $\omega \in B_n(L)$. If $p \in L$, then by Lemma 5.23c

$$\omega_\alpha = \lambda_\alpha\omega_{1\alpha} + (1-\lambda_\alpha)\omega_{2\alpha}$$

where $\omega_{1\alpha} \in \hat{p}$ if $\lambda_\alpha \neq 0$, $\omega_{2\alpha} \in \hat{p}'$ if $\lambda_\alpha \neq 1$, and $\omega_{i\alpha} \in B_n(L)$, $i = 1, 2$. Suppose that $\omega(p) \neq 0$. Since $\lambda_\alpha = \omega_\alpha(p) \to \omega(p)$, $\lambda_\alpha \neq 0$ eventually. Since $B_n(L)$ is assumed compact, a subnet $\omega_{1\beta}$ converges in $B_n(L)$ and hence $\omega_{2\beta}$ also converges in $B_n(L)$. Suppose $\omega_{i\beta} \to \omega_i \in B_n(L)$, $i = 1, 2$, and $\lambda = \omega(p)$. Then $\omega = \lambda\omega_1 + (1-\lambda)\omega_2$. Now $\lambda \neq 0$ and $\omega_1(p) =$

$\lim \omega_{1\beta}(p) = 1$. If $\lambda \neq 1$, then $\omega_2(p') = \lim \omega_{2\beta}(p') = 1$. Similar reasoning holds if $\omega(p') \neq 0$. Applying Lemma 5.23c, we conclude that $\omega \in B_{n+1}(L)$ and hence $B_{n+1}(L)$ is closed.

We have shown in the above that each $B_n(L)$ is convex and compact, $n = 0, 1, \ldots$. It follows that $B_\infty(L)$ is convex and compact. Since by Lemma 5.23a, $\Delta(L) \subseteq B_\infty(L)$, we have $\overline{\text{conv}}[\Delta(L)] \subseteq B_\infty(L)$.

Let $\omega \in B_\infty(L)$ and suppose $\omega \notin \Delta(L)$. Then there is a $p \in L$ with $0 < \omega(p) < 1$. Since $\omega \in B_\infty(L)$, we have

$$\omega = \omega(p)\omega_{1i} + \omega(p')\omega_{2i}$$

where $\omega_{1i} \in B_i(L) \cap (\omega \mid p)$, $\omega_{2i} \in B_i(L) \cap (\omega \mid p')$, $i = 1, 2, \ldots$. Since $\Omega(L)$ is compact, there are convergent subsequences ω_{1j}, ω_{2j} and suppose $\omega_{1j} \to \omega_1$, $\omega_{2j} \to \omega_1$. Since by Lemma 5.23b, $B_n(L)$ forms a decreasing sequence, given an n, $\omega_{1j}, \omega_{2j} \in B_n(L)$ eventually. Since $B_n(L)$ is closed, $\omega_1, \omega_2 \in B_n(L)$. Since n was arbitrary, $\omega_1, \omega_2 \in B_\infty(L)$, so ω is not extremal in $B_\infty(L)$. It follows that $\text{ext}[B_\infty(L)] \subseteq \Delta(L)$. Applying the Krein-Milman theorem gives $B_\infty(L) \subseteq \overline{\text{conv}}[\Delta(L)]$. $\square$

Corollary 5.28. *The following statements are equivalent.*

(a) $B_\infty(L) = \Omega(L)$.

(b) $\Omega(L) = \overline{\text{conv}}[\Delta(L)]$.

(c) $\text{ext}[\Omega(L)] = \Delta(L)$.

The next result gives our characterization of hidden variables.

Corollary 5.29. *L admits hidden variables if and only if $B_\infty(L)$ is order-determining.*

Proof. If $\Delta(L)$ is order-determining, so is $B_\infty(L)$ since $\Delta(L) \subseteq B_\infty(L)$. Suppose $\Delta(L)$ is not order-determining. Then there exist $p, q \in L$ with $p \not\leq q$ but $\omega(p) \leq \omega(q)$ for all $\omega \in \Delta(L)$. Hence, $\nu(p) \leq \nu(q)$ for every $\nu \in \text{conv}[\Delta(L)]$. If $\nu_\alpha(p) \leq \nu_\alpha(q)$ for a net ν_α with $\nu_\alpha \to \nu$, then $\nu(p) \leq \nu(q)$. Hence, $\nu(p) \leq \nu(q)$ for all $\nu \in \overline{\text{conv}}[\Delta(L)] = B_\infty(L)$. Thus, $B_\infty(L)$ is not order-determining. $\square$

5.6 Notes and References

Generalized probability and measure theory began with the work of P. Suppes [1966]. Later developments and more details are reported in [Gudder, 1969, 1979,1984a]. This field is fairly new and there is considerable room for further investigation. A good discussion about the axioms of probability theory which is relevant to this chapter is given in [Fine, 1973]. Further details concerning the embedding of gms's can be found in [Gudder and Marchand, 1980; Prather, 1980; Zanghi and Gudder, 1984].

Some papers selected from the vast literature on the theoretical and experimental aspects of Bell's inequaltities are [Belifante, 1973; Bell, 1966; Beltrametti and Cassinelli, 1981; Pitowski, 1982, 1983b]. Accardi [1981] and Accardi and Fedullo [1982] have made important investigations on probability models. The results in Section 5.2 together with further material can be found in [Zanghi and Gudder, 1984]. It was our intention to just give the reader a brief taste of measurable functions, compatibility and integration on a gms in Section 5.3. We refer those interested in a larger meal to [Gudder, 1969, 1979, 1984a].

Additivity of the generalized integral was an unsolved problem for almost twenty years [Gudder, 1969]. For an outline of the proof see [Gudder, 1984a]; the proof presented here is contained in [Zerbe and Gudder, 1985]. For more details on conditional states we refer the reader to [Beltrametti and Cassinelli, 1981]. The material in Section 5.5 is contained in [Gudder and Armstrong, 1985].

6
Probability Manifolds

This chapter shows that amplitude functions within the framework of a probability manifold give a hidden variables description of quantum mechanics. In particular it will be shown that the important concepts of nonrelativistic quantum mechanics including its dynamics can be derived from probabilistic concepts on a classical phase space. A probability manifold is closely related to a generalized probability space. In fact they generate an important class of such spaces which are not classical probability spaces. Roughly speaking, we can say that traditional quantum mechanics is just classical mechanics which has been averaged over a generalized probability space using amplitude functions. The present chapter unifies and utilizes many of the concepts presented in previous chapters.

6.1 Transfinite Induction

One of the basic tools that we shall need in this chapter is transfinite induction. We now develop the highlights of this useful technique.

Let $(A, \leq)$ be a poset. Two elements $x, y \in A$ are *comparable* if either $x \leq y$ or $y \leq x$; otherwise they are *incomparable*. If any two elements of a subset $B \subseteq A$ are comparable, then B is a *chain* (or *totally ordered subset*) of A. If A itself is a chain, we call A a *totally ordered* set. If $a \in A$, the *initial segment of A determined by a* is the set $S_a = \{x \in A : x < a\}$. It is easy to prove that if P is an initial segment of A, and Q is an initial segment of P, then Q is an initial segment of A. Thus, an initial segment of an initial segment of A is an initial segment of A.

If A and B are posets, a function $f: A \to B$ is *increasing* (or *order-preserving*) if $x \leq y$ implies $f(x) \leq f(y)$. A function $f: A \to B$ is an *isomorphism* if f is bijective and satisfies $x \leq y$ if and only if $f(x) \leq f(y)$. If there exists an isomorphism from A to B, then we call A and B *isomorphic* and write $A \approx B$. It is easy to show that a bijection f is an isomorphism if and only if f and f^{-1} are increasing. Moreover, $\approx$ is an equivalence relation. It is also easy to show that if $f: A \to B$ is bijective and increasing and A is totally ordered, then f is an isomorphism.

Example 6.1. Let $A = \{a, b, c\}$, $B = \{u, v, w\}$ and define partial orders on A and B as follows: $c < a, b$, $w < v < u$ and a, b are incomparable. Define $f: A \to B$ by $f(a) = u$, $f(b) = v$, $f(c) = w$. Then f is bijective and increasing but is not an isomorphism since $f(b) < f(a)$ but a and b are incomparable. □

A poset A is *well ordered* if every nonempty subset of A has a least element. Notice that a well ordered set is totally ordered. The converse of this statement does not hold as the real numbers with their usual order shows. Let A be a poset and let $a \in A$. An element $b \in A$ is an *immediate successor* of a if $a < b$ and there is no element $c \in A$ such that $a < c < b$. If A is well ordered, then every element of A (with the exception of the greatest element of A, if it exists) has a unique immediate successor. Indeed, if $x \in A$ and x is not the greatest element of A, then the set $T = \{y \in A : y > x\}$ is nonempty, so T has a least element which is clearly the immediate successor of x. We define a to be an *immediate predecessor* of b if b is an immediate successor of a.

Example 6.2. Let $A = \{0, \frac{1}{2}, \frac{3}{4}, \frac{7}{8}, \ldots, 1, 1+\frac{1}{2}, 1+\frac{3}{4}, 1+\frac{7}{8}, \ldots\}$. If we order the elements of A in the usual way, A becomes a well ordered set. Notice that every element of A has an immediate successor. However, 0 and 1 do not have immediate predecessors. □

If A is a poset, a *section* of A is any subset $B \subseteq A$ satisfying $y \in B$ implies $S_y \subseteq B$. We now characterize sections of well ordered sets.

Lemma 6.1. *If A is well ordered, then B is a section of A if and only if $B = A$ or B is an initial segment of A.*

Proof. If $B = A$ or B is an initial segment of A, then clearly B is a section of A. Conversely, suppose B is a section of A and $B \neq A$. Since A

TRANSFINITE INDUCTION

is well ordered, $A \setminus B$ has a least element m. We now show that $B = S_m$. If $x \in S_m$, then $x < m$ and since m is the least element of $A \setminus B$, we have $x \in B$. Conversely, suppose $x \in B$. If $m \leq x$, since B is a section, we conclude that $m \in B$. But $m \in A \setminus B$, which is a contradiction. Hence, $x < m$, so $x \in S_m$. □

The following theorem gives an important property of well ordered sets. This property is called the *transfinite induction principle*.

Theorem 6.2. *Let A be a well ordered set and let $P(x)$ be a statement that is either true or false for every $x \in A$. If $P(x)$ is true whenever $P(y)$ is true for every $y < x$, then $P(x)$ is true for every $x \in A$.*

Proof. Suppose that the set $\{x \in A : P(x) \text{ is false}\}$ is nonempty and hence has a least element m. Now $P(x)$ is true for every $x < m$, but $P(m)$ is false. Hence, the contrapositive of the statement to be proved holds. □

Another important property of well ordered sets is, roughly speaking, that they do not differ from one another except in their size. This is the content of Theorem 6.7, but first we need some preparatory lemmas.

Lemma 6.3. *If A is well ordered and f is an isomorphism from A to a subset of A, then $x \leq f(x)$ for all $x \in A$.*

Proof. Suppose $P = \{x \in A : x > f(x)\} \neq \emptyset$, and let a be the least element of P. Then $f(a) < a$. We then have $f(f(a)) < f(a) < a$ so $f(a) \in P$. But this is impssible since a is the least element of P. Hence, $P = \emptyset$. □

Lemma 6.4. *If A is well ordered there is no isomorphism from A to a subset of an initial segment of A.*

Proof. Suppose f is an isomorphism from A to a subset of an initial segment S_a of A. By Lemma 6.3, $a \leq f(a)$ so $f(a) \notin S_a$. Since this is a contradiction, such an f cannot exist. □

Corollary 6.5. *No well ordered set is isomorphic with an initial segment of itself.*

Lemma 6.6. *Let A and B be well ordered sets. If A is isomorphic with an initial segment of B, then B is not isomorphic with any subset of A.*

Proof. Let $f: A \to S_b$ be an isomorphism where $b \in B$. Assume there exists an isomorphism $g: B \to C$ where $C \subseteq A$. Now the composite function

$f \circ g: B \to S_b$ is injective and increasing so by a previous remark, $f \circ g$ is an isomorphism from B to a subset of S_b. But by Lemma 6.4, this is impossible. □

Theorem 6.7. *If A and B are well ordered, then exactly one of the following three cases hold:* (i) $A \approx B$, (ii) $A \approx S_b$ *for some* $b \in B$, (iii) $B \approx S_a$ *for some* $a \in A$.

Proof. Let $C = \{x \in A : S_x \approx S_r \text{ for some } r \in B\}$. If $x \in C$, there is a unique $r \in B$ such that $S_x \approx S_r$. Indeed, suppose $S_x \approx S_r$ and $S_x \approx S_t$ where $r \ne t$, say $r < t$. Then $S_x \approx S_t$ and S_r is an initial segment of S_t. But this is impossible by Corollary 6.5. Denoting the unique $r \in B$ corresponding to $x \in C$ by $F(x)$ provides a function $F: C \to B$. If $D \subseteq B$ denotes the range of F, we now show that $F: C \to D$ is an isomorphism. If $F(u) = F(v) = r$, then $S_u \approx S_r \approx S_v$. If $u \ne v$, say $u < v$, then as before S_u is an initial segment of S_v. But this is impossible by Corollary 6.5. Hence, $u = v$ and F is injective. Suppose $u \le v$, where $F(u) = r$ and $F(v) = t$. Then $S_u \approx S_r$ and $S_v \approx S_t$. Assume that $t < r$ so S_t is isomorphic to an initial segment of S_r and S_r is isomorphic to a subset of S_v. But this is impossible by Lemma 6.6. Hence, $F(u) = r \le t = F(v)$ so F is increasing. It follows that F is an isomorphism.

We next show that C is a section of A. Let $c \in C$ and $x < c$. If $F(c) = r$, then $S_c \approx S_r$ so there exists an isomorphism $g: S_c \to S_r$. It is now easy to show that $g \mid S_x$ is an isomorphism from S_x to $S_{g(x)}$ so $S_x \approx S_{g(x)}$. Hence, $x \in C$ and C is a section of A. An analogous argument shows that S is a section of B.

Now suppose that $C = S_x$ and $D = S_r$ are both initial segments. Then $S_x \approx S_r$ so $x \in C = S_x$ which is impossible. Applying Lemma 6.1 proves that one of the conditions (i), (ii), or (iii) holds. That no two of these conditions hold simultaneously follows from Corollary 6.5 and Lemma 6.6. □

If (ii) of Theorem 6.7 holds we write $A < B$. It then follows that for any well ordered sets A and B we have $A \approx B$ or $A < B$ or $B < A$. The next Corollary follows from Lemma 6.6 and Theorem 6.7.

Corollary 6.8. *If A is well ordered, then every subset of A is isomorphic with A or an initial segment of A.*

Now that we have developed this machinery for well ordered sets we might ask whether it is useful for anything. In particular, are there any interesting well ordered sets? It is clear that any finite set can be well

TRANSFINITE INDUCTION   211

ordered. But are there any infinite well ordered sets? This question does not have an obvious answer. Of course, the set of positive integers N is well-ordered, but this is only because well ordering is postulated as an axiom for N (frequently disguised in terms of mathematical induction). Surprisingly enough, it turns out that any set can be well ordered if we assume the axiom of choice. Because of this fact, the transfinite induction principle has universal applicability.

Let A be a nonempty set with power set $P(A)$ and let $P'(A) = P(A) \setminus \{\emptyset\}$. A *choice function* for A is a function $f: P'(A) \to A$ such that $f(B) \in B$ for every $B \in P'(A)$. The *axiom of choice* postulates that every nonempty set has a choice function. A poset A is *inductive* if every chain of A has an upper bound in A. *Zorn's axiom* states that every nonempty inductive set has a maximal element. A standard result is that the axiom of choice and Zorn's axiom are equivalent [Dunford and Schwartz, 1958; Pinter, 1971]. In the sequel we shall assume the axiom of choice and hence Zorn's axiom.

Theorem 6.9. *Any set A can be well ordered.*

Proof. Let $\mathcal{A}$ be the family of all pairs (B, G) where $B \subseteq A$ and G is an order relation in B that well orders B. We define $(B, G) \leq (B', G')$ if (a) $B \subseteq B'$, (b) $G \subseteq G'$, and (c) if $x \in B$, $y \in B' \setminus B$, then $(x, y) \in G'$. It is easy to verify that $\leq$ is a partial order relation on $\mathcal{A}$. Now let $C = \{(B_i, G_i) : i \in I\}$ be a chain in $\mathcal{A}$ and let $B = \cup B_i$, $G = \cup G_i$. It is straightforward to show that G is a partial order relation on B. We now show that B is well ordered by G. Suppose $D \subseteq B$ is nonempty. Then $D \cap B_i \neq \emptyset$ for some $i \in I$. Since $D \cap B_i \subseteq B_i$, $D \cap B_i$ has a least element b in (B_i, G_i). Thus, for every $y \in D \cap B_i$ we have $(b, y) \in G_i$. Now let $x \in D$. If $x \in B_i$, then $(b, x) \in G_i \subseteq G$. If $x \notin B_i$, then $x \in B_j$ for some $j \in I$ where $B_j \not\subseteq B_i$. Since $(B_j, G_j) \not\leq (B_i, G_i)$ we have $(B_i, G_i) \leq (B_j, G_j)$. Now $b \in B_i$, $x \in B_j \setminus B_i$ so by (c), $(b, x) \in G_j \subseteq G$. Hence, b is the least element of D in (B, G) and B is well ordered by G. It follows that $(B, G) \in \mathcal{A}$. We now show that (B, G) is an upper bound for C. If $(B_i, G_i) \in C$, then clearly $B_i \subseteq B$ and $G_i \subseteq G$. Now suppose that $x \in B_i$ and $y \in B \setminus B_i$. Then $y \in B_j$ for some $j \in I$. Since $B_j \not\subseteq B_i$, $(B_j, G_j) \not\leq (B_i, G_i)$ so $(B_i, G_i) \leq (B_j, G_j)$. Since $x \in B_i$ and $y \in B_j \setminus B_i$, by (c) we have $(x, y) \in G_j \subseteq G$. Therefore, $(B_i, G_i) \leq (B, G)$. Since every chain in $\mathcal{A}$ has an upper bound in $\mathcal{A}$ we can apply Zorn's axiom to conclude that $\mathcal{A}$ has a maximal element (B, G). If $B \neq A$ there exists an $x \in A \setminus B$. But then $G^* = G \cup \{(a, x) : a \in B\}$ is an extension of G that well orders $B \cup \{x\}$. This contradicts the maximality of (B, G). Hence, $B = A$ so A is well ordered by G. □

We call Theorem 6.9 the *well ordering theorem*. It is clear that the axiom of choice can be derived from the well ordering theorem. Indeed, let A be any nonempty set. By the well ordering theorem, A can be well ordered. If $B \in P'(A)$, let $f(B)$ be the least element of B. Then f is a choice function on A. We thus see that the axiom of choice and the well ordering theorem are equivalent.

Another topic that needs discussing in this section is the continuum hypothesis. We say that two sets A and B have the *same cardinality* and write $|A| = |B|$ if there exists a bijection $f: A \to B$. If $|A| \neq |B|$ and there exists an injection $f: A \to B$ we say that A has *smaller cardinality* than B and write $|A| < |B|$. It is well known that $|\mathbb{N}| < |\mathbb{R}|$. The *continuum hypothesis* states that there is no set A such that $|\mathbb{N}| < |A| < |\mathbb{R}|$.

An important application of the previous work is the theory of ordinal numbers. This theory will not be neded in the sequel so the remainder of this section can be omitted without loss of continuity. We can not resist presenting it here since we have already developed much of the needed background material. In axiomatic set theory one must distinguish between sets and classes in order to avoid certain technical inconsistencies. It is not our intention to present the details here and we shall take the naive position that a class is a generalization of a set. Except for the axiom of choice and its equivalents, all of our previous definitions and results hold for classes as well as for sets. In particular, one can define partially ordered and well ordered classes in the usual way.

Ordinal numbers can be used to rank well ordered sets according to their "ordinality". More precisely, since $\approx$ is an equivalence relation on the class W of well ordered sets, we can form the equivalence class $[A]$ for each well ordered set A. The "ordinal numbers" are well ordered sets, one from each equivalence class. The only problem with this definition is that it requires the axiom of choice on W, but W is a class and the axiom of choice only holds for sets. However, this problem can be overcome and the ordinal numbers can be "constructed" using the axioms of set theory [Pinter, 1971]. We then have the following theorem.

Theorem 6.10. *There is a class OR of well ordered sets, called ordinal numbers such that for any $A \in W$ there exists a unique $\alpha \in OR$ with $A \approx \alpha$.*

It is clear that $\leq$ is a partial order on OR. In fact, the following result shows that $(OR, \leq)$ is well ordered.

TRANSFINITE INDUCTION   213

Theorem 6.11. $(OR, \leq)$ *is a well ordered class.*

Proof. Let $\mathcal{A}$ be a nonempty class of ordinal numbers and let $\alpha \in \mathcal{A}$. If α is the least element of $\mathcal{A}$ we are done; otherwise, let $\mathcal{B} = \{\beta \in \mathcal{A} : \beta < \alpha\}$. It follows from the definition of $<$ that every $\beta \in \mathcal{B}$ is isomorphic to an initial segment of α. For each $\beta \in \mathcal{B}$, let $\phi(\beta)$ be the element $x \in \alpha$ such that $b \approx S_x$. Now the set $\{\phi(\beta) : \beta \in \mathcal{B}\}$ has a least element $\phi(\delta)$ since it is a subset of α. If $\beta \in \mathcal{B}$, then $\phi(\delta) \leq \phi(\beta)$ so $S_{\phi(\delta)} \subseteq S_{\phi(\beta)}$. Hence, by Corollary 6.8 $S_{\phi(\delta)} \leq S_{\phi(\beta)}$. Therefore, $\delta \approx S_{\phi(\delta)} \leq S_{\phi(\beta)} \approx \beta$ and it follows that $\delta \leq \beta$. Hence, δ is the least element of $\mathcal{B}$, so δ is the least element of $\mathcal{A}$. □

Notice that OR does not have a largest element. This is because we can always adjoin a new element x to any well ordered set A and extend the well ordering to $A \cup \{x\}$ by defining $y \leq x$ for every $y \in A$. It follows from Theorem 6.11 that every $\alpha \in OR$ has an immediate successor α^+. In fact, α^+ is the least element of $\{\beta \in OR : \alpha < \beta\}$. Of course, if A is a finite set we can identify its ordinal number with $|A|$. If we denote the ordinal number of $\mathbb{N}$ by ω, the first ordinal numbers become: $0, 1, 2, \ldots, \omega, \omega^+, (\omega^+)^+, \ldots$. If a nonzero ordinal number β has no immediate predecessor; that is, $\beta \neq \alpha^+$ for any $\alpha \in OR$, then β is called a *limit ordinal*. For example, ω is the first limit ordinal. If β has an immediate predecessor we call β a *nonlimit ordinal*.

The last result of this section is called the *transfinite recursion theorem*. One of the applications of this theorem is in developing the "arithmetic" of ordinal numbers.

Theorem 6.12. *Let A be a well ordered set (or class), let $f : A \to A$ be a function, and let $a \in A$. Then there exists a unique function $u : OR \to A$ such that* (i) $u(0) = a$, (ii) $u(\alpha^+) = f(u(\alpha))$ *for every $\alpha \in OR$*, (iii) $u(\alpha) = \vee\{u(\gamma) : \gamma < \alpha\}$ *if α is a limit ordinal.*

Proof. If $G \subseteq OR \times A$, we call G *perfect* if it satisfies: (i') $(0, a) \in G$, (ii') for every $\alpha \in OR$, if $(\alpha, x) \in G$, then $(\alpha^+, f(x)) \in G$, (iii') for every limit ordinal α, if $(\gamma, x_\gamma) \in G$ for all $\gamma < \alpha$ and if $x = \vee\{x_\gamma : \gamma < \alpha\}$, then $(\alpha, x) \in G$. There are perfect subclasses of $OR \times A$, for example $OR \times A$ itself. Now let u be the intersection of all the perfect subclasses of $OR \times A$. Thus,

$$u = \{(\alpha, x) : G \text{ perfect} \Rightarrow (\alpha, x) \in G\}.$$

It is trivial to show that u is perfect. We now prove that u is a function from OR to A. Let

$$S = \{\gamma \in OR : (\gamma, x) \in u \text{ for exactly one } x \in A\}.$$

Using transfinite induction (Theorem 6.2) we shall prove that $S = OR$ and the result will follow. Suppose that for every $\gamma < \beta$, $\gamma \in S$; that is, for every $\gamma < \beta$, $(\gamma, x_\gamma) \in u$ for exactly one element $x_\gamma \in A$. We consider two cases.

(a) β is a nonlimit ordinal, $\beta = \delta^+$. Then $(\delta, x_\delta) \in u$, so by (ii$'$)

$$(\delta^+, f(x_\delta)) = (\beta, f(x_\delta)) \in u$$

Suppose now that $(\beta, y) \in u$ where $y \neq f(x_\delta)$. We shall show that $u' = u \setminus \{(\beta, y)\}$ is perfect, which is impossible by the way we defined u, hence $(\beta, y) \notin u$. Note that u' is obtained from u by deleting only one ordered pair (β, y). Now $\beta \neq 0$, so $(\beta, y) \neq (0, a)$. Hence, $(0, a) \in u'$ and u satisfies (i$'$). Suppose $(\alpha, x) \in u'$. Then $(\alpha, x) \in u$, so $(\alpha^+, f(x)) \in u$. If $\alpha^+ \neq \beta$, then $(\alpha^+, f(x)) \neq (\beta, y)$ so $(\alpha^+, f(x)) \in u'$. If $\alpha^+ = \beta = \delta^+$, then $\alpha = \delta$ so $(\alpha, x) = (\delta, x_\delta)$ and hence $(\alpha^+, f(x)) = (\alpha^+, f(x_\delta))$. Since $y \neq f(x_\delta)$, $(\beta, y) \neq (\alpha^+, f(x_\delta))$. Hence, $(\beta, y) \neq (\alpha^+, f(x))$ so $(\alpha^+, f(x)) \in u'$ and u' satisfies (ii$'$). If α is a limit ordinal and $(\gamma, x_\gamma) \in u'$ for every $\gamma < \alpha$, then $(\gamma, x_\gamma) \in u$ for every $\gamma < \alpha$ so by (iii$'$), $(\alpha, \vee x_\gamma) \in u$. But $(\alpha, \vee x_\gamma) \neq (\beta, y)$ since β is a limit ordinal. Hence, $(\alpha, \vee x_\gamma) \in u'$ and u' satisfies (iii$'$).

(b) β is a limit ordinal. Since $(\gamma, x_\gamma) \in u$ for every $\gamma < \beta$ by (iii$'$), $(\beta, \vee\{x_\gamma : \gamma < \beta\}) \in u$. Suppose that $(\beta, y) \in u$ where $y \neq \vee x_\gamma$. We shall show that $u' = u \setminus \{(\beta, y)\}$ is perfect, which again is impossible, so $(\beta, y) \notin u$. Certainly $\beta \neq 0$ so $(\beta, y) \neq (0, a)$. Hence, $(0, a) \in u'$ and u' satisfies (i$'$). If $(\alpha, x) \in u'$, then $(\alpha, x) \in u$, so by (ii$'$), $(\alpha^+, f(x)) \in u$. But $\alpha^+ \neq \beta$ since β is a limit ordinal. Hence, $(\alpha^+, f(x)) \neq (\beta, y)$ so $(\alpha^+, f(x)) \in u'$. Thus, u' satisfies (ii$'$). If α is a limit ordinal and $(\gamma, x_\gamma) \in u'$ for every $\gamma < \alpha$, then $(\gamma, x_\gamma) \in u$ for every $\gamma < \alpha$ so $(\alpha, \vee\{x_\gamma : \gamma < \alpha\}) \in u$. If $\alpha \neq \beta$, then $(\alpha, \vee\{x_\gamma : \gamma < \alpha\}) \neq (\beta, y)$ so $(\alpha, \vee\{x_\gamma : \gamma < \alpha\}) \in u'$. If $\alpha = \beta$, then

$$(\alpha, \vee\{x_\gamma : \gamma < \alpha\}) = (\beta, \vee\{x_\gamma : \gamma < \beta\}).$$

But $y \neq \vee\{x_\gamma : \gamma < \beta\}$ so $(\alpha, \vee\{x_\gamma : \gamma < \alpha\}) \neq (\beta, y)$. Hence, $(\alpha, \vee x_\gamma) \in u'$ and u' satisfies (iii$'$).

We have proved that u is a function from OR to A satisfying conditions (i), (ii) and (iii). The proof of the uniqueness of u is straightforward. $\square$

6.2 Basic Results

This section presents the basic definitions and results for probability manifolds. Let X be a nonempty set, and let (Y, Σ_Y, ν) be a probability space. Suppose that for every $(x, y) \in X \times Y$ there exists a probability space $(X(x, y), \Sigma(x, y), \mu_{x,y})$ such that (a) $X = \cup_{y \in Y} X(x, y)$ for every $x \in X$, and (b) $X(x, y_1) \cap X(x, y_2) = \emptyset$, $y_1 \neq y_2$, for every $x \in X$. We then call the triple
$$M = (X, Y, \{X(x, y) : (x, y) \in X \times Y\})$$
a *probability manifold* and we call $\{X(x, y) : (x, y) \in X \times Y\}$ the *local structure* of M.

Roughly speaking, X is covered by local probability spaces which join together in a stochastic fashion. In general, X itself does not form a probability space, but as we shall later see, it is closely associated with a generalized probability space. Figure 6.1 gives a rough picture of a probability manifold in which two points are illustrated together with part of their corresponding local structures.

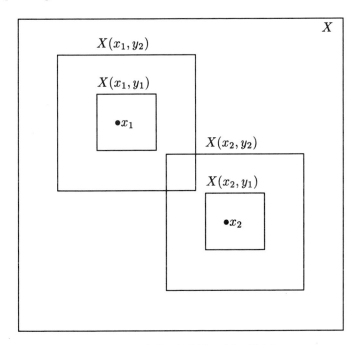

Figure 6.1. A Probability Manifold.

A set $A \subseteq X$ is *locally measurable* if $A \cap X(x,y) \in \Sigma(x,y)$ for every $(x,y) \in X \times Y$ and in this case we call

$$\mu(A \mid x,y) = \mu_{x,y}[A \cap X(x,y)]$$

the *probability of A along $X(x,y)$*. A set $A \subseteq X$ is *globally measurable* if A is locally measurable and for every $x \in X$, the function $y \mapsto \mu(A \mid x,y)$ is Σ_Y-measurable. In this case, we call

$$\mu(A \mid x) = \int \mu(A \mid x,y) \nu(dy)$$

the *probability of A at x*. Finally, a set $A \subseteq X$ is *totally measurable* if A is globally measurable and $\mu(A \mid x)$ is independent of x. In this case, we call $\mu(A) = \mu(A \mid x)$ for every $x \in X$ the *total probability* of A. We define $\Sigma_L, \Lambda_G, \Lambda_T$, respectively, as the collections of locally, globally, and totally measurable subsets of X. It is easy to show that Σ_L is a σ-algebra and $\mu(\cdot \mid x,y)$ is a probability measure on Σ_L for every $(x,y) \in X \times Y$. Hence, $(X, \Sigma_L, \mu(\cdot \mid x,y))$ is a probability space for every $(x,y) \in X \times Y$. In general, we have $\Lambda_T \subseteq \Lambda_G \subseteq \Sigma_L$. However, Λ_T and Λ_G need not be σ-algebras, although, as we now show, they are σ-additive classes.

Theorem 6.13. *If $(X, Y, \{X(x,y): (x,y) \in X \times Y\})$ is a probability manifold, then $(X, \Lambda_G, \mu(\cdot \mid x))$ and (X, Λ_T, μ) are gps's for every $x \in X$.*

Proof. We first show that $(X, \Lambda_G, \mu(\cdot \mid x))$ is a gps for $x \in X$. Clearly $X \in \Lambda_G$ and $\mu(X \mid x) = 1$. Suppose $A \in \Lambda_G$. Then

$$A^c \cap X(x,y) = X(x,y) \setminus A \cap X(x,y) \in \Sigma(x,y)$$

for every $y \in Y$ so $A^c \in \Sigma_L$. Also,

$$\mu(A^c \mid x,y) = 1 - \mu(A \mid x,y)$$

so $y \mapsto \mu(A^c \mid x,y)$ is Σ_Y-measurable. Hence, $A^c \in \Lambda_G$. Let $A_i \in \Lambda_G$, $i = 1, 2, \ldots$, be mutually disjoint, and let $A = \cup A_i$. Then

$$A \cap X(x,y) = \cup[A_i \cap X(x,y)] \in \Sigma(x,y)$$

for every $(x,y) \in X \times Y$ so $A \in \Sigma_L$. Moreover,

$$\mu(A \mid x,y) = \sum \mu(A_i \mid x,y)$$

BASIC RESULTS 217

so $y \mapsto \mu(A \mid x, y)$ is the limit of Σ_Y-measurable functions. Hence, $y \mapsto \mu(A \mid x, y)$ is Σ_Y-measurable. By the monotone convergence theorem,

$$\mu(A \mid x) = \int \mu(A \mid x, y) \nu(dy)$$
$$= \int \mu_{x,y}[A \cap X(x,y)] \nu(dy)$$
$$= \sum \int \mu_{x,y}[A_i \cap X(x,y)] \nu(dy)$$
$$= \sum \mu(A_i, x).$$

It follows that $(X, \Lambda_G, \mu(\cdot \mid x))$ is a gps for every $x \in X$. It now easily follows that (X, Λ_T, μ) is a gps. □

The set of integrable random variables on a gps (X, Λ, μ) is denoted by $L^1(X, \Lambda, \mu)$. As usual, the *expectation* $E_\mu(f)$ of $f \in L^1(X, \Lambda, \mu)$ is defined by $E_\mu(f) = \int f d\mu$. The next result gives the relationship between global and local integrability.

Theorem 6.14. (a) If $f \in L^1(X, \Sigma_L, \mu(\cdot \mid x, y))$, then

$$f \mid X(x, y) \in L^1(X(x, y), \Sigma(x, y), \mu_{x,y}).$$

(b) If $f \in L^1(X, \Lambda_G, \mu(\cdot \mid x))$ then $f \in L^1(X, \Sigma_L, \mu(\cdot \mid x, y))$ for every $y \in Y$,

$$y \mapsto \int_{X(x,y)} f(w) \mu_{x,y}(dw)$$

is integrable on (Y, Σ_Y, ν) and

$$\int_Y \left[\int_{X(x,y)} f(w) \mu_{x,y}(dw) \right] \nu(dy) = \int_X f(w) \mu(dw \mid x).$$

(c) If $f \in L^1(X, \Lambda_T, \mu)$ then $f \in L^1(X, \Lambda_G, \mu(\cdot \mid x))$ for every $x \in X$ and

$$\int_X f(w) \mu(dw \mid x) = \int_X f(w) \mu(dw)$$

for every $x \in X$.

Proof. Part (a) is straightforward and (c) easily follows from (b). We therefore only prove part (b). Let $f \in L^1(X, \Lambda_G, \mu(\cdot \mid x))$ be a simple function, $f = \sum_{i=1}^{n} c_i \chi_{A_i}$, $A_i \in \Lambda_G$. Then $A_i \cap X(x,y) \in \Sigma(x,y)$ so $f \mid X(x,y)$ is $\Sigma(x,y)$-measurable. Since

$$y \mapsto \mu_{x,y}\left[A_i \cap X(x,y)\right]$$

is Σ_Y-measurable,

$$y \mapsto \int_{X(x,y)} f(w)\mu_{x,y}(dw) = \Sigma c_i \mu_{x,y}\left[A_i \cap X(x,y)\right]$$

is Σ_Y-measurable. Now

$$\int_Y \left[\int_{X(x,y)} f(w)\mu_{x,y}(dw)\right]\nu(dy) = \Sigma c_i \int_Y \mu_{x,y}\left[A_i \cap X(x,y)\right]\nu(dy)$$

$$= \Sigma c_i \mu(A_i \mid x) = \int_X f(w)\mu(dw \mid x).$$

Now let $f \in L^1(X, \Lambda_G, \mu(\cdot \mid x))$ be nonnegative. Then there exists an increasing sequence of simple functions $f_i \in L^1(X, \Lambda_G, \mu(\cdot \mid x))$ such that $f_i \to f$ almost everywhere. Since $f \mid X(x,y) = \lim f_i \mid X(x,y)$, we have $f \mid X(x,y)$ is $\Sigma(x,y)$-measurable. By the monotone convergence theorem

$$\int_{X(x,y)} f(w)\mu_{x,y}(dw) = \lim \int_{X(x,y)} f_i(w)\mu_{x,y}(dw).$$

Since $y \mapsto \int_{X(x,y)} f_i(w)\mu_{x,y}(dw)$ is Σ_Y-measurable, so is its limit

$$y \mapsto \int_{X(x,y)} f(w)\mu_{x,y}(dw).$$

Again, by the monotone convergence theorem

$$\int_X f(w)\mu(dw \mid x) = \lim \int_X f_i(w)\mu(dw \mid x)$$

$$= \lim \int_Y \left[\int_{X(x,y)} f_i(w)\mu_{x,y}(dw)\right]\nu(dy)$$

$$= \int_Y \left[\int_{X(x,y)} f(w)\mu_{x,y}(dw)\right]\nu(dy).$$

BASIC RESULTS 219

Finally, if $f \in L^1(X, \Lambda_G, \mu(\cdot \mid x))$ is arbitrary, we can write $f = f_+ - f_-$ where $f_+, f_- \geq 0$, $f_+, f_- \in L^1(X, \Lambda_G, \mu(\cdot \mid x))$ and the ranges of f_+ and f_- are contained in a common σ-subalgebra of Λ_G. The result now follows from the additivity of the integral in this case. □

If $f \mid X(x,y) \in L^1(X(x,y), \Sigma(x,y), \mu_{x,y})$, we call

$$E(f \mid x, y) = \int_{X(x,y)} f(w)\mu(dw \mid x, y)$$

the *local expectation of f along $X(x,y)$*. If $f \in L^1(X, \Lambda_G, \mu(\cdot \mid x))$ we call

$$E(f \mid x) = \int_X f(w)\mu(dw \mid x)$$

the *global expectation of f at x*. It follows from Theorem 6.14 that

$$E(f \mid x) = \int_Y E(f \mid x, y)\nu(dy).$$

If $f \in L^1(X, \Lambda_T, \mu)$ we call $E(f) = \int_X f d\mu$ the *total expectation of f*.

We now give some simple examples that illustrate the above framework. As we shall see, this framework includes product probability spaces and conditional expectations relative to countable measurable partitions.

Example 6.3. Let μ_x be a probability measure on a measurable space (X, Σ) for each $x \in X$. Let $Y = \{y\}$ be a singleton set with the trivial probability measure $\nu(y) = 1$. If we let $X(x,y) = X$, $\Sigma(x,y) = \Sigma$, and $\mu_{x,y} = \mu_x$ for every $x \in X$, then $(X, Y, \{X(x,y): (x,y) \in X \times Y\})$ becomes a probability manifold. Such a probability manifold has been called a *stochastic* (or *fuzzy*) *phase space* [Prugovecki, 1984]. In such a space, a point x is replaced by its corresponding "fuzzy" point μ_x. The probability measure μ_x is usually concentrated near the point x and it gives a measure of the confidence that the system is near x. Clearly, every locally measurable set is globally measurable, but it is not totally measurable in general. Stochastic phase spaces form the basis for the stochastic quantum mechanics approach [Prugovecki, 1984]. □

Example 6.4. Let (Y, Σ_Y, ν) and (Z, Σ_Z, w) be probability spaces and let $X = Y \times Z$. For $(x,y) \in X \times Y$, let $X(x,y) = y \times Z$, $\Sigma(x,y) =$

$\{y \times B : B \in \Sigma_Z\}$, $\mu_{x,y}(y \times B) = w(B)$. Then $M = (X, Y, \{X(x,y) : (x,y) \in X \times Y\})$ is a probability manifold. It is clear that

$$\Sigma_Y \times \Sigma_Z \subseteq P(Y) \times \Sigma_Z \subseteq \Sigma_L.$$

Let $C \times D \in \Sigma_Y \times \Sigma_Z$ be a measurable rectangle. Then $C \times D \in \Sigma_L$ and

$$\mu_{x,y}[C \times D \cap X(x,y)] = \mu_{x,y}[C \times D \cap (y \times Z)] = \chi_C(y)w(D).$$

Since this is a measurable function of y, $C \times D \in \Lambda_G$. Since the intersection of two measurable rectangles is a measurable rectangle, it can be shown [Gudder, 1969] that

$$\Sigma_Y \times \Sigma_Z \subseteq \Lambda_G \subseteq \Sigma_L.$$

Moreover,

$$\mu(C \times D \mid x) = \int_Y \mu_{x,y}[C \times D \cap X(x,y)] \nu(dy)$$

$$= \nu(C)w(D)$$

is independent of x, so $\Lambda_T = \Lambda_G$. Hence,

$$(X, \Sigma_y \times \Sigma_Z, \nu \times w) \subseteq (X, \Lambda_T, \mu)$$

and we may think of M as a generalization of the Cartesian product of two probability spaces. The probability of $C \times D$ along $X(x, y)$ becomes $\mu(C \times D \mid x) = \chi_C(y)w(D)$ and the probability of $C \times D$ at x becomes

$$\mu(C \times D \mid x) = \mu(C \times D) = \nu(C)w(D)$$

$$= \nu \times w(C \times D). \qquad \square$$

Example 6.5. Let $X = Y = \mathbb{R}$, $\Sigma_y = B(\mathbb{R})$ and $\nu(B) = \pi^{-1/2} \int_B e^{-y^2} dy$ for $B \in \Sigma_y$. For $(x, y) \in X \times Y$, let $X(x,y) = \{x+y\}$, $\Sigma(x,y) = \{\emptyset, \{x+y\}\}$ and $\mu_{x,y}(\{x+y\}) = 1$. Then

$$M = \{X, Y, \{X(x,y) : (x,y) \in X \times Y\}\}$$

is a probability manifold. It is clear that $\Sigma_L = P(X)$. If $A \in \Sigma_L$ then

$$\mu(A \mid x, y) = \mu_{x,y}[A \cap X(x,y)] = \phi_{A-x}(y).$$

It follows that $\Lambda_G \subseteq B(\mathbb{R})$. For $A \in \Lambda_G$, we have

$$\mu(A \mid x) = \int \mu(A \mid x, y) \nu(dy)$$

$$= \nu(A - x) = \pi^{-\frac{1}{2}} \int_{A-x} e^{-y^2} dy.$$

Hence, $\Lambda_T = \{\emptyset, \mathbb{R}\}$. $\square$

BASIC RESULTS   221

Example 6.6. Let (X, Σ, μ) be a probability space and let B_i, $i = 1, 2, \ldots$, be a countable measurable partition of X, with $\mu(B_i) \neq 0$, $i = 1, 2, \ldots$. Let $Y = \{y_1, y_2, \ldots\}$ be a countable set and define $\Sigma_y = P(Y)$, $\nu(\{y_i\}) = \mu(B_i)$. For any $x \in X$ define $X(x, y_i) = B_i$, $\Sigma(x, y_i) = \{A \cap B_i, A \in \Sigma\}$ and $\mu_{x,y_i}(B) = \mu(B)/\mu(B_i)$ for $B \in \Sigma(x, y_i)$, $i = 1, 2, \ldots$. It is clear that $\Sigma_L = \Sigma$. For $A \in \Sigma$, we have

$$\mu(A \mid x, y_i) = \frac{\mu(A \cap B_i)}{\mu(B_i)}$$

which is a Σ_Y-measurable function, so $\Lambda_G \subseteq \Sigma$. Moreover, $\mu(A \mid x, y_i)$ is the usual conditional probability $\mu(A \mid B_i)$, $i = 1, 2, \ldots$. Also for $A \in \Sigma$ we have

$$\mu(A \mid x) = \int_Y \mu(A \mid x, y) \nu(dy)$$
$$= \sum \mu(A \cap B_i) = \mu(A).$$

It follows that $\Lambda_T = \Sigma$. □

We now present a simple global limit theorem that follows from local properties. We restrict our attention to the simplest version of the strong law of large numbers (see Section 1.5). One can prove more general versions of the law of large numbers and also the central limit theorem in the present context. Moreover, we shall only consider what we call "g-harmonic" functions. The result we present generalizes some theorems due to I. Pitowski [1983a] that are considered in the next section.

Let $(X, Y, \{X(x, y) : (x, y) \in X \times Y\})$ be a probability manifold and let $f_i : X \to \mathbb{R}$, $i = 1, 2, \ldots$, be a sequence of locally measurable functions. If $f_i \mid X(x, y)$ are stochastically independent for every $(x, y) \in X \times Y$, we say that f_i, $i = 1, 2, \ldots$, are *locally independent*. If $f_i \mid X(x, y)$ are identically distributed, we call f_i, $i = 1, 2, \ldots$, *locally identically distributed*. Let $g \in L^1(Y, \Sigma_Y, \nu)$ be a function that is not identically zero. A locally integrable function $f : X \to \mathbb{R}$ is *g-harmonic* if $E(f \mid x, y) = f(x)g(y)$ for every $(x, y) \in X \times Y$.

Notice that if f is g-harmonic, then f is globally integrable and

$$E(f \mid x) = \int_Y E(f \mid x, y) \nu(dy)$$
$$= \int_Y f(x) g(y) \nu(dy) \qquad (6.1)$$
$$= f(x) \int_Y g(y) \nu(dy).$$

It follows that f is totally measurable if and only if $\int g(y)\nu(dy) = 0$ or f is a constant function. For a fixed $x \in X$, each $w \in X$ is contained in a unique set $X(x, y_w)$. Define $\pi_x: X \to Y$ by $\pi_x(w) = y_w$.

Theorem 6.15. *Let f_i, $i = 1, 2, \ldots$, be a sequence of locally independent, locally indentically distributed, g-harmonic functions.*

(a) *If $f_i(x_0) = \alpha$ $i = 1, 2, \ldots$, then*

$$\mu\left[\{w \in X: \left(\frac{1}{n}\right) \sum_{i=1}^{n} f_i(w) \to \alpha g(\pi_{x_0}(w))\} \mid x_0\right] = 1.$$

(b) *If $g(y) = 0$, then for every $x_0 \in X$*

$$\mu\left[\{w \in X: \left(\frac{1}{n}\right) \sum_{i=1}^{n} f_i(w) \to 0\} \mid x_0, y\right] = 1.$$

(c) *If $n^{-1} \sum_{i=1}^{n} f_i(x_0) \to 0$, then*

$$\mu\left[\{w \in X: \left(\frac{1}{n}\right) \sum_{i=1}^{n} f_i(w) \to 0\} \mid x_0\right] = 1.$$

Proof. (a) Since $f_i(x_0) = \alpha$, we have $E(f_i \mid x_0, y) = \alpha g(y)$ for $i = 1, 2, \ldots$. Applying the law of large numbers to $X(x_0, y)$ we obtain

$$\left(\frac{1}{n}\right) \sum_{i=1}^{n} f_i(w) \to \alpha g(y) = \alpha g[\pi_{x_0}(w)]$$

for almost every $w \in X(x_0, y)$. Hence,

$$\mu\left[\{w \in X: \left(\frac{1}{n}\right) \sum_{i=1}^{n} f_i(w) \to \alpha g(\pi_{x_0}(w))\} \mid x_0, y\right] = 1 \qquad (6.2)$$

for every $y \in Y$. Integrating (6.2) over Y gives the result.

(b) This follows from the law of large numbers on $X(x_0, y)$ using the fact that

$$E(f_i \mid x_0, y) = f_i(x_0)g(y) = 0.$$

(c) The proof is similar to part (a) and uses the fact that

$$\frac{1}{n} \sum_{i=1}^{n} E(f_i \mid x_0, y) = \frac{g(y)}{n} \sum_{i=1}^{n} f_i(x_0) \to 0. \quad \square$$

BASIC RESULTS

Given an integrable function g, how do we know that g-harmonic functions exist? The next result shows that they do indeed exist under fairly general conditons. A probability space is called a *Lebesgue space* if it is isomorphic to $[0,1]$ with Lebesgue measure.

Theorem 6.16. *Let $(X, Y, \{X(x,y): (x,y) \in X \times Y\})$ be a probability manifold, with $\text{card}(X) = \text{card}(\mathbb{R})$, and let $g: Y \to \mathbb{R}$ be Σ_Y-measurable with $|g(y)| \leq 1$ for every $y \in Y$. If $X(x,y)$ are Lebesgue spaces such that $X(x_1, y_1) \cap X(x_2, y_2)$ is at most countable whenever $x_1 \neq x_2$, then there exists a g-harmonic function f.*

Proof. We assume the axiom of choice and the continuum hypothesis. Well order the family of sets

$$F = \{\{x\} \cup X(x,y): (x,y) \in X \times Y\}.$$

Since $\text{card}(F) = \text{card}(\mathbb{R})$, it follows from the continuum hypothesis that there is such a well ordering in which every element has only countably many predecessors. We define f by transfinite induction on this ordering.

Let $\{x_1\} \cup X(x_1, y_1)$ be the first element of F in the well ordering. Since $X(x_1, y_1)$ is a Legesgue space, we can partition $X(x_1, y_1)$ into two measurable sets A_1, A_2 where A_1 has measure $[1 - g(y_1)]/2$ and A_2 has measure $[1 + g(y_1)]/2$. If $x_1 \in X(x_1, y_1)$ we can assume that $x_1 \in A_2$, and we define f to be -1 on A_1 and 1 on A_2. If $x_1 \notin X(x_1, y_1)$ we again define f to be -1 on A_1 and 1 on A_2 and define $f(x_1) = 1$. Then

$$\int_{X(x_1,y_1)} f(w) \mu_{x_1,y_1}(dw) = \frac{1+g(y_1)}{2} - \frac{1-g(y_1)}{2} \qquad (6.3)$$

$$= g(y_1) = f(x_1)g(y_1).$$

Now suppose we have defined f on all the elements of F up to but not including $\{x_\alpha\} \cup X(x_\alpha, y_\alpha)$ in the order. If x_α does not belong to any preceding element of F, define $f(x_\alpha) = 1$. Since $\{x_\alpha\} \cup X(x_\alpha, y_\alpha)$ is preceded by at most countably many elements and since the intersection of two different $X(x,y)$'s is at most countable, f has already been defined on at most countably many points of $X(x_\alpha, y_\alpha)$. Hence, we can proceed as in the original step. Otherwise, x_α belongs to a preceding element. If $f(x_\alpha) = 1$, proceed as before. If $f(x_\alpha) = -1$, reverse the measures of the two sets. As before, f satisfies (6.3) with (x_1, y_1) replaced by (x_α, y_α). Hence, f is g-harmonic. □

6.3 Deterministic Spin Model

In this section we give an application to a deterministic spin model due to Pitowski [1983a]. Although this model is deterministic, it gives the same probabilistic predictions as quantum mechanics. Unlike Pitowski, we shall pose this model in the framework of a probability manifold.

Let X be the unit sphere in $\mathbb{R}^3$. We shall denote the usual inner product in $\mathbb{R}^3$ by $\langle x, y \rangle$ so $X = \{x \in \mathbb{R}^3 : \langle x, x \rangle = 1\}$. For $x \in X$, $0 \le \theta \le \pi$, let

$$X(x, \theta) = \{u \in X : \langle u, x \rangle = \cos \theta\}.$$

Thus, $X(x, \theta)$ is a circle in X consisting of the points that form an angle θ to x. The radius of this circle is $\sin \theta$. Let $m_{x,\theta}$ be Lebesgue measure on $X(x, \theta)$ and let $\mu_{x,\theta} = (2\pi \sin \theta)^{-1} m_{x,\theta}$ be the uniform distribution on $X(x, \theta)$ when $0 < \theta < \pi$. For $q = 0$ or π, let $\mu_{x,\theta}$ be the point probability measures concentrated at x and $-x$, respectively. Define $Y = [0, \pi]$, $\Sigma_Y = B([0, \pi])$ and let ν be the distribution $d\nu/d\theta = \frac{1}{2} \sin \theta$ on Y. Then it is easy to check that

$$M = (X, Y, \{X(x, \theta) : (x, \theta) \in X \times Y\})$$

is a probability manifold. The next result shows that every Borel set is totally measurable and that μ restricted to $B(X)$ is just normalized Lebesgue measure μ_0.

Lemma 6.17. $B(X) \subseteq \Lambda_T$ and $\mu \mid B(X) = \mu_0$.

Proof. Let $A \in B(X)$. Then clearly $A \cap X(x, \theta)$ is $\Sigma(x, \theta)$-measurable and by Fubini's theorem $\theta \mapsto \mu_{x,\theta}[A \cap X(x, \theta)]$ is Σ_Y-measurable. Now let (r, θ, ϕ) be a set of spherical coordinates and let $x = (1, 0, 0)$. Applying Fubini's theorem we obtain

$$\mu(A) = \frac{1}{4\pi} \int_0^\pi \int_0^{2\pi} \phi_A(\theta, \phi) \sin \theta \, d\phi d\theta$$

$$= \frac{1}{2} \int_0^\pi \left[\frac{1}{2\pi} \int_0^{2\pi} \phi_A(\theta, \phi) \, d\phi \right] \sin \theta \, d\theta$$

$$= \frac{1}{2} \int_0^\pi \mu_{x,\theta}[A \cap X(x, \theta)] \sin \theta \, d\theta$$

$$= \int \mu(A \mid x, \theta) \nu(d\theta) = \mu(A \mid x). \quad \square$$

DETERMINISTIC SPIN MODEL 225

Even though $B(X) \subseteq \Lambda_T$, we shall later show that Λ_T is not a σ-algebra. Hence, $B(X) \neq \Lambda_T$ so Λ_T contains sets that are not Lebesgue measurable. Of course, we have shown in Theorem 6.13 that Λ_T is a σ-additive class and (X, Λ_T, μ) is a gps.

We now give the main definition in this model. A *spin-$\frac{1}{2}$ function* is a function $f \in L^1(X, \Lambda_T, \mu)$ that satisfies the following.

(a) $f: X \to \{-\frac{1}{2}, \frac{1}{2}\}$.
(b) $f(-w) = -f(w)$ for every $w \in X$.
(c) $E(f \mid x, \theta) = f(x) \cos \theta$ for every $(x, \theta) \in X \times Y$.

Notice that f is g-harmonic where $g(\theta) = \cos \theta$. Spin-$\frac{1}{2}$ functions may be interpreted as giving a deterministic spin model. In fact, (a), (b) and (c) are properties of a quantum mechanical spin-$\frac{1}{2}$ system where (c) corresponds to Equation (2.27) (see Section 2.4). If a system is described by a spin-function f, then $f(x)$ specifies the spin in the x-direction. Unlike quantum mechanics, which can only predict the probability of the spin in a specified direction, this model is deterministic since f gives the precise spin value in any given direction.

It is not at all clear that spin-$\frac{1}{2}$ functions exist. Although we have shown in Theorem 6.16 that cos-harmonic functions exist, we still need conditions (a) and (b). To do this, we must refine the proof of Theorem 6.16.

Theorem 6.18. *There exists a spin-$\frac{1}{2}$ function.*

Proof. Again we assume the axiom of choice and the continuum hypothesis. For $x \in X$ and $0 < \theta \leq \pi/2$ define

$$R(x, \theta) = X(x, \theta) \cup X(-x, \theta) \cup \{-x, x\}$$

and let

$$F = \left\{ R(x, \theta) : x \in X, 0 < \theta \leq \frac{\pi}{2} \right\}.$$

Since $\text{card}(F) = \text{card}(\mathbb{R})$ there is a well ordering of F in which every element $R(x, \theta)$ has only countably many predecessors. We shall define a spin-$\frac{1}{2}$ function f on X by transfinite induction on such an ordering.

Let $R(x_1, \theta_1)$ be the first element in the order. Define $f(x_1) = -f(-x_1) = \frac{1}{2}$. Partition $X(x_1, \theta_1)$ into two measurable sets A_1, A_2 where A_1 has measure $\cos^2(\theta_1/2)$ and A_2 has measure $\sin^2(\theta_1/2)$. Define f to be $\frac{1}{2}$ on A_1 and $-\frac{1}{2}$ on A_2. For $w \in X(-x_1, \theta_1)$ define $f(w) = -f(-w)$. (In the case $\theta = \pi/2$, partition $X(x_1, \pi/2)$ into subsets A_1, A_2 of measure $\frac{1}{2}$ such that the reflection of A_2 is A_1.)

Suppose we have defined f on all the elements of f up to but not including $R(x_\alpha, \theta_\alpha)$ in the ordering. There are now two cases.

(a) x_α (and thus $-x_\alpha$) do not belong to any preceding element in the ordering. In this case, define $f(x_\alpha) = \frac{1}{2} = -f(-x_\alpha)$. Since $R(x_\alpha, \theta_\alpha)$ is preceded by at most countably many elements and since the intersection of two different circles contains at most two points it follows that f has already been defined on at most countably many points of $X(x_\alpha, \theta_\alpha)$ hence only on a subset of $X(x_\alpha, \theta_\alpha)$ of measure zero. Thus, we can partition the rest of $X(x_\alpha, \theta_\alpha)$ into two subsets of measure $\cos^2(\theta_\alpha/2)$ and $\sin^2(\theta_\alpha/2)$, respectively, and define f to be $\frac{1}{2}$ on the first and $-\frac{1}{2}$ on the second. Moreover, for $w \in X(-x_\alpha, \theta_\alpha)$ define $f(w) = -f(-w)$.

(b) x_α (and thus $-x_\alpha$) belong to some preceding element in the ordering. In this case, $f(x_\alpha)$ and $f(-x_\alpha)$ have already been defined. If $f(x_\alpha) = \frac{1}{2}$, proceed as in Case a. If $f(x_\alpha) = -\frac{1}{2}$, follow the same procedure reversing the roles of x_α and $-x_\alpha$.

The resulting function f clearly satisfies Conditions a and b for a spin-$\frac{1}{2}$ function. Also, (c) is satisfied since for all $x \in X$, $0 < \theta < \pi$, we have

$$E(f \mid x, \theta) = f(x)\left(\cos^2 \frac{\theta}{2} - \sin^2 \frac{\theta}{2}\right)$$

$$= f(x) \cos \theta. \quad \square$$

We now show that a spin-$\frac{1}{2}$ function f gives the same statistical predictions as quantum mechanics. Of course, it follows from (6.1) that $f \in L^1(X, \Lambda_T, \mu)$ and $E(f) = 0$. In this specific case, we have

$$E(f) = \frac{1}{2} \int_0^\pi E(f \mid x, \theta) \sin \theta \, d\theta$$

$$= \frac{1}{2} f(x) \int_0^\pi \cos \theta \sin \theta \, d\theta = 0.$$

Corresponding to f define the sets $A^+ = f^{-1}(\frac{1}{2})$ and $A^- = f^{-1}(-\frac{1}{2})$. Since $\phi_{A^+} = f + \frac{1}{2}$ we have

$$\mu(A^+ \mid x, \theta) = E(\phi_{A^+} \mid x, \theta)$$

$$= E\left(f + \frac{1}{2} \mid x, \theta\right) \quad (6.4)$$

$$= \frac{1}{2} + f(x) \cos \theta$$

DETERMINISTIC SPIN MODEL   227

for all $x \in X$, $0 < \theta < \pi$. Equation (6.4) gives

$$\mu(A^+ \mid x, \theta) = \cos^2\left(\frac{\theta}{2}\right) \quad \text{if } f(x) = \frac{1}{2} \tag{6.5}$$

$$\mu(A^+ \mid x, \theta) = \sin^2\left(\frac{\theta}{2}\right) \quad \text{if } f(x) = -\frac{1}{2} \tag{6.6}$$

for all $x \in X$, $0 < \theta < \pi$. Equations (6.5), (6.6) coincide with the quantum mechanical formulas (2.25), (2.26), respectively. Similar formulas hold for A^-. By definition $A^+, A^- \in \Lambda_T$ and moreover

$$\mu(A^+) = \frac{1}{2}\int_0^\pi \mu(A^+ \mid x, \theta) \sin\theta \, d\theta = \frac{1}{2}$$

with a similar formula holding for A^-.

If $\alpha \in SO(3)$ and f is a spin-$\frac{1}{2}$ function, it is easy to check that

$$E(f \circ \alpha \mid x, \theta) = f(\alpha(x)) \cos\theta$$

for all $x \in X$, $0 < \theta < \pi$. It follows that $f \circ \alpha$ is again a spin-$\frac{1}{2}$ function. In this way we obtain an infinite set of spin-$\frac{1}{2}$ functions [Pitowski, 1983b]. The proof of the following theorem is similar to that of Theorems 6.16 and 6.18.

Theorem 6.19. *Let $\alpha \in SO(3)$, $\alpha \neq \pm I$. Then there exists a spin-$\frac{1}{2}$ function f such that for all $x \in X$, $0 < \theta < \pi$, we have*

$$\begin{aligned} E(ff \circ \alpha \mid x, \theta) &= E(f \mid x, \theta) E(f \circ \alpha \mid x, \theta) \\ &= f(x) f(\alpha(x)) \cos^2\theta. \end{aligned} \tag{6.7}$$

Now each of the functions $f, f \circ \alpha$ in Theorem 6.19 is total measurable and it follows from (6.7) that their product $ff \circ \alpha$ is globally measurable. Moreover

$$\begin{aligned} E(ff \circ \alpha \mid x) &= \frac{1}{2}\int_0^\pi E(ff \circ \alpha \mid x, \theta) \sin\theta \, d\theta \\ &= \frac{1}{2}f(x)f(\alpha(x)) \int_0^\pi \cos^2\theta \sin\theta \, d\theta \\ &= \frac{1}{3}f(x)f(\alpha(x)). \end{aligned} \tag{6.8}$$

However, (6.8) shows that $ff \circ \alpha$ is not totally measurable. Indeed, from (6.8) we have $E(ff \circ \alpha \mid x) = 1/12$ when $f(x) = f(\alpha(x))$ and $E(ff \circ \alpha \mid x) = -1/12$ when $f(x) \neq f(\alpha(x))$. Since the product of measurable functions is always measurable in a σ-algebra we conclude that Λ_T is not a σ-algebra. Hence, $\Lambda_T \neq B(X)$. In the same way, one can show that

$$f^{-1}\left(\frac{1}{2}\right) \cap (f \circ \alpha)^{-1}\left(\frac{1}{2}\right) \notin \Lambda_T$$

It then follows that $f^{-1}(\frac{1}{2}) \notin B(X)$. We thus see that a deterministic spin model is not possible within the framework of classical probability theory.

The next theorem gives another way of verifying the quantum mechanical formula (2.27) in the present model. The proof of this result follows from Theorem 6.15.

Theorem 6.20. *Let $f_1, f_2, \ldots$ be a locally independent sequence of spin-$\frac{1}{2}$ functions satisfying $f_i(x_0) = \frac{1}{2}$. Then*

$$\mu\left[\left\{w \in X \colon \left(\frac{1}{n}\right) \sum_{i=1}^{n} f_i(w) \to \frac{1}{2}\langle w, x_0 \rangle\right\} \mid x_0\right] = 1$$

One can prove many more interesting results for this model. Moreover, the model can be generalized to systems with higher order spins [Pitowski, 1983a]. In a sense, this scheme gives a hidden variables model for spin-$\frac{1}{2}$ systems. The hidden variables are the points of X and the states are given by spin-$\frac{1}{2}$ functions. Once the state f is known, the spin is determined in every direction. If f is averaged over the relevant hidden variables, then the quantum probabilities are obtained.

6.4 Phase Space Model

Einstein, Schrödinger, DeBroglie and others believed that quantum mechanics gives an incomplete description of reality. They believed in the existence of hidden variables, that if known, would give exact values for observables and measurement results. The quantum mechanical predictions are probabilities that are only averages over the hidden variables. In a sense, these investigators were right. The reason that they were unable to convince the rest of the physics community was that they did not give the appropriate averaging procedure. If we average correctly, that is, if we give the appropriate probability theory on the space of hidden variables, then we do obtain the usual quantum probabilities. To see how to do this, we must look more closely at quantum probability theory.

As mentioned before, Feynman and others suggested that probabilities in quantum mechanics are computed by summing an amplitude function and taking the modulus squared. But what do we sum (or integrate) over, and how do we find the amplitude function? The answer to this question has actually been known for a long time. The key lies in Schrödinger's equation and the strange appearance of the imaginary unit i in that equation. Because of the i, the wave function $\psi(x,t)$ cannot represent a physical property or wave. In fact, $\psi(x,t)$ is closely related to the probability amplitude we are seeking. The main problem with ψ is that it is only a function of position x and time t and is not a function of momentum p. There are two reasons why this is a problem. First, it appears that x and p should have an equal footing, so an appropriate probability amplitude should be a function of them both. Second, the most obvious choice for the hidden variables space is phase space itself. After all, we know from classical mechanics that given a point in phase space, the values of all dynamical variables are known precisely. Hence, the amplitude function should be a function on phase space.

In this section we present a phase space model for quantum mechanics. There is a large body of literature on such phase space models [Belifante, 1973; Cohen, 1966; Groblicki, 1984; Schulman, 1981; Wigner, 1932]. However, to our knowledge, this is the first that incorporates the concept of an amplitude density. In this model, the hidden variables are the points in a phase space X, the observables are represented by real-valued functions on X, the events by subsets of X and the states by amplitude densities on X. Once a hidden variable is known, the observables and events are determined precisely. However, when the expectation of an observable or the probability of an event are computed relative to an amplitude density, the usual quantum results are obtained. Even the quantum dynamics can be derived in this way. In particular, if Hamilton's equations of classical mechanics are averaged with an amplitude density, then Schrödinger's equation is obtained. In a sense, quantum mechanics is just classical mechanics averaged over amplitude densities.

In the sequel we shall consider a two-dimensional phase space $\mathbb{R}^2 = \{(q,p): q,p \in \mathbb{R}\}$. The generalization to systems with any finite number of degrees of freedom is straightforward. A function $f: \mathbb{R}^2 \to \mathbb{C}$ is an *amplitude density* if (a) for every p, $f(\cdot,p)$ is Lebesgue integrable, $p \mapsto \int_{\mathbb{R}} f(q,p)\,dq$ is

Lebesgue measurable, and

$$\int_{\mathbf{R}} \left| \int_{\mathbf{R}} f(q,p)\,dq \right|^2 dp = 1 \qquad (6.9)$$

(b) for every q, $f(q,\cdot)$ is Lebesgue integrable, $q \mapsto \int_{\mathbf{R}} f(q,p)\,dp$ is Lebesgue measurable, and

$$\int_{\mathbf{R}} \left| \int_{\mathbf{R}} f(q,p)\,dp \right|^2 dq = 1. \qquad (6.10)$$

Although f is measurable relative to each of its arguments separately, in general f need not be Lebesgue measurable on $\mathbf{R}^2$.

The idea behind this definition of an amplitude density f is the following. We think of $(\mathbf{R}^2, f)$ as a quantum probability space, where f is now an amplitude density instead of being a probability amplitude. The elements of $\mathbf{R}^2$ are the sample points of the space and instead of summing the values of f over sets of sample points we now integrate. A set of the form $\mathbf{R} \times \{p\} \subseteq \mathbf{R}^2$ is a momentum outcome in which the momentum has value p, and $\int_{\mathbf{R}} f(q,p)\,dq$ is the amplitude of that outcome. The probability density for this outcome becomes $\left|\int_{\mathbf{R}} f(q,p)\,dq\right|^2$ and (6.9) shows that the collection of these outcomes is an operation, namely the momentum measurement operation. Equation 6.10 has a similar interpretation in terms of position outcomes and the position operation. We shall later consider other operations corresponding to other observables such as energy.

Let f be an amplitude density and let $A \in B(\mathbf{R})$. We then interpret $A \times \{p\}$ as the partial outcome in which we first measure position and secure a value in A and then measure momentum and secure the value p. We then call

$$f_P(A)(p) = \int_A f(q,p)\,dq \qquad (6.11)$$

the P-amplitude density following an A-measurement of Q. Similarly, we call

$$f_Q(A)(q) = \int_A f(q,p)\,dp \qquad (6.12)$$

the Q-amplitude density following an A-measurement of P. For $A, B \in B(\mathbf{R})$ we call

$$\mu_P(B,A) = \int_B |f_P(A)(p)|^2 dp \qquad (6.13)$$

the P-probability of B following an A-measuremert of Q. Notice that $\mu_P(B,A)$ gives the probability of first securing a position value in A and subsequently securing a momentum value in B. Similarly, we call

$$\mu_Q(B,A) = \int_B |f_Q(A)(q)|^2 dq \tag{6.14}$$

the *Q-probability of B following an A-measurement of P*. Notice that these probabilities correspond to the classical expression $\mu(A \cap B)$ for the probability of the simultaneous occurrence of A and B except now we have two measurements and we must specify which is executed first. If we think of $\mathbf{R} \times B$ as the event in which the momentum has a value in B, then the probability of this event is

$$\mu_P(B,\mathbf{R}) = \int_B |f_P(\mathbf{R})(p)|^2 dp$$

$$= \int_B \left| \int_\mathbf{R} f(q,p)\, dq \right|^2 dp$$

Similarly, we have

$$\mu_Q(B,\mathbf{R}) = \int_B |f_Q(\mathbf{R})(q)|^2 dq$$

$$= \int_B \left| \int_\mathbf{R} f(q,p)\, dp \right|^2 dq.$$

It is clear that $\mu_Q(B) = \mu_Q(B,\mathbf{R})$ and $\mu_P(B) = \mu_P(B,\mathbf{R})$ are probability measures on $\mathcal{B}(\mathbf{R})$. If we let $Y = Z = \mathbf{R}$, $\Sigma_Y = \Sigma_Z = \mathcal{B}(\mathbf{R})$, then (Y, Σ_Y, μ_Q) and (Z, Σ_Z, μ_P) become probability spaces. Let $X = Y \times Z = \mathbf{R}^2$. As in Example 6.4, for $(x,y) \in X \times Y$, let $X(x,y) = Y \times Z$, $\Sigma(x,y) = \{y \times B : B \in \Sigma_Z\}$, $\mu_{x,y}(y \times B) = \mu_P(B)$. Then

$$M_1 = \bigl(X, Y, \{X(x,y) : (x,y) \in X \times Y\}\bigr)$$

is a probability manifold. In a similar way, for $(x,z) \in X \times Z$, let $X(x,z) = Y \times Z$, $\Sigma(x,z) = \{B \times z : B \in \Sigma_Y\}$, $\mu_{x,z}(B \times z) = \mu_Q(B)$. Then

$$M_2 = \bigl(X, Z, \{X(x,z) : (x,z) \in X \times Z\}\bigr)$$

is a probability manifold. In this way, the phase space $X = \mathbf{R}^2$ can be thought of as a probability manifold with two different local structures

M_1, M_2 where M_1 corresponds to a position measurement and M_2 corresponds to a momentum measurement. The amplitude density f combines the two local structures.

We can now define conditional probabilities in the natural way. If $\mu_P(\mathbb{R}, A) \neq 0$ then the *P-probability of B given an A-measurement of Q* is

$$\mu_P(B \mid A) = \frac{\mu_P(B, A)}{\mu_P(\mathbb{R}, A)}. \tag{6.15}$$

If $\mu_Q(\mathbb{R}, A) \neq 0$ then the *Q-probability of B given an A-measurement of P* is

$$\mu_Q(B \mid A) = \frac{\mu_Q(B, A)}{\mu_Q(\mathbb{R}, A)}. \tag{6.16}$$

Notice that if the denominators are nonzero, then $B \mapsto \mu_P(B \mid A)$ and $B \mapsto \mu_Q(B \mid A)$ are probability measures on $B(\mathbb{R})$.

In order to find expectations of observables (i.e., functions on $\mathbb{R}^2$) we need to extend our amplitude definitions (6.11), (6.12) to functions. We call a function $F: \mathbb{R}^2 \to \mathbb{C}$ an *observable* and write $F \in L^2(f)$ if F is Lebesgue measurable on $\mathbb{R}^2$,

$$\int F(q,p) f(q,p) \, dp \in L^2(\mathbb{R}, dq)$$

and

$$\int F(q,p) f(q,p) \, dq \in L^2(\mathbb{R}, dp).$$

Note that the function I which is identically 1 on $\mathbb{R}^2$ is an observable. For $F \in L^2(f)$ we define the P and Q amplitude densities of F, respectively by

$$f_P(F)(p) = \int F(q,p) f(q,p) \, dq \tag{6.17}$$

$$f_Q(F)(q) = \int F(q,p) f(q,p) \, dp. \tag{6.18}$$

Notice that (6.17), (6.18) are extensions of (6.11) and (6.12), respectively, since

$$f_P(\phi_{A \times \mathbb{R}}) = f_P(A) \tag{6.19}$$

$$f_Q(\phi_{\mathbb{R} \times A}) = f_Q(A). \tag{6.20}$$

In fact, (6.17), (6.18) are the natural continuous linear extensions of (6.19), (6.20), respectively.

PHASE SPACE MODEL

Just as we have defined probabilities in terms of $f_P(A)$ and $f_Q(A)$ we would like to define expectations in terms of $f_P(F)$ and $f_Q(F)$. First of all, the P-expectation E_P should satisfy

$$E_P(\phi_{\mathbf{R}\times A}) = \mu_P(A, \mathbf{R}) \qquad (6.21)$$

and similarly the Q-expectation E_Q should satisfy

$$E_Q(\phi_{A\times \mathbf{R}}) = \mu_Q(A, \mathbf{R}). \qquad (6.22)$$

It follows from (6.21) that

$$\begin{aligned} E_P(\phi_{\mathbf{R}\times A}) &= \int_A \left| \int_{\mathbf{R}} f(q,p)\,dq \right|^2 dp \\ &= \int_{\mathbf{R}} \left[\int_{\mathbf{R}} \bar{f}(q,p)\,dq \int \phi_{\mathbf{R}\times A}(q,p) f(q,p)\,dq \right] dp \qquad (6.23) \\ &= \int_{\mathbf{R}} \overline{f_P(I)} f_P(\phi_{\mathbf{R}\times A})\,dp. \end{aligned}$$

In a similar way, we have

$$E_Q(\phi_{A\times \mathbf{R}}) = \int_{\mathbf{R}} \overline{f_Q(I)} f_Q(\phi_{A\times \mathbf{R}})\,dq. \qquad (6.24)$$

Now the natural continuous linear extensions of (6.23) and (6.24) are

$$E_P(F) = \int \overline{f_P(I)}(p) f_P(F)(p)\,dp \qquad (6.25)$$

$$E_Q(F) = \int \overline{f_Q(I)}(q) f_Q(F)(q)\,dq \qquad (6.26)$$

for all $F \in L^2(f)$. We therefore use (6.25), (6.26) as the defining equations for the expectations E_P and E_Q respectively.

6.5 Quantum Mechanics Derived

Section 6.4 presented the basic concepts and definitions for a phase space model of quantum mechanics. The present section shows that this model gives the usual quantum probabilities, expectations and dynamics. In this sense, we shall derive traditional quantum mechanics from the model.

We first need to know the relationship between amplitude densities and quantum mechanical pure states. In the traditional quantum mechanics of the present context pure states are usually given by unit vectors in $L^2(\mathbf{R}, dq)$. Since $f_Q(I) = f_Q(\mathbf{R})$ is a unit vector in $L^2(\mathbf{R}, dq)$ we use the notation $\psi_f = f_Q(I)$ and call ψ_f the *state corresponding* to the amplitude density f. Unfortunately, the map $f \mapsto \psi_f$ is not injective. However, if we restrict attention to certain well-behaved amplitude densities, this problem is remedied.

For $\psi \in L^2(\mathbf{R}, dq)$, we denote the Fourier transform by

$$\hat{\psi}(p) = (2\pi\hbar)^{-\frac{1}{2}} \int \psi(q) e^{-iqp/\hbar}\, dq.$$

The inverse Fourier transform is denoted $\check{\psi}(q)$ and of course

$$\check{\psi}(q) = (2\pi\hbar)^{-\frac{1}{2}} \int \psi(p) e^{iqp/\hbar}\, dp.$$

We say that $F \in L^2(f)$ is a *function of one variable* if $F(q, p) = G(q)$ or $F(q, p) = G(p)$ for some function G, and all $q, p \in \mathbf{R}$. We say that an amplitude density f is *regular* if $f_Q(F)\hat{} = f_P(F)$ for all $F \in L^2(f)$ that are functions of one variable. It should not be surprising that regular amplitude densities have a close connection to traditional quantum mechanics. First, in traditional quantum mechanics, the Fourier transform of position is momentum (in a certain sense). Second, the regularity condition allows Planck's constant $\hbar$ to be introduced into the formalism. The next result characterizes regular amplitude densities in terms of quantum mechanical states.

Theorem 6.21. *An amplitude density f is regular if and only if the following two conditions hold.*

(a) *For every $p \in \mathbf{R}$*

$$f(q, p) = (2\pi\hbar)^{-\frac{1}{2}} \psi_f(q) e^{-iqp/\hbar} \qquad \text{a.e. } [q].$$

QUANTUM MECHANICS DERIVED

(b) *For every* $q \in \mathbb{R}$

$$f(q,p) = (2\pi\hbar)^{-\frac{1}{2}} \hat{\psi}_f(p) e^{iqp/\hbar} \quad \text{a.e. } [p].$$

Proof. Let f be an amplitude density satisfying (a) and (b). Then for any function $F(q,p) = G(p) \in L^2(f)$ we have

$$f_Q(F) = \int G(p) f(q,p) \, dp$$

$$= (2\pi\hbar)^{-\frac{1}{2}} \int G(p) \hat{\psi}_f(p) e^{iqp/\hbar} \, dp$$

$$= (G\hat{\psi}_f)\check{}.$$

Hence, $f_Q(F)\hat{} = G\hat{\psi}_f$. We then have

$$f_P(F) = \int G(p) f(q,p) \, dq$$

$$= G(p)(2\pi\hbar)^{-\frac{1}{2}} \int \psi_f(q) e^{-iqp/\hbar} \, dq$$

$$= G\hat{\psi}_f = f_Q(F)\hat{}.$$

Moreover, for any function $F(q,p) = G(q) \in L^2(f)$ we have

$$f_P(F) = \int G(q) f(q,p) \, dq$$

$$= (2\pi\hbar)^{-\frac{1}{2}} \int G(q) \psi_f(q) e^{-iqp/\hbar} \, dq$$

$$= (G\psi_f)\hat{}$$

and

$$f_Q(F) = \int G(q) f(q,p) \, dq = G\psi_f.$$

Hence, $f_Q(F)\hat{} = (G\psi_f)\hat{} = f_P(F)$ and f is regular.

Conversely, suppose f is regular. Then for $F(q,p) = G(q) \in L^2(f)$ we have

$$\int G(q) f(q,p) \, dq = f_P(F) = f_Q(F)\hat{}$$

$$= (2\pi\hbar)^{-\frac{1}{2}} \int f_Q(F) e^{-iqp/\hbar} \, dq$$

$$= (2\pi\hbar)^{-\frac{1}{2}} \int G(q) \psi_f(q) e^{-iqp/\hbar} \, dq.$$

It follows that Condition a holds. Applying (a) gives

$$\int f(q,p)\, dq = \hat{\psi}_f(p). \tag{6.27}$$

If $F(q,p) = G(p) \in L^2(f)$, then from (6.27) we have

$$\int G(p)f(q,p)\, dp = f_Q(F) = f_P(F)\tilde{\ }$$

$$= (2\pi\hbar)^{-\frac{1}{2}} \int f_P(F) e^{iqp/\hbar}\, dp$$

$$= (2\pi\hbar)^{-\frac{1}{2}} \int G(p)\hat{\psi}_f(p) e^{iqp/\hbar}\, dp.$$

It follows that Condition b holds. □

Theorem 6.21 shows that corresponding to a regular amplitude density f there is a quantum state ψ_f satisfying (a) and (b). But what about the converse? If $\psi \in L^2(\mathbb{R}, dq)$ is a unit vector, does there exist a regular amplitude density f such that $\psi_f = \psi$? The answer is yes, but to prove it we must again assume the axiom of choice and the continuum hypothesis.

Theorem 6.22. *If $\psi \in L^2(\mathbb{R}, dq)$ with $\|\psi\| = 1$, then there exists a regular amplitude density f such that $\psi = \psi_f$.*

Proof. The proof is similar to that of Theorems 6.16 and 6.18. Define the family of sets

$$K = \{q \times \mathbb{R}, \mathbb{R} \times p : q, p \in \mathbb{R}\}.$$

Now K can be well ordered as in the proofs of those theorems. We now define f by transfinite induction. If the first set in this ordering has the form $q_1 \times \mathbb{R}$, define f on this set by

$$f(q_1, p) = (2\pi\hbar)^{-\frac{1}{2}} \hat{\psi}(p) e^{iq_1 p/\hbar}$$

and if the first set has the form $\mathbb{R} \times p_1$, define f on this set by

$$f(q, p_1) = (2\pi\hbar)^{-\frac{1}{2}} \psi(q) e^{-iqp_1/\hbar}.$$

The proof now proceeds as in Theorem 6.18 to obtain a function $f \colon \mathbb{R}^2 \to \mathbb{C}$ satisfying the following conditions.

(a') For every $p \in \mathbb{R}$
$$f(q,p) = (2\pi\hbar)^{-\frac{1}{2}}\psi(q)e^{-iqp/\hbar} \qquad \text{a.e. } [q].$$
(b') For every $q \in \mathbb{R}$
$$f(q,p) = (2\pi\hbar)^{-\frac{1}{2}}\hat{\psi}(p)e^{iqp/\hbar} \qquad \text{a.e. } [p].$$

Applying (a') gives
$$\int \left| \int f(q,p)\, dq \right|^2 dp = \int |\hat{\psi}(p)|^2 dp = \|\hat{\psi}\| = \|\psi\| = 1.$$

Applying (b') gives
$$\int \left| \int f(q,p)\, dp \right|^2 dq = \int |\psi(q)|^2 dq = \|\psi\| = 1.$$

Hence, f is an amplitude density. Moreover, by (b')
$$\psi_f(q) = \int f(q,p)\, dp = \psi(q).$$

It follows from Theorem 6.21 that f is regular. □

The next two results show that the regular amplitude density corresponding to a quantum state produces the usual quantum marginal distributions and conditional probabilities. We denote the spectral measures for the usual position and momentum operators Q and P by E^Q and E^P, respectively.

Lemma 6.23. *Let ψ be a unit vector in $L^2(\mathbb{R}, dq)$ and let f be the corresponding regular amplitude density. Then*

(a) $f_P(A) = (\psi\phi_A)\hat{\ }, \quad f_P(\mathbb{R}) = \hat{\psi}.$

(b) $f_Q(A) = (\hat{\psi}\phi_A)\check{\ }, \quad f_Q(\mathbb{R}) = \psi.$

(c) $\mu_P(B, A) = \int_B |(\psi\phi_A)\hat{\ }(p)|^2 dp$

$\mu_P(\mathbb{R}, A) = \int_A |\psi(q)|^2 dq, \quad \mu_P(A, \mathbb{R}) = \int_A |\hat{\psi}(p)|^2 dp.$

(d) $\mu_Q(B, A) = \int_B |(\hat{\psi}\phi_A)\check{\ }(q)|^2 dq$

$\mu_Q(\mathbb{R}, A) = \int_A |\hat{\psi}(p)|^2 dp, \quad \mu_Q(A, \mathbb{R}) = \int_A |\psi(q)|^2 dq.$

Proof. Straightforward application of definitions. □

Theorem 6.24. *If f is the regular amplitude density corresponding to $\psi \in L^2(\mathbb{R}, dq)$ and P_ψ denotes the one-dimensional projection onto ψ, then relative to f we have*

(a) $\mu_P(B \mid A) = \dfrac{\operatorname{tr}\left[E^P(B)E^Q(A)P_\psi E^Q(A)\right]}{\operatorname{tr}[E^Q(A)P_\psi]}.$

(b) $\mu_Q(B \mid A) = \dfrac{\operatorname{tr}\left[E^Q(B)E^P(A)P_\psi E^P(A)\right]}{\operatorname{tr}[E^P(A)P_\psi]}.$

Proof. (a) Denote the Fourier transform by F. Using the fact that $E^Q(B)$ is multiplication by ϕ_B and $E^P(B) = F^* E^Q(B) F$ we obtain from Lemma 6.23,

$$\mu_P(B, A) = \int_B |(\psi\phi_A)\hat{\ }(p)|^2 dp$$

$$= \|\phi_B(\psi\phi_A)\hat{\ }\|^2 = \|F^*\phi_B F(\psi\phi_A)\|^2 = \|E^P(B)(\psi\phi_A)\|^2$$

$$= \langle E^P(B)\psi\phi_A, \psi\phi_A \rangle = \langle E^P(B)E^Q(A)\psi, E^Q(A)\psi \rangle$$

$$= \langle E^Q(A)E^P(B)E^Q(A)P_\psi\psi, \psi \rangle$$

$$= \operatorname{tr}\left[E^Q(A)E^P(B)E^Q(A)P_\psi\right] = \operatorname{tr}\left[E^P(B)E^Q(A)P_\psi E^Q(A)\right].$$

Hence,

$$\mu_P(B \mid A) = \frac{\mu_P(B, A)}{\mu_P(\mathbb{R}, A)}$$

$$= \frac{\operatorname{tr}\left[E^P(B)E^Q(A)P_\psi E^Q(A)\right]}{\operatorname{tr}[E^Q(A)P_\psi]}.$$

(b) As in Part a we have

$$\mu_Q(B, A) = \int_B |(\hat{\psi}\phi_A)\check{\ }(q)|^2 dq$$

$$= \|\phi_B(\hat{\psi}\phi_A)\check{\ }\|^2 = \|\phi_B F^* E^Q(A) F\psi\|^2 = \|E^Q(B)E^P(A)\psi\|^2$$

$$= \langle E^Q(B)E^P(A)\psi, E^P(A)\psi \rangle = \langle E^P(A)E^Q(B)E^P(A)P_\psi\psi, \psi \rangle$$

$$= \operatorname{tr}\left[E^P(A)E^Q(B)E^P(A)P_\psi\right] = \operatorname{tr}\left[E^Q(B)E^P(A)P_\psi E^P(A)\right].$$

Hence,

$$\mu_Q(B \mid A) = \frac{\mu_Q(B, A)}{\mu_Q(\mathbb{R}, A)}$$

$$= \frac{\operatorname{tr}\left[E^Q(B)E^P(A)P_\psi E^P(A)\right]}{\operatorname{tr}[E^P(A)P_\psi]}. \quad \square$$

QUANTUM MECHANICS DERIVED

Notice that Theorem 6.24a and b are the usual von Neumann formulas for conditional probabilities. We now consider expectations of observables. To assure that ψ is in the domain of the relevant operators, we asume that ψ is a Schwartz test function. Moreover, we define the usual quantum operators $(Q\psi)(q) = q\psi(q)$, $(P\psi)(q) = -i\hbar(d\psi/dq)(q)$, $(P\hat{\psi})(p) = p\hat{\psi}(p)$. In the next theorem, the first two equations of Part a derive a version of the Bohr correspondence principle and the remaining results show that expectations of functions of Q and P reproduce the usual quantum formulas.

Theorem 6.25. *Let f be the regular amplitude density corresponding to $\psi \in L^2(\mathbf{R}, dq)$.*

(a) *If $F(q,p) = \Sigma a_{ij} q^i p^j$ is a polynomial, $a_{ij} \in \mathbf{R}$, then*
$$f_Q(F) = \Sigma a_{ij} Q^i P^j \psi, \qquad f_P(F) = \Sigma a_{ij} P^j (Q^i \psi)\hat{}$$
$$E_Q(F) = \langle \Sigma a_{ij} Q^i P^j \psi, \psi \rangle, \qquad E_P(F) = \langle \Sigma a_{ij} P^j Q^i \psi, \psi \rangle.$$

(b) *For any functions of one variable $F(q), G(p)$ for which the expressions are defined*
$$E_Q[F(q) + G(p)] = E_P[F(q) + G(p)] = \langle [F(Q) + G(P)]\psi, \psi \rangle.$$

Proof. (a) We shall prove this result for a function of the form $F(q,p) = qp$. The proof for more general f is similar.

$$f_Q(F) = \int qp f(q,p)\, dp$$
$$= (2\pi\hbar)^{-\frac{1}{2}} q \int p\hat{\psi}(p) e^{iqp/\hbar}\, dp$$
$$= q\left(-i\hbar \frac{d\psi}{dq}\right) = QP\psi$$
$$f_P(F) = \int qp f(q,p)\, dq$$
$$= (2\pi\hbar)^{-\frac{1}{2}} p \int q\psi(q) e^{-iqp/\hbar}\, dp = p(q\psi)\hat{} = P(Q\psi)\hat{}$$
$$E_Q(F) = \int \overline{f_Q(I)} f_Q(F)\, dq$$
$$= \int \bar{\psi}(q)(QP\psi)(q)\, dq = \langle QP\psi, \psi \rangle$$
$$E_P(F) = \int \overline{f_P(I)} f_P(F)\, dp$$
$$= \int \bar{\hat{\psi}}(p) p(Q\psi)\hat{}(p)\, dp$$
$$= \langle pF(Q\psi), F\psi \rangle = \langle F^* pFQ\psi, \psi \rangle = \langle PQ\psi, \psi \rangle.$$

(b) The proof is similar to (a). □

Notice that in computing $E_Q(F)$, the momentum operators act first and the position operators act second. This is to be expected since in the E_Q average momentum is measured first and position second. A similar observation holds for $E_P(F)$. Although the formulas for $E_Q(F)$ and $E_P(F)$ are essentially the same as in traditional quantum mechanics, there is an important difference in interpretation. In traditional quantum mechanics Q and P are thought of as being measured simultaneously, while in the present formalism one is measured after the other so the order is relevant. This is why we have two expectations $E_Q(F)$ and $E_P(F)$ and these can be different. One of the problems in traditional quantum mechanics is that products such as $Q^2 P^2$ must be symmetrized to obtain a self-adjoint operator. Unfortunately, there are many ways to do this and no canonical method is known. The present formalism has no such problem since the order is determined by the type of expectation performed. But there is a tradeoff. We now have the problem that $E_Q(F)$ (or $E_P(F)$) may not be a real number. We are again faced with an artificial symmetrization or the realization that the expectation does not exist for certain states.

The next corollary shows that, in a certain sense, the Heisenberg commutation relations and uncertainty principle hold in this theory.

Corollary 6.26. *If f is a regular amplitude density corresponding to $\psi \in L^2(\mathbb{R}, dq)$, then*

(a) $E_Q(qp) - E_P(qp) = \langle [Q,P]\psi, \psi \rangle = i\hbar$

(b) $\Delta p \Delta q \geq \hbar/2$ *where* $\Delta q = [E(q^2) - E^2(q)]^{\frac{1}{2}}$ *and* $\Delta p = [E(p^2) - E^2(p)]^{\frac{1}{2}}$.

Notice that in Part b we need not specify a P or Q expectation since these coincide by Theorem 6.25b. Part b is essentially the same as the usual uncertainty principle while Part a is a little strange. In this case, the function $F(q,p) = qp$ is the same in both expectations but the order of measurement for the two observables is different.

We now consider Schrödinger's equation. Suppose the classical Hamiltonian is

$$H(q,p) = \frac{p^2}{2m} + \vee(q). \qquad (6.28)$$

If the system is closed, then we have conservation of energy

$$H(q,p) = E. \qquad (6.29)$$

QUANTUM MECHANICS DERIVED 241

Now suppose f is a regular amplitude density with corresponding state $\psi \in L^2(\mathbb{R}, dq)$. Taking the Q-amplitude average of (6.29) gives

$$f_Q[H(q,p)] = f_Q(E). \tag{6.30}$$

Using the linearity of f_Q and Theorem 6.25 we have

$$H(Q,P)\psi = E\psi. \tag{6.31}$$

Of course, (6.31) is the time-independent Schrödinger equation. This shows that E, ψ form an eigenpair for the operator $H(Q,P)$.

We have seen in Section 2.1 that the classical dynamics is described by Hamilton's equation

$$\frac{dp}{dt} = -\frac{\partial H}{\partial q}. \tag{6.32}$$

Suppose that for any time $t \in (-\infty, \infty)$, the system is described by a regular amplitude density $f(q,p,t)$ with corresponding state $\psi(q,t)$ and assume that f and ψ are differentiable with respect to t. Moreover, suppose (6.32) holds in the Q-amplitude average. That is,

$$\frac{d}{dt}\int pf(q,p,t)\,dp = -\frac{\partial}{\partial q}\int H(q,p)f(q,p,t)\,dp. \tag{6.33}$$

Then (6.33) has the form

$$\frac{d}{dt}f_Q(p) = -\frac{\partial}{\partial q}f_Q(H). \tag{6.34}$$

Applying Theorem 6.25 to Equation 6.34 gives

$$\frac{d}{dt}\left(-i\hbar\frac{\partial \psi}{\partial q}\right) = -\frac{\partial}{\partial q}H(Q,P)\psi. \tag{6.35}$$

Interchanging the order of differentiation in (6.35) gives

$$\frac{\partial}{\partial q}\left(-i\hbar\frac{\partial \psi}{\partial t}\right) = -\frac{\partial}{\partial q}H(Q,P)\psi. \tag{6.36}$$

Integrating both sides of (6.36) we obtain (except for a constant which we can set equal to 0)

$$i\hbar\frac{\partial \psi}{\partial t} = H(Q,P)\psi. \tag{6.37}$$

Equation 6.37 is the time-dependent Schrödinger equation (see Section 2.2). We conclude that Schrödinger's equation is an amplitude averaged version of Hamilton's equation for classical mechanics.

Instead of (6.32), what if we use Hamilton's other equation, $dq/dt = \partial H/\partial p$? In this case we would take the P-amplitude average to obtain

$$\frac{d}{dt} f_P(q) = \frac{\partial}{\partial p} f_P(H).$$

Proceeding in a similar way, we obtain the Fourier transformed time-dependent Schrödinger equation

$$i\hbar \frac{\partial \hat{\psi}}{\partial t} = \frac{p^2}{2m} \hat{\psi} + (\vee \psi)\hat{}\,.$$

6.6 Notes and References

For further details on transfinite induction and axiomatic set theory, we refer the reader to [Pinter, 1971]. Of course, we have just scratched the surface of this subject and mainly discussed material that was needed later. It is curious that we need such "heavy" set-theoretic machinery as the axiom of choice and the continuum hypothesis in a physical theory. As one might suspect, this brings up various philosophical questions [Pitowski, 1983b].

The material in Section 6.2 appeared in [Gudder, 1984c] which the reader may consult for more details. There appears to be considerable room in probability manifolds for further investigation. In particular it would be interesting to prove more general extensions of classical limit theorems. Section 6.3 presents only a small fraction of Pitowski's interesting work. For more information, the reader can consult [Pitowski, 1982, 1983a, 1983b].

Sections 6.4 and 6.5 have appeared in [Gudder, 1984c, 1985; Pitowski, 1984]. Section 6.5 was mainly concerned with regular amplitude densities since these correspond to the usual quantum states and gave the usual quantum predictions. It would be interesting to investigate nonregular amplitude densities. Although it is not clear whether they have physical relevance, they might provide an amusing alternative to traditional quantum mechanics. It was not our intention in 6.5 to argue that this is a better way to do quantum mechanics. Our purpose was to show that traditional quantum mechanics is encompassed in the more general framework of quantum probability. This framework has its roots in classical probability theory. However, there are two main deviations from this classical theory. The first is that probabilities are computed using amplitude functions or amplitude densities. Secondly, in order to treat more than one measurement, a more general measure theory involving σ-additive classes is required. As a consequence, the cherished role of Lebesgue measurable sets is called to question.

Einstein, Podolsky and Rosen [1935] wrote a famous paper entitled "Can quantum mechanical description of physical reality be considered complete?". Since that time there have been many books and articles discussing the pros and cons of that paper [Belifante, 1973; Bohm, 1951; Bohr, 1935; Lahti and Mittlestaedt, 1985; Pitowski, 1982]. The idea of the paper was to show that the description of a system by the formalism of quantum mechanics failed to tell us everything about what the system was really like. They attempted to show that not all the elements of physical reality associated with a system entered into its description by quantum mechanics.

Section 6.5 gives another way to answer the question posed by Einstein, Podolsky and Rosen. The answer given by Section 6.5 is no. Then how can we obtain a complete description, and how is quantum mechanics related to such a description? The answer to the first part of this question has been known for hundreds of years. The complete description (at least for nonrelativistic systems with a finite number of degrees of freedom) is given by classical mechanics. Quantum mechanics itself answers the second part of the question because of the unusual way that quantum probabilities are computed. Quantum mechanics is an amplitude average of classical mechanics.

Does it do us any good to know that quantum mechanics possesses an underlying complete description of physical reality? At the present time, the answer is not very clear, although we shall take some preliminary steps toward using this information in the next chapter. The problem is that the (pure) states of the underlying complete description may not be physically accessible. In other words, nature may not allow us to view the world through a window provided by states that are more accurate than the quantum states. Future studies in elementary particle physics may shed more light on this question. Chapter 7 will have a little more to say about this.

7
Discrete Quantum Mechanics

Previous chapters have emphasized that probability amplitudes (or transition amplitudes) are basic to a quantum mechanical description of a physical system. The present chapter gives a more detailed discussion of implementing such a program. Since this is most easily accomplished for discrete quantum mechanics, we shall work within that framework. Moreover, discreteness may not be a severe limitation. First, it can be shown that the usual continuum quantum mechanics can be approximated arbitrarily closely by a discrete quantum mechanics [Gudder and Naroditsky, 1981]. Second, there are reasons for believing that natural phenomena are intrinsically discrete and should not be based upon a continuum spacetime model [Gudder, 1984b, 1986a, 1988; Russell, 1954; Shale, 1979, 1982]. Such continuum models are frequently plagued with infinities and singularities and they break down at small distances and high energies. An example of this was seen in our previous discussion of the path integral formalism. There is, in fact, speculation that there exists in nature an elementary length and an elementary time and that all length and time measurements must be integer multiples of these [Gudder, 1984b]. Since the existence of an elementary length does not automatically lead to a discrete spacetime, it is not clear whether discrete quantum mechanics will play the role of a fundamental physical theory. In any case, the present chapter illustrates the techniques and methods developed in previous chapters

In Sections 7.1 and 7.2 we shall discuss quantum Markov chains. These are the quantum counterparts of classical Markov chains and they can be used to describe a discrete quantum dynamics. These sections are based in the framework of a quantum probability space (Section 4.8). Sections 7.3–7.7 discuss quantum mechanical models in terms of one-step transition amplitudes. Such models can be used to describe a quantum dynamics in discrete spacetime. Moreover, they may be useful for describing the internal dynamics of elementary particles. The motion induced by such dynamics corresponds to a quantum random walk.

7.1 Markov Amplitude Chains

Throughout this section, (Ω, f) will denote a quantum probability space (Section 4.8) and $S = \{a_1, \ldots, a_r\}$ will be a fixed finite subset of $\mathbb{R}$. If $h_j \colon D \subseteq \Omega \to S$, $j = 0, \ldots, n$ are Σ_0 measurable, we call

$$\{h_j\}_0^n = \{h_j \colon j = 0, \ldots, n\}$$

an *amplitude n-chain* with *value space* S. We interpret S as the set of possible locations of a randomly moving physical system, and $h_0, \ldots, h_n$ give the location of the system at the times $0, 1, \ldots, n$, respectively. The numbers in S need not correspond to position values, but could be interpreted as values for any physical quantity such as energy, momentum, spin, charge, etc. The possible n-step paths of the system are given by the set of *sample paths* $(h_0(\omega), \ldots, h_n(\omega))$, $\omega \in D$. The kth *vector* of the n-chain is the vector $\phi_k \in \mathbb{C}^r$ given by $\phi_k(j) = \hat{f}[h_k^{-1}(a_j)]$, $j = 1, \ldots, r$; $k = 0, \ldots, n$. Notice that ϕ_k is a unit vector in $\mathbb{C}^r$ if and only if h_k is observable. We call ϕ_0 and ϕ_n the *initial* and *final vectors*, respectively.

We define the *amplitude matrix* of $\{h_j\}_0^n$ by

$$A(k, j) = \hat{f}\big[h_n^{-1}(a_k) \cap h_0^{-1}(a_j)\big], \qquad j, k = 1, \ldots, r.$$

The matrix element $A(k, j)$ gives the amplitude that a system moves from a_j to a_k in n steps. This amplitude is the sum of the amplitudes for all possible n-step paths from a_j to a_k. Indeed, since $\hat{f}$ is a complex measure, it follows that

$$A(k, j) = \sum_{i_1, \ldots, i_{n-1}} \hat{f}\big[h_0^{-1}(a_j) \cap h_1^{-1}(a_{i_1}) \cap \cdots \cap h_{n-1}^{-1}(a_{i_{n-1}}) \cap h_n^{-1}(a_k)\big].$$

We also define the *conditional amplitude matrix*

$$A'(k, j) = \hat{f}\big[h_n^{-1}(a_k) \mid h_0^{-1}(a_j)\big], \qquad j, k = 1, \ldots, r.$$

MARKOV AMPLITUDE CHAINS 247

If the initial vector is nonvanishing (i.e., $\phi_0(j) \neq 0$, $j = 1, \ldots, r$), then $A(k,j) = \phi_0(j)A'(k,j)$ and hence

$$\phi_n(k) = \hat{f}[h_n^{-1}(a_k)]$$
$$= \sum_j \hat{f}[h_n^{-1}(a_k) \cap h_0^{-1}(a_j)]$$
$$= \sum_j A(k,j) = \sum_j A'(k,j)\phi_0(j) = (A'\phi_0)(k).$$

Thus, $\phi_n = A'\phi_0$.

We call an $r \times r$ complex matrix T a *quasistochastic matrix* if $\Sigma_k T(k,j) = 1$, $j = 1, \ldots, r$. An $r \times r$ complex matrix T is called a *stochastic amplitude matrix* if $|T(k,j)|^2$ is quasistochastic (that is, $\Sigma_k |T(k,j)|^2 = 1$, $j = 1, \ldots, r$). Notice that h_n is h_0 observable if and only if ϕ_0 is nonvanishing and A' is a stochastic amplitude matrix. Now suppose A is a stochastic amplitude matrix. This holds if and only if the collection of sets

$$\{h_n^{-1}(a_k) \cap h_0^{-1}(a_j) : k = 1, \ldots, r\}$$

is an operation for each $j = 1, \ldots, r$. We then call $\{h_j\}_0^n$ a *constrained n-chain*. In this case, we interpret $|A(k,j)|^2$ as the probability that a system which is initially constrained at a_j, then moves to a_k in n steps. Another important case is when h_0 and h_n are jointly observable. This holds if and only if $\Sigma_{j,k}|A(k,j)|^2 = 1$, and we then call $\{h_j\}_0^n$ a *definite n-chain*. We interpret $|A(k,j)|^2$, in this case, as the probability that a system begins at a_j and moves to a_k in n steps.

We call $\{h_j\}_0^n$ a *Markov amplitude n-chain* if

$$\hat{f}[h_j^{-1}(a_{i_j}) \mid h_0^{-1}(a_{i_0}) \cap h_{j-1}^{-1}(a_{i_{j-1}})] = \hat{f}[h_j^{-1}(a_{i_j}) \mid h_{j-1}^{-1}(a_{i_{j-1}})]$$

for all $i_0, \ldots, i_j$; $j = 1, \ldots, n$. We say that $\{h_j\}_0^n$ is *stationary* if

$$\hat{f}[h_m^{-1}(a_k) \mid h_{m-1}^{-1}(a_j)] = \hat{f}[h_1^{-1}(a_k) \mid h_0^{-1}(a_j)]$$

for all $j, k = 1, \ldots, r$; $m = 1, \ldots, n$.

Theorem 7.1. *Let $\{h_j\}_0^n$ be a stationary, Markov, amplitude n-chain for which ϕ_0 is nonvanishing and define the matrix*

$$T(k,j) = \hat{f}[h_1^{-1}(a_k) \mid h_0^{-1}(a_j)].$$

(a) *Then T is a quasistochastic matrix and*

$$\hat{f}[h_0^{-1}(a_{i_0}) \cap \cdots \cap h_n^{-1}(a_{i_n})] = \phi_0(i_0)T(i_1,i_0)\ldots T(i_n,i_{n-1}) \quad (7.1)$$

$$A(k,j) = \phi_0(j)T^n(k,j) \quad (7.2)$$

$$A'(k,j) = T^n(k,j) \quad (7.3)$$

$$\phi_k = T^k \phi_0, \quad k = 0,1,\ldots,n. \quad (7.4)$$

(b) *If in adddition, h_0 is observable and h_1 is h_0 observable and h_0 orthogonal, then T is unitary and any subset of $\{h_j: j = 0,\ldots,n\}$ is jointly observable. In particular, each h_j is observable and $\{h_j\}_0^n$ is definite.*

Proof. (a) Since $\hat{f}[h_0^{-1}(a_j)] \ne 0$ and $\hat{f}[\cdot \mid h_0^{-1}(a_j)]$ is a complex measure, we have

$$\Sigma_k T(k,j) = \hat{f}[\cup h_1^{-1}(a_k) \mid h_0^{-1}(a_j)]$$
$$= \hat{f}[D \mid h_0^{-1}(a_j)] = 1.$$

Hence, T is a quasistochastic matrix. Equation 7.1 follows from Markovicity, stationarity, and (4.25). Equation 7.2 follows from letting $i_0 = j$, $i_n = k$ in (7.1) and summing over $i_1,\ldots,i_{n-1}$ Equation 7.3 follows directly from (7.2). Replace n by k in (7.1) and sum over $i_0,\ldots,i_{k-1}$ to obtain (7.4).

(b) Since h_1 is h_0 observable and h_0 orthogonal, we have

$$\Sigma_k T^*(j',k)T(k,j) = \Sigma_k T(k,j)\bar{T}(k,j') = \delta_{jj'}$$

and hence, T is unitary. We next show that T^m is quasistochastic for any $m = 0,1,\ldots$. Indeed, T^0 and T are quasistochastic and suppose T^m is quasistochastic. Then

$$\Sigma_k T^{m+1}(k,j) = \Sigma_k \Sigma_i T^m(k,i)T(i,j)$$
$$= \Sigma_i T(i,j)\Sigma_k T^m(k,j) = \Sigma_i T(i,j) = 1.$$

Hence, the result follows by induction. Consider a subset $\{h_0, h_{j_1},\ldots, h_{j_k}\}$ of $\{h_j: j = 0,\ldots,n\}$ where $0 < j_1 < \cdots < j_k < n$. Summing (7.1) over the other indices gives

$$\hat{f}[h_0^{-1}(a_{i_0}) \cap h_{j_1}^{-1}(a_{i_1}) \cap \cdots \cap h_{j_k}^{-1}(a_{i_k})]$$
$$= \phi_0(i_0)T^{j_1}(i_1,i_0)T^{j_2-j_1}(i_2,i_1)\cdots T^{j_k-j_{k-1}}(i_k,i_{k-1})\Sigma_i T^{n-j_k}(i,i_k).$$

Since T^{n-j_k} is quasistochastic, the last sum is 1. Moreover, since T^k is unitary for $k = 0, 1, \ldots$, and $\|\phi_0\| = 1$, taking the modulus squared and summing over $i_0, i_1, \ldots, i_k$ again gives 1. It follows that $\{h_0, h_{j_1}, \ldots, h_{j_k}\}$ is jointly observable. The other cases follow in a similar manner. □

The above theorem shows that stationary, Markov, amplitude n-chains can be very well behaved. In particular, for the situation described in Part b, we can not only observe any trajectory (sample path), but we can observe a sequence of "snap shots" in which time is stopped whenever we wish. We shall see later that n-chains which describe quantum systems need not be stationary although they frequently are Markov. We then lose the ability to observe individual trajectories. However, frequently h_n is observable so we can observe where the system ends, or h_0 and h_n are jointly observable so we can observe the beginning and the end. We shall also show that (7.1)–(7.4) hold in slightly altered form.

The reader may wonder if it is possible for a nontrivial matrix to be both unitary and quasistochastic as in Theorem 7.1b. Of course, the identity matrix has this property and it is easy to construct $0 - 1$ matrices with this property. However, there are also nontrivial unitary quasistochastic matrices as the following examples show.

$$\begin{bmatrix} \frac{1}{2} + \frac{1}{2}i & \frac{1}{2} - \frac{1}{2}i \\ \frac{1}{2} - \frac{1}{2}i & \frac{1}{2} + \frac{1}{2}i \end{bmatrix}$$

$$\begin{bmatrix} \frac{1}{3} + \left(\frac{1}{\sqrt{3}}\right)i & \frac{1}{3} & \frac{1}{3} - \left(\frac{1}{\sqrt{3}}\right)i \\ \frac{1}{3} - \left(\frac{1}{\sqrt{3}}\right)i & \frac{1}{3} + \left(\frac{1}{\sqrt{3}}\right)i & \frac{1}{3} \\ \frac{1}{3} & \frac{1}{3} - \left(\frac{1}{\sqrt{3}}\right)i & \frac{1}{3} + \left(\frac{1}{\sqrt{3}}\right)i \end{bmatrix}$$

We have seen that a stationary, Markov, amplitude n-chain has a joint amplitude satisfying (7.1). We now use this equation to define an important class of n-chains. An n-chain $\{h_j\}_0^n$ is called a *quantum n-chain* if there exists a vector $0 \neq \psi_0 \in \mathbb{C}^r$ and an $r \times r$ complex matrix T such that

$$\hat{f}[h_0^{-1}(a_{i_0}) \cap \cdots \cap h_n^{-1}(a_{i_n})] = \psi_0(i_0) T(i_1, i_0) \cdots T(i_n, i_{n-1}). \quad (7.5)$$

We call ψ_0 and T the *vector* and *matrix* of $\{h_j\}_0^n$, respectively. We denote the vector of $\{h_j\}_0^n$ by ψ_0 instead of ϕ_0 since the vector may not equal the initial vector. The relationship between ψ_0 and ϕ_0 is given in Theorem 7.2. If $\|\psi_0\| = 1$, we call ψ_0 the *initial state*. If T is unitary we call $\{h_j\}_0^n$ a *closed* quantum n-chain.

Theorem 7.2. *Let $\{h_j\}_0^n$ be a quantum n-chain with vector ψ_0 and matrix T. (a) The amplitude matrix, conditional amplitude matrix, initial vector, and final vector satisfy*

$$A(k,j) = \psi_0(j) T^n(k,j) \tag{7.6}$$

$$A'(k,j) = \frac{T^n(k,j)}{\Sigma_k T^n(k,j)} \quad \text{if } \phi_0(j) \neq 0 \tag{7.7}$$

$$\phi_0(j) = \psi_0(j) \Sigma_k T^n(k,j) \tag{7.8}$$

$$\phi_n = T^n \psi_0. \tag{7.9}$$

(b) *If $\|\psi_0\| = 1$ and T^n is a stochastic amplitude matrix, then $\{h_j\}_0^n$ is definite (and hence, h_0 and h_n are jointly observable).* (c) *If $\psi_0 \equiv 1$, then T^n is a stochastic amplitude matrix if and only if $\{h_j\}_0^n$ is constrained.* (d) *If $\{h_j\}_0^n$ is closed, then $\|\psi_0\| = 1$ implies it is definite and h_n is observable, and $\psi_0 \equiv 1$ implies it is constrained.*

Proof. (a) The proof of (7.6) is the same as the proof of (7.2). Sum (7.6) over k to obtain (7.8). Equation 7.7 follows from (7.6) and (7.8). Sum (7.6) over j to obtain (7.9). (b) Take the square of the modulus of (7.6) and sum over j and k. (c) Same proof as (b). (d) The first part follows from (b) and (7.9). The second part follows from (c). □

Notice that Formulas 7.5, 7.6 and 7.9 are analogous to (4.24) and (4.22) of Section 4.8. In that case the system was closed and T was given by the unitary matrix U. Observe from Theorem 7.2 that $\psi_0 = \phi_0$ if T (or T^n) is a quasistochastic matrix.

The next theorem summarizes some of the important properties of quantum n-chains. Among other things, it shows that if certain amplitudes are nonzero, then a quantum n-chain is Markov. Corresponding to an $r \times r$ matrix T we define vectors $T_m \in \mathbb{C}^r$, $m = 0, \ldots, n$ by

$$T_m(j) = \Sigma_k T^{n-m}(k,j), \qquad j = 1, \ldots, r.$$

Notice that T is quasistochastic if and only if $T_m \equiv 1$, $m = 0, \ldots, n$.

Theorem 7.3. Let $\{h_j\}_0^n$ be a quantum n-chain with vector ψ_0 and matrix T. (a) If ϕ_k is nonvanishing and

$$\hat{f}[h_0^{-1}(a_{i_0}) \cap \cdots \cap h_k^{-1}(a_{i_k})] \neq 0$$

for all $k, i_0, \ldots, i_k$, then $\{h_j\}_0^n$ is Markov. (b) If the terms are nonzero then

$$\hat{f}[h_m^{-1}(a_k) \mid h_{m-1}^{-1}(a_j)] = \frac{T(k,j)T_m(k)}{T_{m-1}(j)}.$$

(c) If the terms are nonzero, then

$$\hat{f}[h_m^{-1}(a_k) \mid h_{m-1}^{-1}(a_j)] = \hat{f}[h_i^{-1}(a_k) \mid h_{i-1}^{-1}(a_j)] \frac{T_m(k)T_{i-1}(j)}{T_{m-1}(j)T_i(k)}.$$

Proof. (a) Summing over various indices in (7.5) gives the following equations:

$$\hat{f}[h_0^{-1}(a_{i_0}) \cap \cdots \cap h_m^{-1}(a_{i_m})] = \psi_0(i_0)T(i_1,i_0)\cdots T(i_m,i_{m-1})T_m(i_m) \tag{7.10}$$

$$\hat{f}[h_m^{-1}(a_j)] = (T^m \psi_0)(j) T_m(j) \tag{7.11}$$

$$\hat{f}[h_m^{-1}(a_j) \cap h_{m-1}^{-1}(a_k)] = (T^{m-1}\psi_0)(k) T(j,k) T_m(j). \tag{7.12}$$

It follows from (7.10) that

$$\hat{f}[h_m^{-1}(a_{i_m}) \mid h_0^{-1}(a_{i_0}) \cap \cdots \cap h_{m-1}^{-1}(a_{i_{m-1}})] = \frac{T(i_m, i_{m-1})T_m(i_m)}{T_{m-1}(i_{m-1})}.$$

Applying (7.11) and (7.12) we obtain

$$\hat{f}[h_m^{-1}(a_{i_m}) \mid h_{m-1}^{-1}(a_{i_{m-1}})] = \frac{T(i_m, i_{m-1})T_m(i_m)}{T_{m-1}(i_{m-1})}. \tag{7.13}$$

Hence, $\{h_j\}_0^n$ is Markov. (b) This follows from (7.13). (c) This follows from Part b. □

In Theorem 7.3a we gave a sufficient condition for a quantum n-chain to be Markov. In the following corollary we strengthen this to a characterization.

Corollary 7.4. *A quantum n-chain with vector ψ_0 and matrix T is Markov if and only if the following two conditions hold for all $m = 1, 2, \ldots, n$.*

(a) $(T^m \psi_0)(i_m) T_m(i_m) = 0$ *whenever*

$$\psi_0(i_0) T(i_1, i_0) \cdots T(i_m, i_{m-1}) T_m(i_m) = 0$$

for some $i_0, \ldots, i_{m-1}$.

(b) $\psi_0(i_0) T(i_1, i_0) \cdots T(i_m, i_{m-1}) T_m(i_m) = 0$ *for every $i_0, \ldots, i_{m-1}$ whenever*

$$(T^m \psi_0)(i_m) T_m(i_m) = 0.$$

Proof. It follows from the proof of Theorem 7.3 that $\{h_j\}_0^n$ being Markov is equivalent to the two condtioning sets in the Markov definition having zero amplitude simultaneously. The result then follows from (7.10) and (7.11). □

We see from Theorem 7.3b and c that a quantum n-chain is stationary if its matrix T is quasistochastic. However, unlike the Markov condition, this is a strong requirement, and in general we would not expect a quantum n-chain to be stationary. Nevertheless, the conditions in Theorem 7.3b and c are weak stationary conditions. Although $\hat{f}\bigl[h_m^{-1}(a_k) \mid h_{m-1}^{-1}(a_j)\bigr]$ is not independent of m, we can separate out some of the m dependence. We now make this idea precise. We call an n-chain $\{h_j\}_0^n$ *almost stationary* if there exist vectors $\delta_m \in \mathbb{C}^r$, $m = 0, \ldots, n$ with $\delta_n \equiv 1$ such that

$$\begin{aligned}
\delta_{m-1}(j) \delta_i(k) \hat{f}\bigl[h_m^{-1}(a_k) \mid h_{m-1}^{-1}(a_j)\bigr] \\
= \delta_m(k) \delta_{i-1}(j) \hat{f}\bigl[h_i^{-1}(a_k) \mid h_{i-1}^{-1}(a_j)\bigr]
\end{aligned} \quad (7.14)$$

for all applicable i, j, k, m.

Lemma 7.5. (a) *If $\{h_j\}_0^n$ is a quantum n-chain with matrix T and ϕ_m, T_m are nonvanishing for $m = 0, \ldots, n$, then $\{h_j\}_0^n$ is almost stationary.*
(b) *If $\{h_j\}_0^n$ is an almost stationary, Markov n-chain with ϕ_m and δ_m nonvanishing for $m = 0, \ldots, n$, then $\{h_j\}_0^n$ is quantum with matrix*

$$T(j, k) = \delta_0(k) \delta_1(j)^{-1} \hat{f}\bigl[h_1^{-1}(a_j) \mid h_0^{-1}(a_k)\bigr]$$

vector $\psi_0(j) = \delta_0(j)^{-1} \phi_0(j)$, and $\delta_m = T_m$, $m = 0, \ldots, n$.

Proof. (a) This follows from Theorem 7.3c. (b) Using Markovicity and (7.14), the joint amplitude becomes

$$\hat{f}[h_0^{-1}(a_{i_0}) \cap \cdots \cap h_n^{-1}(a_{i_n})]$$
$$= \hat{f}[h_0^{-1}(a_{i_0})]\hat{f}[h_1^{-1}(a_{i_1}) \mid h_0^{-1}(a_{i_0})] \cdots \hat{f}[h_n^{-1}(a_{i_n}) \mid h_{n-1}^{-1}(a_{i_{n-1}})]$$
$$= \delta_0(i)^{-1}\hat{f}[h_0^{-1}(a_{i_0})]T(i_1,i_0)\cdots T(i_n,i_{n-1})$$

where
$$T(i_j,i_{j-1}) = \delta_0(i_{j-1})\delta_1(i_j)^{-1}\hat{f}[h_1^{-1}(a_{i_j}) \mid h_0^{-1}(a_{i_{j-1}})]$$

This shows that $\{h_j\}_0^n$ is a quantum n-chain with the prescribed matrix T and vector ψ_0. We now prove that $\delta_m = T_m$, $m = 0,\ldots,n$ by reverse induction. Clearly, $\delta_n = T_n \equiv 1$. Suppose that $\delta_m = T_m$, for some integer m where $0 < m \leq n$. By (7.14) we have

$$\hat{f}[h_m^{-1}(a_k) \mid h_{m-1}^{-1}(a_j)]$$
$$= \delta_m(k)\delta_0(j)[\delta_{m-1}(j)\delta_1(k)]^{-1}\hat{f}[h_1^{-1}(a_k) \mid h_0^{-1}(a_j)]$$
$$= \delta_m(j)T(k,j)[\delta_{m-1}(j)]^{-1}.$$

Summing over k gives

$$\delta_{m-1}(j) = \sum_k \delta_m(k)T(k,j)$$
$$= \sum_k \sum_i T^{n-m}(i,k)T(k,j)$$
$$= \sum_i T^{n-m+1}(i,j) = T_{m-1}(j). \quad \square$$

We close this section by showing that quantum n-chains exist with any vector and matrix. Let $S = \{a_1,\ldots,a_r\} \subseteq \mathbb{R}$ and let $P_n = \{p\colon \{0,1,\ldots,n\} \to S\}$. We call the functions in P_n n-paths. Notice that $|P_n| = r^{n+1}$. Let $\psi_0 \in \mathbb{C}^r$ and let T be an $r \times r$ complex matrix. We then call (P_n, ψ_0, T) a *path space*. Define the map $I\colon S \to 1,\ldots,r$ by $Ia_j = j$. For $p \in P_n$ define

$$f(p) = \psi_0[Ip(0)]T[Ip(1),Ip(0)]\cdots T[Ip(n),Ip(n-1)]. \qquad (7.15)$$

Then (P_n, f) becomes a quantum probability space. If $\|\psi_0\| = 1$ and T is a stochastic amplitude matrix, then (P_n, f) is point closed. Define the *path*

n-chain $\{h_j\}_0^n$ on (P_n, f) by $h_j(p) = p(j)$, $j = 0, \ldots, n$. Then the joint amplitude of $\{h_j\}_0^n$ becomes

$$\hat{f}[h_0^{-1}(a_{i_0}) \cap \cdots \cap h_n^{-1}(a_{i_n})] = f[(a_{i_0}, \ldots, a_{i_n})]$$
$$= \psi_0(i_0) T(i_1, i_0) \ldots T(i_n, i_{n-1})$$

where $(a_{i_0}, \ldots, a_{i_n})$ denotes the path $p(j) = a_{i_j}$, $j = 0, \ldots, n$. Hence, $\{h_j\}_0^n$ is a quantum n-chain with vector ψ_0 and matrix T. We thus see that quantum n-chains with arbitrary vectors and matrices exist. One can characterize quantum n-chains that are isomorphic to path n-chains [Gudder, 1984b]. Path chains will be one of the basic constituents for the discrete quantum models that we shall construct in later sections.

7.2 Random Phases and Quantum Processes

We begin by studying changes that result in a quantum probability space and its n-chains due to random phase changes. In this section, $\{h_j\}_0^n$ will be an n-chain in a quantum probability space (Ω, f) with value space $S = \{a_1, \ldots, a_r\}$. For simplicity, we shall assume that the functions h_j, $j = 0, \ldots, n$, are defined on all of Ω. In this case, we have $\Omega \in \Sigma_0$. We say that $\{h_j\}_0^n$ is *separating* if $h_j(\omega) = h_j(\omega')$ for $j = 0, \ldots, n$ implies that $\omega = \omega'$. Define

$$L^1(\Omega, f) = \{\beta \colon \Omega \to \mathbb{C} \colon \sum_{\omega \in \Omega} |\beta(\omega) f(\omega)| < \infty\}.$$

For $\beta \in L^1(\Omega, f)$ define

$$\sum \beta f \equiv \sum \beta(\omega) f(\omega) = \sum_{\omega \in \Omega} \beta(\omega) f(\omega)$$

and

$$\sum_{a_j}^{a_k} \beta f \equiv \sum_{a_j}^{a_k} \beta(\omega) f(\omega) = \sum \{\beta(\omega) f(\omega) \colon h_0(\omega) = a_j, h_n(\omega) = a_k\}$$

Notice that

$$A(k, j) = \sum_{a_j}^{a_k} f.$$

For $\alpha\colon \Omega \to \mathbb{R}$, define $f^\alpha\colon \Omega \to \mathbb{C}$, by $f^\alpha(\omega) = e^{-i\alpha(\omega)}f(\omega)$. We call $f \mapsto f^\alpha$ a *random phase transformation*. Notice that $f^0 = f$. If α is a constant function, then f^α and f are essentially the same since the probabilities are unchanged and $\hat{f}^\alpha(\cdot \mid \cdot) = \hat{f}(\cdot \mid \cdot)$. For a general α, the quantum probability space (Ω, f^α) has an entirely different structure than (Ω, f). If (Ω, f) is point closed, then it is clear that (Ω, f^α) is also. In this case, the collection of singleton outcome sets is an operation in both (Ω, f) and (Ω, f^α). In general however, (Ω, f) and (Ω, f^α) have different operations and outcomes. If $\{h_j\}_0^n$ is separating, then it is jointly observable in (Ω, f) if and only if it is jointly observable in (Ω, f^α). Indeed, we then have

$$\hat{f}^\alpha\bigl[h_0^{-1}(a_{i_0}) \cap \cdots \cap h_n^{-1}(a_{i_n})\bigr] = e^{-i\alpha(\omega)}\hat{f}\bigl[h_0^{-1}(a_{i_0}) \cap \cdots \cap h_n^{-1}(a_{i_n})\bigr]$$

where $h_j(\omega) = a_{i_j}$, $j = 0, \ldots, n$ (if such an ω does not exist, then both f and f^α are zero on $h_0^{-1}(a_{i_0}) \cap \cdots \cap h_n^{-1}(a_{i_n})$). However, even when $\{h_j\}_0^n$ is separating, we can have h_j observable in (Ω, f) but not observable in (Ω, f^α).

If $\{h_j\}_0^n$ is Markov, stationary, almost stationary, quantum, etc., relative to f, it need not be relative to f^α. The most regular nontrivial case is when α has the following form. Let $v\colon S \to \mathbb{R}$ and define $\alpha_v\colon \Omega \to \mathbb{R}$ by $\alpha_v(\omega) = \sum_{j=0}^{n} v[h_j(\omega)]$. Define $f^v = f^{\alpha_v}$.

Theorem 7.6. *Let $\{h_j\}_0^n$ be a quantum n-chain on (Ω, f) with vector ψ_0 and matrix T, and let $v\colon S \to \mathbb{R}$. Let $\{h'_j\}_0^n$ be the same n-chain considered on (Ω, f^v).* (a) *$\{h'_j\}_0^n$ is a quantum n-chain with vector $\psi'_0(j) = e^{-iv(a_j)}\psi_0(j)$ and matrix $T' = e^{-iv(Q)}T$, where Q is the $r \times r$ diagonal matrix $Q = \operatorname{diag}(a_1, \ldots, a_r)$.* (b) *The amplitude matrix A^v for $\{h'_j\}_0^n$ becomes*

$$A^v(k, j) = \psi'_0(j)T'^n(k, j)$$
$$= \sum_{a_j}^{a_k} \exp\left[-i\sum_{m=0}^{n} v(h_m(\omega))\right]f(\omega).$$

(c) *If $\{h_j\}_0^n$ is closed, then so is $\{h'_j\}_0^n$. If $\{h_j\}_0^n$ is closed and definite, then so is $\{h'_j\}_0^n$.*

Proof. (a) The joint amplitude of $\{h'_j\}_0^n$ becomes

$$\hat{f}^v[h_0^{-1}(a_{i_0}) \cap \cdots \cap h_n^{-1}(a_{i_n})]$$
$$= \sum \{f^v(\omega): h_0(\omega) = a_{i_0}, \ldots, h_n(\omega) = a_{i_n}\}$$
$$= \sum \{e^{-i\alpha_v(\omega)} f(\omega): h_0(\omega) = a_{i_0}, \ldots, h_n(\omega) = a_{i_n}\}$$
$$= \exp i\left[-\sum_{j=o}^n v(a_{i_j})\right] \hat{f}[h_0^{-1}(a_{i_0}) \cap \cdots \cap h_n^{-1}(a_{i_n})]$$
$$= \prod_{j=0}^n e^{-iv(a_{i_j})} \psi_0(i_0) T(i_1, i_0) \cdots T(i_n, i_{n-1})$$
$$= e^{-iv(a_{i_0})} \psi_0(i_0) \prod_{j=1}^n e^{-iv(a_{i_j})} T(i_j, i_{j-1})$$
$$= e^{-iv(a_{i_0})} \psi_0(i_0) \prod_{j=1}^n [e^{-iv(Q)} T](i_j, i_{j-1}).$$

The proofs of (b) and (c) are straightforward. □

Notice that Theorem 7.6b gives a result that is analogous to a path integral. In this sense, we can interpret that result as a discrete path integral. If $\{h_j\}_0^n$ is a closed quantum n-chain, then its matrix T is unitary so it has the form $T = e^{-iH_0}$ where H_0 is a (not necessarily unique) self-adjoint matrix. It follows from Theorem 7.6a that $T' = e^{-iv(Q)} e^{-iH_0}$. We can interpret H_0 as a free Hamiltonian, Q as the position observable, and $v(Q)$ as the potential energy observable. The Hamiltonian H for the closed quantum n-chain $\{h'_j\}_0^n$, then satisfies $e^{-iH} = e^{-iv(Q)} e^{-iH_0}$. This is different than traditional quatnum mechanics where the Hamiltonian has the form $H = H_0 + v(Q)$ and we shall have more to say about this later.

Let (Ω, f) be a point closed quantum probability space. As in Section 4.8 we define the probability distribution $\mu_f(A) = \sum_{\omega \in A} |f(\omega)|^2$, $A \in P(\Omega)$. Notice that $\mu_f(\omega) = \bar{f}(\omega) f(\omega)$ so μ_f results from a random change in f although it is not a random phase change. If $\{h_j\}_0^n$ is an n-chain on (Ω, f) we call $\{h_j\}_0^n$ on (Ω, μ_f) the *parallel classical n-chain*. Of course, the latter is just a set of ordinary random variables on a classical probability space.

Lemma 7.7. *Let* $\{h_j\}_0^n$ *be a separating, definite, closed quantum n-chain with vector ψ_0 and matrix T on a point closed, quantum probability* space (Ω, f). Then the parallel classical n-chain is a stationary Markov n-chain with initial probability vector $\psi_0'(j) = |\psi_0(j)|^2$ and transition matrix $W(j,k) = |T(j,k)|^2$.

Proof. The result easily follows from

$$\mu_f\bigl[h_0^{-1}(a_{i_0}) \cap \cdots \cap h_n^{-1}(a_{i_n})\bigr] = \bigl|\hat{f}\bigl[h_0^{-1}(a_{i_0}) \cap \cdots \cap h_n^{-1}(a_{i_n})\bigr]\bigr|^2$$
$$= |\psi_0(i_0)|^2 |T(i_1, i_0)|^2 \cdots |T(i_n, i_{n-1})|^2. \quad \Box$$

Thus, a quantum n-chain satisfying the conditions of Lemma 7.7 has a corresponding classical Markov description of a physical system. Does this mean that the system behaves classically? No it does not; the parallel classical n-chain gives entirely different probabilities than the original amplitude n-chain. For example, in the notation of Lemma 7.7 we have

$$\bigl|\hat{f}[h_n^{-1}(a_k)]\bigr|^2 = \bigl|(T^n \psi_0)(k)\bigr|^2$$

and

$$\mu_f[h_n^{-1}(a_k)] = (W^n \psi_0')(k).$$

In particular, if ψ_0 is the standard basis element e_j, then

$$\bigl|\hat{f}[h_n^{-1}(a_k)]\bigr|^2 = |T^n(k,j)|^2$$

and

$$\mu_f[h_n^{-1}(a_k)] = W^n(k,j).$$

For 1-chains these agree. This indicates that quantum mechanical effects do not have time to occur during just one time step. More precisely, in one time step, paths are unique so there is no summation over various paths to cause interference. However, for 2-chains we have

$$\bigl|\hat{f}[h_2^{-1}(a_k)]\bigr|^2 = \Bigl|\sum_i T(k,i)T(i,j)\Bigr|^2$$

while

$$\mu_f[h_2^{-1}(a_k)] = \sum_i |T(k,i)T(i,j)|^2$$

which are quite different in general.

There is another important difference. As we have seen, a random phase transformation $f \mapsto f^\alpha$ makes a profound change in an amplitude n-chain. However, since $\mu_f = \mu_{f^\alpha}$, random phase transformations cannot be distinguished in the parallel classical n-chain.

We now present a lemma that is interesting in its own right and is needed to prove Theorem 7.9.

Lemma 7.8. *Let $\{h_j\}_0^n$ be a quantum n-chain with vector ψ_0 and matrix T. Let s be a positive integer and let F be a map from $S \times S \times \cdots \times S$ (s times) to $\mathbb{R}$. If $0 \le i_1 \le \cdots \le i_s \le n$, then*

$$\sum_{a_j}^{a_k} F(h_{i_1}(\omega), \ldots, h_{i_s}(\omega))f(\omega)$$
$$= \sum_{j_1,\ldots,j_s} F(a_{j_1}, \ldots, a_{j_s})\psi_0(j)T^{i_1}(j_1,j)T^{i_2-i_1}(j_2,j_1)\ldots T^{n-i_s}(k,j_s).$$

Proof. Using the notation $\sum' = \sum_{j_1,\ldots,j_s}$ we have

$$\sum_{a_j}^{a_k} F(h_{i_1}(\omega), \ldots, h_{i_s}(\omega))f(\omega)$$
$$= \sum \{F(h_{i_1}(\omega), \ldots, h_{i_s}(\omega))f(\omega) : h_0(\omega) = a_j, h_n(\omega) = a_k\}$$
$$= \sum{}' \sum \{F(h_{i_1}(\omega), \ldots, h_{i_s}(\omega))f(\omega) : h_0(\omega) = a_j, h_n(\omega)$$
$$= a_k, h_{i_m}(\omega) = a_{j_m}, m = 1, \ldots, s\}$$
$$= \sum{}' F(a_{j_1}, \ldots, a_{j_s}) \sum \{f(\omega) : h_0(\omega) = a_j, h_n(\omega)$$
$$= a_k, h_{i_m}(\omega) = a_{j_m}, m = 1, \ldots, s\}$$
$$= \sum{}' F(a_{j_1}, \ldots, a_{j_s})$$
$$\hat{f}[h_0^{-1}(a_j) \cap h_{i_1}^{-1}(a_{j_1}) \cap \cdots \cap h_{i_s}^{-1}(a_{j_s}) \cap h_n^{-1}(a_k)]$$
$$= \sum{}' F(a_{j_1}, \ldots, a_{j_s})\psi_0(j)T^{i_1}(j_1,j)T^{i_2-i_1}(j_2,j_1)\ldots T^{n-i_s}(k,j_s). \quad \square$$

Let $0 \le i_1, \cdots, i_s \le n$ be integers and place these integers in nondecreasing order. For $1 \le m \le s$, define i'_m to be the mth element of the resulting ordered sequence. The next theorem gives a perturbation expansion for $A^v(k,j)$. Unlike the path integral perturbation expansions [Feynman and Hibbs, 1965; Schulman, 1981], this one is rigorous and convergent for a large class of potentials. In the theorem we use the notation $\sum' = \sum_{j_1,\ldots,j_s}$, $\sum'' = \sum_{i_1,\ldots,i_s}$.

Theorem 7.9. *If $\{h_j\}_0^n$ is a quantum n-chain with vector ψ_0 and matrix T, and $v: S \to \mathbb{R}$, then*

$$A^v(k,j) = A(k,j) + A_1(k,j) + A_2(k,j) + \cdots$$

where

$$A_s(k,j) = \frac{(-i)^s}{s!} \sum{}'' \sum{}' v(a_{j_1})$$
$$\cdots v(a_{j_s})\psi_0(j)T^{i'_1}(j_1,j)T^{i'_2-i'_1}(j_2,j_1)\cdots T^{n-i'_s}(k,j_s).$$

Proof. Expanding the exponential in Theorem 7.6b gives

$$A^v(k,j)$$
$$= \sum_{a_j}^{a_k} \{\{-i\sum_m v(h_m(\omega)) + \cdots + \frac{(-i)^s}{s!}[\sum_m v(h_m(\omega))]^s + \cdots\}f(\omega)$$
$$= \sum_{a_j}^{a_k} f(\omega) - i\sum_{a_j}^{a_k}\sum_m v(h_m(\omega))f(\omega)$$
$$+ \cdots + \frac{(-i)^s}{s!}\sum_{a_j}^{a_k}[\sum_m v(h_m(\omega))]^s f(\omega) + \cdots$$
$$= A(k,j) + \cdots + \frac{(-i)^s}{s!}\sum{}'' \sum_{a_j}^{a_k} v(h_{i_1}(\omega))\cdots v(h_{i_s}(\omega))f(\omega) + \cdots.$$

Applying Lemma 7.8 gives

$$\sum_{a_j}^{a_k} v(h_{i_1}(\omega))\cdots v(h_{i_s}(\omega))f(\omega)$$
$$= \sum{}' v(a_{j_1})\cdots v(a_{j_s})\psi_0(j)T^{i'_1}(j_1,j)T^{i'_2-i'_1}(j_2,j_1)\cdots T^{n-i'_s}(k,j_s)$$

The result now follows. □

To get a better understanding of the perturbation expansion, let us write out $A_1(k,j)$ and $A_2(k,j)$ in detail:

$$A_1(k,j) = -i \sum_{m=0}^{n} \sum_{j_1=0}^{r} \psi_0(j) T^m(j_1,j) v(a_{j_1}) T^{n-m}(k,j_1)$$

$$A_2(k,j) = -\sum_{i_1<i_2} \sum_{j_1,j_2} \psi_0(j) T^{i_1}(j_1,j) v(a_{j_1}) T^{i_2-i_1}(j_2,j_1) v(a_{j_2}) T^{n-j_2}(k,j_2)$$

$$-\frac{1}{2} \sum_{i_1=0}^{n} \sum_{j_1=1}^{r} \psi_0(j) T^{i_1}(j_1,j) v(a_{j_1})^2 T^{n-i_1}(k,j_1).$$

We follow Feynman and Hibbs [1965] in interpreting these expansions. The amplitude matrix element $A^v(k,j)$ corresponds to a sum of alternative ways of going from a_j to a_k: not scattered at all $[A(k,j)]$, scattered once $[A_1(k,j)]$, scattered twice $[A_2(k,j)],\ldots$. Each of these alternatives is a sum of further alternatives. For example, $A_1(k,j)$ is a sum of terms

$$\psi_0(j) T^m(j_1,j) v(a_{j_1}) T^{n-m}(k,j_1).$$

The system moves "freely" from a_j to a_{j_1} in m steps, is scattered by the potential $v(a_{j_1})$, and then moves "freely" from a_{j_1} to a_k in $n-m$ steps. The term $A_2(k,j)$ has a similar interpretation except something new happens in the second summation of this term. Here the system moves "freely" from a_j to a_{j_1} in i_1 steps, is doubly scattered at a_{j_1}, and then moves "freely" from a_{j_1} to a_k. This second summand does not appear in the continuum valued case [Feynman and Hibbs, 1965].

We now consider quantum processes. While a quantum n-chain describes a quantum system as it evolves in n-steps, a quantum process describes a system as it evolves in any number of steps. To be precise, a *quantum process* is a sequence of chains $\{h_j^n\}_{j=0}^n$, $n=1,2,\ldots$ on quantum probability spaces (Ω_n, f_n) with the same value space $S = \{a_1,\ldots,a_r\} \subseteq \mathbb{R}$ such that the *product rule*

$$\hat{f}_n\big[(h_0^n)^{-1}(a_{i_0}) \cap \cdots \cap (h_n^n)^{-1}(a_{i_n})\big]$$
$$= \hat{f}_j\big[(h_0^j)^{-1}(a_{i_0}) \cap \cdots \cap (h_j^j)^{-1}(a_{i_j})\big]$$
$$\hat{f}_{n-j}\big[(h_0^{n-j})^{-1}(a_{i_j}) \cap \cdots \cap (h_{n-j}^{n-j})^{-1}(a_{i_n})\big]$$

holds for every n, $i_0,\ldots,i_n$, $1 \leq j \leq n$. Notice that the product rule is the same as Feynman's Rule 3 in Section 3.6. It states that the amplitude

RANDOM PHASES AND QUANTUM PROCESSES 261

that a system moves from a_{i_0} to a_{i_n} in n steps along a particular path is the amplitude that it moves from a_{i_0} to a_{i_j} along the first part of the path in j steps times the amplitude that it moves from a_{i_j} to a_{i_n} along the second part of the path in $n-j$ steps. The next theorem shows that a quantum process consists of a sequence of quantum chains. We denote the amplitude matrix corresponding to the chain $\{h_j^n\}_{j=0}^n$ by $A_n(k,j)$. (By convention $A_0(k,j) = \delta_{kj}$.)

Theorem 7.10. *If $\{h_j^n\}_{j=0}^n$, $n = 1, 2, \ldots$, is a quantum process, then the n-chain $\{h_j^n\}_{j=0}^n$ is a quantum n-chain with vector $\psi_0 \equiv 1$ and matrix $A_1(k,j)$. Moreover,*

$$A_n(k,j) = \sum_i A_m(i,j) A_{n-m}(k,i) \tag{7.16}$$

for every $j, k = 1, \ldots, r$, and $m = 0, \ldots, n$. Conversely, if $\{h_j^n\}_{j=0}^n$, $n = 1, 2, \ldots$, is a sequence of quantum chains with the same value space, vector $\psi_0 \equiv 1$, and matrix T, then this sequence is a quantum process.

Proof. We prove this using induction on n. For $n = 1$, we have

$$\hat{f}_1\left[(h_0^1)^{-1}(a_{i_0}) \cap (h_1^1)^{-1}(a_{i_1})\right] = A_1(i_1, i_0).$$

Assume the result holds for $n = k - 1$. Then

$$\hat{f}_k\left[(h_0^k)^{-1}(a_{i_0}) \cap \cdots \cap (h_k^k)^{-1}(a_{i_k})\right]$$
$$= \hat{f}_{k-1}\left[(h_0^{k-1})^{-1}(a_{i_0}) \cap \cdots \cap (h_{k-1}^{k-1})^{-1}(a_{i_{k-1}})\right]$$
$$\quad \hat{f}_1\left[(h_0^1)^{-1}(a_{i_{k-1}}) \cap (h_1^1)(a_{i_k})\right]$$
$$= A_1(i_1, i_0) \cdots A_1(i_{k-1}, i_{k-2}) A_1(i_k, i_{k-1}).$$

This completes the proof by induction. For the next part, we let A_1 be the matrix with components $A_1(k,j)$ and apply (7.6) to obtain

$$A_n(k,j) = A_1^n(k,j)$$
$$= \sum_i A_1^{n-m}(k,i) A_1^m(i,j)$$
$$= \sum_i A_m(i,j) A_{n-m}(k,i).$$

The converse follows from

$$\hat{f}_n\left[(h_0^n)^{-1}(a_{i_0}) \cap \cdots \cap (h_n^n)^{-1}(a_{i_n})\right]$$
$$= \left[T(i_1, i_0) \cdots T(i_j, i_{j-1})\right]\left[T(i_{j+1}, i_j) \cdots T(i_n, i_{n-1})\right]$$
$$= \hat{f}_j\left[(h_0^j)^{-1}(a_{i_0}) \cap \cdots \cap (h_j^j)^{-1}(a_{i_j})\right]$$
$$\hat{f}_{n-j}\left[(h_0^{n-j})^{-1}(a_{i_j}) \cap \cdots \cap (h_{n-j}^{n-j})^{-1}(a_{i_n})\right].$$

This completes the proof. □

Notice that (7.16) is the amplitude version of the Chapman-Kolmogorov condition which holds for classical Markov chains.

Let $\{h_j^n\}_{j=0}^n$, $n = 1, 2, \ldots$ be a quantum process with the value space S and let $v: S \to \mathbf{R}$. Let e_j, $j = 1, \ldots, r$, be the standard basis for $\mathbf{C}^r$. Applying Theorems 7.6 and 7.10, we have

$$A_n^v(k, j) = \sum_{a_j}^{a_k} \exp\left[-i \sum_m v(h_m^n(\omega))\right] f_n(\omega)$$
$$= e^{-iv(a_j)} \left\langle [e^{-iv(Q)} A_1]^n e_j, e_k \right\rangle.$$

The phase factor $e^{-iv(a_j)}$ is inessential since it does not affect the probability $|A_n^v(k, j)|^2$. For a closed system, we have $A_1 = e^{-iH_0}$, where H_0 is a "free" Hamiltonian, and then

$$\left\langle [e^{-iv(Q)} e^{-iH_0}]^n e_j, e_k \right\rangle = e^{iv(a_j)} A_n^v(k, j). \tag{7.17}$$

Let us now make a scale change in time. We replace $A_1 = e^{-iH_0}$ by $A_1(\tau) = e^{-i\tau H_0}$ and replace $e^{-iv(Q)}$ by $e^{-i\tau v(Q)}$, where $\tau \in \mathbf{R}$ is fixed. We denote the corresponding quantum process by $\{h_j^n(\tau)\}_{j=0}^n$, $n = 1, 2, \ldots$. This process is interpreted as giving the location of the system at the time steps $0, \tau, 2\tau, \ldots$. Equation (7.17) then becomes

$$\left\langle [e^{-i\tau v(Q)} e^{-i\tau H_0}]^n e_j, e_k \right\rangle = e^{i\tau v(a_j)} \sum_{a_j}^{a_k} \exp\left\{-i\tau \sum_m v\left[h_m^n(\tau)(\omega)\right]\right\} f_n(\omega).$$
(7.18)

For continuous time t, the perturbed Hamiltonian is taken to be $H_0 + v(Q)$ and the evolution is given by

$$\exp\left[-it(H_0 + v(Q))\right].$$

The next result gives a Feynman formula for the finite-dimensional case (see Section 2.6).

Theorem 7.11. *For every $t \in \mathbb{R}$ we have*

$$\left\langle e^{-it(H_0+v(Q))} e_j, e_k \right\rangle = \lim_{n\to\infty} \sum_{a_j}^{a_k} \exp\left\{-i\frac{t}{n}\sum_m v\left[h_m^n\left(\frac{t}{n}\right)(\omega)\right]\right\} f_n(\omega).$$

Proof. Applying (7.18) with $\tau = t/n$ and Trotter's formula [Dunford and Schwartz, 1958] gives

$$\lim_{n\to\infty} \sum_{a_j}^{a_k} \exp\left\{-i\frac{t}{n}\sum_m v\left[h_m^n\left(\frac{t}{n}\right)(\omega)\right]\right\} f_n(\omega)$$

$$= \lim_{n\to\infty} e^{-itv(a_j)/n} \left\langle [e^{-itv(Q)/n} e^{-itH_0/n}]^n e_j, e_k \right\rangle$$

$$= \left\langle e^{-it(H_0+v(Q))} e_j, e_k \right\rangle. \quad \square$$

7.3 Quantum Graphic Dynamics

In the previous two sections, a main role was played by a quantum n-chain or quantum process and probabilities were computed from the corresponding amplitude matrices. Moreover, the possible n-paths were determined by the quantum n-chain. In the sequel, these roles will be reversed and the n-paths and transition matrices will dominate. The n-chains and quantum process can then be easily retrieved. The method we shall use is similar to the one we began at the end of Section 7.1. This method turns out to be more basic and convenient for constructing various discrete quantum graphic dynamics models.

As its name implies, quantum graphic dynamics (QGD) consists of two main ingredients, a graph (or multigraph) and a quantum dynamics. The graph (or multigraph) is interpreted as a generalized discrete phase space in which the vertices represent discrete positions that a particle can occupy and the edges represent discrete directions that a particle can propagate. The quantum dynamics is induced by a (one-step) transition amplitude that generates a quantum random walk on the graph.

Our development of QGD has two purposes. First, QGD can be used to describe the motion of a particle in discrete spacetime. In this case, the graph represents an actual discrete phase space and the discrete time is given by the time steps of the random walk. Second, QGD might be

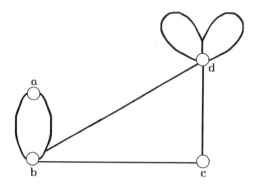

Figure 7.1 (A Multigraph)

useful for describing the internal dynamics of "elementary particles". In this case, the vertices represent quark-like constituents of a particle and edges represent interaction paths for gluons that are emitted or absorbed by the vertices. A vertex can emit or absorb a gluon at each time step of the random walk. It is too early to tell whether QGD has any practical importance. Nevertheless, it illustrates the type of rigorous mathematical models that can be constructed using quantum probability theory.

We begin with some basic definitions. A *multigraph* is a pair $G = (V, E)$ where V is a nonempty set and E is a collection of one and two element subsets of V. The elements of V are called *vertices*, the two element sets in E are called *edges*, and the one element sets in E are called *loops*. The sets in E may be repeated so there may be more than one loop containing a vertex or more than one edge containing a pair of vertices. The *degree* of a vertex is the number of edges containing (or *incident* to) it. As far as the degree is concerned, a loop is considered to be two edges incident to a vertex. Thus, each loop incident to a vertex adds two to its degree. Figure 7.1 illustrates a multigraph $G = (V, E)$ where $V = \{a, b, c, d\}$ and

$$E = \{\{a,b\}_1, \{a,b\}_2, \{b,c\}, \{b,d\}, \{c,d\}, \{d\}_1, \{d\}_2\}.$$

The degrees of the vertices are, $\deg a = 2$, $\deg b = 4$, $\deg c = 2$, $\deg d = 6$.

Suppose that $V = \{v_j : j \in J\}$ and that the edges and loops containing v_j are denoted by e_{jk}, $k \in K(j)$. If $\{v_r, v_s\} \in E$ we write $v_r \perp v_s$ and say that v_r, v_s are *adjacent* (r may equal s). The *phase space* on G is

$$S = \{(v_j, e_{jk}) : j \in J, k \in K(j)\} \subseteq V \times E.$$

If $q = (v_j, e_{jk})$, $q' = (v_{j'}, e_{j'k'}) \in S$ and $v_j \perp v_{j'}$, we write $q \perp q'$. For $n \in \mathbb{N}$, an *n-path* is a sequence of not necessarily distinct elements $q_0, \ldots, q_n \in S$ with $q_j \perp q_{j+1}$, $j = 0, \ldots, n-1$. We call q_0, q_n the *initial* and *final* elements of the n-path, respectively. Denote the set of n-paths with initial element q_0 and final element q by $P_n(q_0, q)$.

We now interpret V as a set of discrete positions for a particle and adjacent vertices correspond to "nearest neighbor" positions. An edge (or loop) corresponds to a direction that a particle can move so it represents a direction of momentum. If $e = \{v_1, v_2\} \in E$, then a particle located at v_1 can move along e to v_2 in one time step. An n-path $p \in P_n(q_0, q)$ is a possible trajectory for a particle propagating in a discrete phase space from q_0 to q in n time steps.

A function $A: S \times S \to \mathbb{C}$ is a *one-step transition amplitude* if $A(q, q') = 0$ if $q \not\perp q'$ and for every $q_1, q_2 \in S$ we have

$$\sum_q A(q_1, q)\bar{A}(q_2, q) = \sum_q A(q, q_1)\bar{A}(q, q_2) = \delta_{q_1 q_2}. \tag{7.19}$$

Equation (7.19) is a type of unitarity condition. Notice the difference between a one-step transition amplitude and the (instantaneous) transition amplitude discussed in Sections 4.4 and 4.5. In the present case, the instantaneous transition amplitude A_0 is the trivial one $A_0(q_1, q_2) = \delta_{q_1 q_2}$. The one step transition amplitude corresponds to the matrix A_1 discussed in Section 7.2 (except that we have reversed the arguments for convenience).

If A is a one-step transition amplitude and

$$p = (q_0, \ldots, q_n) \in P_n(q_0, q_n)$$

the *amplitude* of p is

$$A(p) = A(q_0, q_1) A(q_1, q_2) \cdots A(q_{n-1}, q_n). \tag{7.20}$$

Notice that (7.20) is the Markov property or product rule discussed in Sections 7.1 and 7.2. For $q_0, q \in S$, the *n-step transition amplitude* from q_0 to q is

$$A_n(q_0, q) = \sum \{A(p): p \in P_n(q_0, q)\} \tag{7.21}$$

and by convention $A_0(q_0, q) = \delta_{q_0 q}$. Of course, $A_1(q_0, q) = A(q_0, q)$ so the terminology is consistent. We see from (7.21) that A_n corresponds to the amplitude matrix discussed in Section 7.1. The *n-step transition probability* from q_0 to q is $\mu_n(q_0, q) = |A_n(q_0, q)|^2$.

Let us examine how the present framework fits in with the quantum probability space formalism (Section 4.8). Let Ω be the set of all paths; that is,

$$\Omega = \{p \in P_n(q_0, q) : q_0, q \in S, n \in \mathbb{N}\}.$$

Then (Ω, A) becomes a quantum probability space, and we can consider the paths as sample points. As we shall see, for any fixed $q_0, q \in S, n \in \mathbb{N}$, the set of sample points $P_n(q_0, q)$ is an outcome. In fact, $P_n(q_0, q)$ is an outcome of the operation $\{P_n(q_0, q) : q \in S\}$. Moreover,

$$\hat{A}[P_n(q_0, q)] = A_n(q_0, q)$$

and $\hat{A}$ becomes an amplitude function on the corresponding quasimanual $\mathcal{A}$. The state μ corresponding to $\hat{A}$ satisfies

$$\mu[P_n(q_0, q)] = \mu_n(q_0, q).$$

If we think of n as n units of time, then each discrete time period and each $q_0 \in S$ generate an operation

$$E(n, q_0) = \{P_n(q_0, q) : q \in S\}.$$

Roughly speaking, we can think of $E(n, q_0)$ as the operation which records the position-momentum of a particle at time n if its initial position-momentum is q_0. Although the $E(n, q_0)$ are important operations in $\mathcal{A}$, we shall also find others.

We now physically motivate our definition of a one-step transition amplitude. First, we interpret $A(q, q')$ as the probability amplitude that a particle moves from q to q' in one time step and we interpret $\bar{A}(q, q')$ as the probability amplitude that a particle moves from q' to q in minus one time steps. Alternatively, if a particle is at q' at time n, then $\bar{A}(q, q')$ is the probability amplitude that it was at q at time $n-1$. Then $A(q, q') = 0$ if $q \not\perp q$ means that a particle can only move to adjacent vertices in one time step. Moreover, we then interpret $A(q_1, q)\bar{A}(q_2, q)$ as the probability amplitude that a particle moves from q_1 to q_2 via q in zero time steps. It follows that $\sum_q A(q_1, q)\bar{A}(q_2, q)$ is the total probability amplitude that a particle moves from q_1 to q_2 in zero time steps. It is clear that this should equal $\delta_{q_1 q_2}$. Similar reasoning applies to the other equality in (7.19).

The following result shows that probability is conserved and that an amplitude Chapman-Kolmogorov equation holds.

Theorem 7.12. (a) $\sum_q |A_n(q_0,q)|^2 = \sum_q |A_n(q,q_0)|^2 = 1$, $n \in \mathbb{N}$.
(b) If $m, n \in \mathbb{N}$, $m \leq n$, then

$$A_n(q_0, q) = \sum_{q'} A_m(q_0, q') A_{n-m}(q', q). \tag{7.22}$$

Proof. It is clear that (7.22) holds if $m = n$, so assume that $m < n$. We prove (a) and (b) by induction on m. For $n = 1$, (a) holds by (7.19). For $n = 2$, we have

$$A_2(q_0, q) = \sum \{A(p) : p \in P_2(q_0, q)\}$$
$$= \sum_{q'} A_1(q_0, q') A_1(q', q)$$

so (b) holds for $n = 2$. Moreover, the sum converges absolutely since by Schwarz's inequality

$$\sum_{q'} |A(q_0, q') A(q', q)| \leq \left[\sum_{q'} |A(q_0, q')|^2 \sum_{q'} |A(q', q)|^2 \right]^{\frac{1}{2}} = 1.$$

It follows from (7.19) that

$$\sum_q |A_2(q_0, q)|^2 = \sum_q \sum_{q'} A(q_0, q') A(q', q) \sum_{q''} \bar{A}(q_0, q'') \bar{A}(q'', q)$$
$$= \sum_{q',q''} A(q_0, q') \bar{A}(q_0, q'') \sum_q A(q', q) \bar{A}(q'', q)$$
$$= \sum_{q',q''} A(q_0, q') \bar{A}(q_0, q'') \delta_{q'q''} = 1.$$

Hence, (a) holds for $n = 2$. Now suppose that (a) and (b) hold for $n = 2, 3, \ldots, r$ and let $n = r + 1$, $m < n$. Then

$$A_{r+1}(q_0, q) = \sum \{A(p) : p \in P_{r+1}(q_0, q)\}$$
$$= \sum_{q'} \sum \{A(p') A(q', q) : p' \in P_r(q_0, q')\}$$
$$= \sum_{q'} A(q', q) \sum \{A(p') : p' \in P_r(q_0, q')\}$$
$$= \sum_{q'} A_1(q', q) A_r(q_0, q').$$

Hence, (b) holds for $m = r$. Moreover, using the induction hypothesis and proceeding as before, the sum converges absolutely and

$$\sum_q |A_{r+1}(q_0, q)|^2 = 1.$$

Hence, (a) holds for $n = r + 1$. This proves (a) by induction. Finally, if $m < r$ we have by the induction hypothesis:

$$\begin{aligned} A_{r+1}(q_0, q) &= \sum_{q'} A_1(q', q) A_r(q_0, q') \\ &= \sum_{q'} A_1(q', q) \sum_{q''} A_m(q_0, q'') A_{r-m}(q'', q') \\ &= \sum_{q''} A_m(q_0, q'') \sum_{q'} A_{r-m}(q'', q') A_1(q', q) \\ &= \sum_{q''} A_m(q_0, q'') A_{r+1-m}(q'', q). \end{aligned}$$

Thus, (b) holds for $n = r + 1$, and the proof by induction is complete. □

Define the Hilbert space of functions

$$\ell^2(S) = \left\{ f \colon S \to \mathbb{C} \colon \sum_q |f(q)|^2 < \infty \right\}$$

with inner product $\langle f, g \rangle = \sum f(q) \bar{g}(q)$. An orthonormal basis for $\ell^2(S)$ is the set $\{\delta_q : q \in S\}$ where $\delta_q(q') = \delta_{qq'}$. Notice that $f(q) = \langle f, \delta_q \rangle$ for every $f \in \ell^2(S)$, $q \in S$. It follows from (7.19) that $A(q_0, \cdot), A(\cdot, q_0) \in \ell^2(S)$ and from Theorem 7.12a that $A_n(q_0, \cdot), A_n(\cdot, q_0) \in \ell^2(S)$. Using the usual quantum terminology, we call a unit vector $\psi \in \ell^2(S)$ a *state* and if ψ_1, ψ_2 are states, we call $\langle \psi_1, \psi_2 \rangle$ the *transition amplitude* from ψ_2 to ψ_1. Moreover, $|\langle \psi_1, \psi_2 \rangle|^2$ is the *transition probability* from ψ_2 to ψ_1. Define the linear operators T, U on $\ell^2(S)$ by

$$(Tf)(q) = \sum_{q'} A(q, q') f(q'), \quad (Uf)(q) = \sum_{q'} \bar{A}(q', q) f(q'). \tag{7.23}$$

We call T and U *propagators* for A. It is clear that $A(q_0, q) = \langle T\delta_q, \delta_{q_0} \rangle$. It will follow from the next theorem that $T\delta_q$ is a state and hence $A(q_0, q)$ can be interpreted as the transition amplitude from δ_{q_0} to $T\delta_q$.

QUANTUM GRAPHIC DYNAMICS

Theorem 7.13. *The operators T and U are unitary and $U = T^*$.*

Proof. To show that T is unitary, we have for $f, g \in \ell^2(S)$

$$\langle Tf, Tg \rangle = \sum_q (Tf)(q)\overline{(Tg)}(q)$$

$$= \sum_{q,q'} A(q,q')f(q') \sum_{q''} \bar{A}(q,q'')\bar{g}(q'')$$

$$= \sum_{q',q''} f(q')\bar{g}(q'') \sum_q A(q,q')\bar{A}(q,q'')$$

$$= \sum_{q',q''} f(q')\bar{g}(q'')\delta_{q'q''} = \langle f, g \rangle.$$

Hence, $T^*T = I$ and similarly $U^*U = I$. To show that $U = T^*$, for $f, g \in \ell^2(S)$ we have

$$\langle f, Tg \rangle = \sum_q f(q)\overline{Tg}(q)$$

$$= \sum_q f(q) \sum_{q'} \bar{A}(q,q')\bar{g}(q')$$

$$= \sum_{q'} \bar{g}(q') \sum_q \bar{A}(q,q')f(q)$$

$$= \sum_{q'} \bar{g}(q')(Uf)(q') = \langle Uf, g \rangle.$$

It follows that T and U are unitary. $\square$

We have seen that if A is a one-step transition amplitude, then its propagator T is unitary. We now show that the converse holds. Let $A \colon S \times S \to \mathbb{C}$ be a function satisfying $A(q, q') = 0$ if $q \not\sim q'$. Define the linear operator T on $\ell^2(S)$ by (7.23) and suppose that T is unitary. Then

$$\sum_{q'} A(q_1, q')\bar{A}(q_2, q') = \sum_{q'} \langle T\delta_{q'}, \delta_{q_1} \rangle \overline{\langle T\delta_{q'}, \delta_{q_2} \rangle}$$

$$= \sum_{q'} \langle T^*\delta_{q_2}, \delta_{q'} \rangle \langle \delta_{q'}, T^*\delta_{q_1} \rangle$$

$$= \langle T^*\delta_{q_2}, T^*\delta_{q_1} \rangle$$

$$= \langle \delta_{q_2}, \delta_{q_1} \rangle = \delta_{q_1 q_2}.$$

In a similar way, the other equality in (7.19) holds so A is a one-step transition amplitude.

For $f \in \ell^2(S)$ we define the *n-step amplitude expectation*
$$A_n(f \mid q_0) = \sum_q f(q) A_n(q_0, q).$$

Note that $A_0(f \mid q_0) = f(q_0)$ and $A_1(f \mid q_0) = Tf(q_0)$. Moreover, $A_n(\cdot \mid q_0)$ is a bounded linear functional on $\ell^2(S)$ and $A_n(\delta_q \mid q_0) = A_n(q_0, q)$. For $f: S \to \mathbf{R}$ we define the *n-step expectation*
$$E_n(f \mid q_0) = \sum_q f(q) \mu_n(q_0, q)$$
provided that the sum converges.

Theorem 7.14. *For $f \in \ell^2(S)$ we have $A_n(f \mid q_0) = T^n f(q_0)$.*

Proof. Proceeding by induction on n, the result holds for $n = 1$. Suppose the result holds for n and apply Theorem 7.12b to obtain
$$\begin{aligned}A_{n+1}(f \mid q_0) &= \sum_q f(q) \sum_{q'} A_1(q_0, q') A_n(q', q) \\ &= \sum_{q'} A(q_0, q') \sum_q f(q) A_n(q', q) \\ &= \sum_{q'} A(q_0, q') A_n(f \mid q') \\ &= T A_n(f \mid q_0) = T^{n+1} f(q_0). \quad \square\end{aligned}$$

Corollary 7.15. $A_n(q_0, q) = \langle \delta_q, U^n \delta_{q_0} \rangle$.

Proof. Applying Theorem 7.14 gives
$$\begin{aligned}A_n(q_0, q) &= A_n(\delta_q \mid q_0) \\ &= T^n \delta_q(q_0) = \langle T^n \delta_q, \delta_{q_0} \rangle \\ &= \langle \delta_q, (T^n)^* \delta_{q_0} \rangle = \langle \delta_q, U^n \delta_{q_0} \rangle. \quad \square\end{aligned}$$

Let P_q be the one-dimensional projection onto the unit vector $\delta_q \in \ell^2(S)$. For $f: S \to \mathbf{R}$ define the self-adjoint operator B_f on $\ell^2(S)$ by $B_f g(q) = f(q) g(q)$. Then B_f has the representation $B_f = \sum_q f(q) P_q$. The next corollary gives the usual quantum probabilities and expectations.

Corollary 7.16. (a) $\mu_n(q_0, q) = \langle P_q U^n \delta_{q_0}, U^n \delta_{q_0} \rangle$. (b) $E_n(f \mid q_0) = \langle B_f U^n \delta_{q_0}, U^n \delta_{q_0} \rangle$.

Proof. (a) Applying Corollary 7.15 gives

$$\mu_n(q_0, q) = |\langle \delta_q, U^n \delta_{q_0} \rangle|^2 = \langle P_q U^n \delta_{q_0}, U^n \delta_{q_0} \rangle.$$

(b) By (a) we have

$$\begin{aligned} E_n(f \mid q_0) &= \sum f(q) \mu_n(q_0, q) \\ &= \sum f(q) \langle P_q U^n \delta_{q_0}, U^n \delta_{q_0} \rangle \\ &= \left\langle \sum f(q) P_q U^n \delta_{q_0}, U^n \delta_{q_0} \right\rangle \\ &= \langle B_f U^n \delta_{q_0}, U^n \delta_{q_0} \rangle. \quad \square \end{aligned}$$

We now see that there is a close correspondence between the present formalism and traditional quantum mechnics. In particular, if we correspond the state δ_q in the Hilbert space $\ell^2(S)$ to the point $q \in S$, then the previous two corollaries show that the usual transition amplitudes, probabilities, and expectations are reproduced. Moreover, the dynamics induced by the n-step transition amplitude A_n is reproduced by the discrete one-parameter unitary group U^n. But which viewpoint is more basic? From the axiomatic standpoint, we have previously stressed that A is more basic than U. This is also frequently true in practice. If we know the Hamiltonian H for the system, then we easily construct $U = e^{-iH}$ and can then find A. However, in practice, H is frequently unknown, and may not be physically derivable. For example, in scattering experiments involving elementary particles, the underlying forces and hence the Hamiltonian are unknown, but the transition amplitudes can be found experimentally.

Our previous discussion notwithstanding, we can carry the Hilbert space analogy further. Suppose that the one-step transition amplitude A corresponds to the free evolution of a particle. Now suppose that a function $v: S \to \mathbf{R}$ represents a potential energy. We then define $A^v: S \times S \in \mathbf{R}$ by $A^v(q, q') = e^{-iv(q)} A(q, q')$. Then clearly, A^v is a new one-step transition amplitude that we can regard as corresponding to a particle evolving under the influence of the potential v. Notice that if $v \equiv 0$, then A^v reduces to the free amplitude A. If

$$p = (q_0, \ldots, q_n) \in P_n(q_0, q)$$

then we easily obtain the Feynman type formula

$$A^v(p) = A(p)\exp\Big[-i\sum_{j=0}^{n-1} v(q_j)\Big].$$

We thus obtain the following perturbed n-step transition amplitude

$$A_n^v(q_0, q) = \sum\{A(p)\exp\Big[-i\sum_{j=0}^{n-1} v(q_j)\Big] : p = (q_0, \ldots, q_n) \in P_n(q_0, q)\}.$$

Note that this agrees with our expression in Section 7.2. Now define the unitary operation e^{-iV} by

$$e^{-iV} f(q) = e^{-iv(q)} f(q).$$

The next corollary shows that the propagator for A^v is the unitary operator $T_v = e^{-iV}T$ and that an expression analogous to (7.17) holds.

Corollary 7.17. (a) If T_v is the propagator for A^v, then $T_v = e^{-iV}T$.
(b) $A_n^v(q_0, q) = \langle \delta_q, (Ue^{iV})^n \delta_{q_0} \rangle$.

Proof. (a) By definition of the propagator for A^v, we have

$$(Tf)(q) = \sum_{q'} A^v(q, q')f(q')$$

$$= \sum_{q'} e^{-iv(q)} A(q, q')f(q')$$

$$= e^{-iv(q)} \sum_{q'} A(q, q')f(q') = (e^{-iV}Tf)(q).$$

(b) Applying Corollary 7.15 and (a) we have

$$A_n^v(q_0, q) = \langle T_n^v \delta_q, \delta_{q_0} \rangle$$
$$= \langle (e^{-iV}T)^n \delta_q, \delta_{q_0} \rangle$$
$$= \langle \delta_q, (e^{-iV}T)^{*n} \delta_{q_0} \rangle$$
$$= \langle \delta_q, (Ue^{iV})^n \delta_{q_0} \rangle. \quad \square$$

QUANTUM GRAPHIC DYNAMICS

We now derive a formula for $A_n(q_0, q)$ that will be useful in the sequel. Suppose the spectral representation for U is given by $U = \int_0^{2\pi} e^{i\theta} P(d\theta)$. It follows from Corollary 7.15 that

$$A_n(q_0, q) = \int_0^{2\pi} e^{in\theta} \langle \delta_q, P(d\theta)\delta_{q_0} \rangle.$$

Hence, there exist measures $\mu(q_0, q)$ on $[0, 2\pi]$ such that

$$A_n(q_0, q) = \int_0^{2\pi} e^{in\theta} \mu(q_0, q)(d\theta) \tag{7.24}$$

for all $n \in \mathbb{N}$, $q_0, q \in S$. In the case that U has a complete set of eigenvectors ϕ_j with corresponding eigenvalues $e^{i\lambda_j}$, we can obtain a discrete version of (7.24). In this case Corollary 7.15 reduces to

$$\begin{aligned} A_n(q_0, q) &= \langle \delta_q, U^n \delta_{q_0} \rangle \\ &= \left\langle \sum_j \langle \delta_q, \phi_j \rangle \phi_j, U^n \sum_{j'} \langle \delta_{q_0}, \phi_{j'} \rangle \phi_{j'} \right\rangle \\ &= \sum_{j,j'} \overline{\phi_j}(q) \phi_{j'}(q_0) \langle \phi_j, e^{in\lambda_{j'}} \phi_{j'} \rangle \\ &= \sum_j e^{-in\lambda_j} \phi_j(q_0) \overline{\phi_j}(q). \end{aligned} \tag{7.25}$$

In this theory, the dynamics of the system is decscribed by the n-step transition amplitude $A_n(q_0, q)$. Given the one-step transition amplitude $A(q_0, q)$ the main problem is to compute $A_n(q_0, q)$ for all $n \in \mathbb{N}$. One approach to this problem is to apply a formula such as (7.25). Another approach is to find and solve a difference equation for $A_n(q_0, q)$. To this end, fix $q_0 \in S$ and let $\psi_n(q) = A_n(q_0, q)$ be the *wave function*. We have already noted that ψ_n is a unit vector in $\ell^2(S)$. Applying Theorem 7.12b we obtain

$$\psi_{n+1}(q) = \sum_{q'} A_n(q_0, q') A(q', q) = \sum_{q'} A(q', q) \psi_n(q'). \tag{7.26}$$

We shall later show that the difference equation (7.26) is a discrete analog of the Dirac equation, which in turn is the relativistic version of Schrödinger's equation for the electron. We call (7.26) the *discrete wave equation*.

7.4 Discrete Spacetime Models

In this section we illustrate the theory developed in Section 7.3 with two discrete spacetime models. We first consider cubic lattices of arbitrary finite dimension. Let $V = Z^m, Z = \{0, \pm 1, \pm 2, \ldots\}$ be the cubic lattice of dimension m. For $v_1, v_2 \in V$, define $v_1 \perp v_2$ if one of their components differ by one and the others are equal. Thus $v_1 \perp v_2$ if and only if v_1, v_2 are nearest neighbors and the edges are represented by lines joining nearest neighbors. Define the following unit vectors in $\mathbf{R}^m$

$$\hat{k}_1 = (1, 0, \ldots, 0), \quad \hat{k}_3 = (0, 1, 0, \ldots, 0), \ldots, \quad \hat{k}_{2m-1} = (0, 0, \ldots, 0, 1)$$
$$\hat{k}_{2n} = -\hat{k}_{2n-1}, \quad n = 1, \ldots, m.$$

We can then represent the phase space S as

$$S = \{(j, k) : j \in V, k \in \{\hat{k}_1, \ldots, \hat{k}_{2m}\}\}.$$

One can now consider various one-step transition amplitudes on S. In this section, we shall consider the simplest nontrivial one that seems to have some physical justification. Define the function $A : S \times S \to \mathbb{C}$ as follows

$$A((j, k), (j + k, k')) = \begin{cases} a & \text{if } k' = k \\ be^{i\theta} & \text{if } k' = -k \\ ce^{i\phi} & \text{if } k' \neq \pm k \end{cases} \quad (7.27)$$

where $a, b, c > 0$, $\theta, \phi \in [0, 2\pi)$ and $A((j, k), (j', k')) = 0$ otherwise. Physical motivations can be given for taking A to have this form. First it is clear that A is translation invariant and that $A(q, q') = 0$ if $q \not\perp q'$. Second, since probabilities are independent of a constant multiplicative phase factor, we can take one of the values of A to be positive; hence, we can take $a > 0$. Of course, as it now stands, A need not be a one-step transition amplitude since the five parameters are unspecified except for the given constraints. We shall later characterize a, b, c, θ, ϕ so that A satisfies (7.19) and becomes a one-step transition amplitude. This one-step transition amplitude will then describe the following motion. Suppose a particle is initially at the phase space point (j, k). We can think of the particle as being at the lattice point $j \in V$ and moving in the direction k. The particle then moves in one time step to the lattice point $j + k$. It continues its motion in the "forward direction" k with amplitude a, in the "backward direction" $-k$ with amplitude $be^{i\theta}$, and in one of the "orthogonal

DISCRETE SPACETIME MODELS   275

directions" $k' \neq \pm k$ with amplitude $ce^{i\phi}$. Since the forward and backward directions are distinguished directions for the particle they each have their own amplitude. Since the particle cannot physically differentiate among the various orthogonal directions, they all have a common amplitude.

We now find conditions on a, b, c, θ, ϕ which characterize when A is a one-step transition amplitude.

Theorem 7.18. *The function A given by (7.27) is a one-step transition amplitude if and only if the following three conditions hold.*

(a) $a^2 + b^2 + 2(m-1)c^2 = 1$
(b) $ab\cos\theta + (m-1)c^2 = 0$
(c) $a\cos\phi + b\cos(\theta - \phi) + (m-2)c = 0$, *if* $m \neq 1$.

Proof.   For this proof we reorder the $\hat{k}_i$'s as follows:

$$\hat{k}_1 = (1, 0, \ldots, 0), \ldots, \quad \hat{k}_m = (0, 0, \ldots, 0, 1)$$

$$\hat{k}_{m+j} = -\hat{k}_j, \quad j = 1, \ldots, m.$$

Let R_0 be the permutation on the set $\{\hat{k}_1, \ldots, \hat{k}_{2m}\}$ defined by $R_0\hat{k}_j = \hat{k}_{j+1}$ (mod $2m$). For $f \in \ell^2(S)$, the propagator T satisfies

$$(Tf)(j,k) = \sum A((j,k),(j',k'))f(j',k')$$

$$= af(j+k,k) + be^{i\theta}f(j+k, R_0^m k)$$

$$+ ce^{i\phi}\sum\{f(j+k, R_0^r k): r = 1, \ldots, 2m-1, r \neq m\}.$$

Let S and R be the linear operators on $\ell^2(S)$ defined by $Sf(j,k) = f(j+k, k)$ and $Rf(j,k) = f(j, R_0 k)$. It is easy to check that the "translation operator" S is unitary and $S^*f(j,k) = f(j-k, k)$. Moreover, the "rotation operator" R is unitary and $R^* = R^{2m-1}$. We then have

$$T = \left[aI + be^{i\theta}R^m + ce^{i\phi}\sum\{R^r : r = 1, \ldots, 2m-1, r \neq m\}\right]S.$$

Letting L be the operator in the square brackets, we have $T = LS$. Since S is unitary, T is unitary if and only if L is unitary. Now

$$LL^* = \left(aI + be^{i\theta}R^m + ce^{i\phi}\sum_{r \neq m} R^r\right)\left(aI + be^{-i\theta}R^m + ce^{-i\phi}\sum_{r \neq m} R^r\right)$$

$$= [a^2 + b^2 + 2(m-1)c^2]I + 2[ab\cos\theta + (m-1)c^2]R^m$$

$$+ 2c[a\cos\phi + b\cos(\theta - \phi) + (m-2)c]\sum_{r \neq m} R^r.$$

Thus, T is unitary if and only if the three conditions hold. We have already shown that T is unitary if and only if A is a one-step transition amplitude. □

Corollary 7.19. *If A is a one-step transition amplitude, then*

$$\left|a + be^{i\theta} + 2(m-1)ce^{i\phi}\right|^2 = 1.$$

Proof. Applying (a), (b), and (c) of Theorem 7.18 we obtain

$$\left|a + be^{i\theta} + 2(m-1)ce^{i\phi}\right|^2$$
$$= a^2 + b^2 + 4(m-1)^2c^2 + 2ab\cos\theta$$
$$\quad + 4(m-1)c[a\cos\phi + b\cos(\theta - \phi)]$$
$$= a^2 + b^2 + 4(m-1)^2c^2 - 2(m-1)c^2 - 4(m-1)(m-2)c^2$$
$$= a^2 + b^2 + 2(m-1)c^2 = 1. \quad \square$$

We interpret Corollary 7.19 as follows. Let B be the outcome that the particle is at the lattice point $j + k \in V$ given that it was at (j, k) the previous time step. Then B occurs with certainty. Notice that this is different from the result given by (a) which says that the sum of the probabilities that the particle reaches $j + k \in V$ from (j, k) in one time step is unity. More generally, we saw in Theorem 7.12a that $\sum_q |A_n(q_0, q)|^2 = 1$. However, when A is given by (7.27) we also have the following.

Corollary 7.20. *If A is a one-step transition amplitude, then*

$$\left|\sum_q A_n(q_0, q)\right|^2 = 1.$$

Proof. The expression $\sum_q A_n(q_0, q)$ is the sum of the amplitudes of all n-paths with initial point q_0. It easily follows by induction on n that this sum equals
$$\left[a + be^{i\theta} + 2(m-1)ce^{i\phi}\right]^n.$$
The result now follows from Corollary 7.19. □

It follows from Theorem 7.18 that if A is a one-step transition amplitude, then only two of the five parameters are left unspecified. These two

DISCRETE SPACETIME MODELS   277

remaining parameters are physical constants corresponding to properties of the particle under consideration.

We now derive some other characterizations for A to be a one-step transition amplitude and these will be useful later. Let B, C be the following 2×2 matrices

$$B = \begin{bmatrix} a & be^{i\theta} \\ be^{i\theta} & a \end{bmatrix}, \quad C = ce^{i\phi}\begin{bmatrix} 1 & 1 \\ 1 & 1 \end{bmatrix}$$

and let M be the $2m \times 2m$ matrix consisting of B's on the main diagonal and C's elsewhere. It is easy to check that $\psi_1 = (1,1), \psi_2 = (1,-1)$ are eigenvectors of B and C with respective eigenvalues $a + be^{i\theta}, a - be^{i\theta}$, and $2ce^{i\phi}, 0$. It follows from a direct computation that $a - be^{i\theta}$ is an eigenvalue of M with multiplicity m and the following are corresponding orthonormal eigenvectors

$$\frac{(\psi_2, 0, \ldots, 0)}{\sqrt{2}}, \ldots, \frac{(0, 0, \ldots, 0, \psi_2)}{\sqrt{2}}. \tag{7.28}$$

Moreover, $a + be^{i\theta} + 2(m-1)ce^{i\phi}$ is a nondegenerate eigenvalue of M with corresponding normalized eigenvector

$$\frac{(\psi_1, \psi_1, \ldots, \psi_1)}{\sqrt{2m}}. \tag{7.29}$$

Finally, $a + be^{i\theta} - 2ce^{i\phi}$ is an eigenvalue with multiplicity $m - 1$ and the following are corresponding orthonormal eigenvectors

$$\begin{aligned} &\frac{(\psi_1, -\psi_1, 0, \ldots, 0)}{2} \\ &\frac{(\psi_1, \psi_1, -2\psi_1, 0, \ldots, 0)}{\sqrt{12}} \\ &\quad \vdots \\ &\frac{(\psi_1, \psi_1, \ldots, \psi_1, -(m-1)\psi_1)}{(2(m-1)m)^{\frac{1}{2}}}. \end{aligned} \tag{7.30}$$

Theorem 7.21. *The following statements are equivalent.*
 (a) *A is a one-step transition amplitude.*
 (b) *$|a - be^{i\theta}| = |a + be^{i\theta} - 2ce^{i\phi}| = |a + be^{i\theta} + 2(m-1)ce^{i\phi}| = 1$*
 (c) *M is a unitary matrix.*

Proof. That (a) implies (b) follows from expanding the squares of the expressions in (b) and applying Theorem 7.18. If (b) holds, then M is a normal matrix with eigenvalues of modulus one. It follows from the spectral theorem that M is unitary and hence (c) holds.

Now suppose M is unitary and denote its entries by M_{jk}, $j, k = 1, \ldots, 2m$. Then

$$\sum_k M_{jk} \bar{M}_{j'k} = \delta_{jj'}, \qquad j, j' = 1, \ldots, 2m.$$

If $j = j'$, we obtain

$$1 = \sum_k |M_{jk}|^2 = a^2 + b^2 + 2(m-1)c^2$$

and hence Condition a of Theorem 7.18 holds. If $j \neq j'$, there are two cases. In the first case $|j - j'| = 1$ and $\min(j, j')$ is odd. For this case we have

$$0 = \sum_k M_{jk} \bar{M}_{j'k} = 2ab \cos \theta + 2(m-1)c^2$$

and hence (b) of Theorem 7.18 holds. Otherwise, we have the second case which gives

$$0 = \sum_k M_{jk} \bar{M}_{j'k} = 2ac \cos \phi + 2bc \cos(\theta - \phi) + 2(m-2)c^2.$$

If $c \neq 0$ and $m \neq 1$, we obtain (c) of Theorem 7.18. Otherwise, $c = 0$ and $m = 1$ so this condition is vacuous. $\square$

We now consider the discrete wave equation (7.26) in this context. Assume that A is a one-step transition amplitude and write (7.26) as follows

$$\psi_{n+1}(j, \hat{k}_r) = \sum_{s=1}^{2m} A((j - \hat{k}_s, \hat{k}_s), (j, \hat{k}_r)) \psi_n(j - \hat{k}_s, \hat{k}_s). \tag{7.31}$$

If we use the notation $\psi^s(n, j) = \psi_n(j, \hat{k}_s)$, $s = 1, \ldots, 2m$, we can write (7.31) in matrix form

$$\begin{bmatrix} \psi^1(n+1, j) \\ \vdots \\ \psi^{2m}(n+1, j) \end{bmatrix} = M \begin{bmatrix} \psi^1(n, j - \hat{k}_1) \\ \vdots \\ \psi^{2m}(n, j - \hat{k}_{2m}) \end{bmatrix} \tag{7.32}$$

DISCRETE SPACETIME MODELS

where M is the unitary matrix defined previously. In a sense, (7.32) is a discrete analog of Dirac's equation

$$i\frac{\partial}{\partial t}\left[\psi^j(t,\mathbf{x})\right] = \Gamma \cdot \nabla\left[\psi^j(t,\mathbf{x})\right]$$

where $[\psi^j(t,\mathbf{x})]$ is a four-component "spinor" and Γ is a matrix-valued three vector. Indeed, in (7.32) the left hand side has the position variable j fixed and the time variable n is incremented while the right hand side has the time variable n fixed and the position variable j is incremented.

For simplicity, let us consider the one-dimensional case, $m = 1$. In this case, the parameters c and ϕ do not appear and the conditions for unitarity in Theorem 7.18 become: $a^2 + b^2 = 1$, $ab\cos\theta = 0$. It follows that $b = (1-a^2)^{\frac{1}{2}}$ and $\theta = \pi/2$ or $3\pi/2$. For concreteness, suppose $\theta = \pi/2$. We then have only one free parameter $0 < a < 1$. In this case, (7.32) is equivalent to the following two equations, where $j \in Z$.

$$\psi^1(n+1,j) = a\psi^1(n,j-1) + ib\psi^2(n,j+1) \qquad (7.33)$$

$$\psi^2(n+1,j) = ib\psi^1(n,j-1) + a\psi^2(n,j+1). \qquad (7.34)$$

From (7.33) we obtain

$$\psi^2(n,j) = -\frac{i}{b}\psi^1(n+1,j-1) + \frac{ia}{b}\psi^1(n,j-2). \qquad (7.35)$$

Substituting (7.35) into (7.34) and simplifying gives

$$\psi^1(n+1,j) + \psi^1(n-1,j) = a[\psi^1(n,j+1) + \psi^1(n,j-1)]. \qquad (7.36)$$

A similar procedure shows that (7.36) also holds for $\psi^2(n,j)$. Notice that (7.36) is a discrete analog of the Klein-Gordon equation

$$\frac{\partial^2}{\partial t^2}\psi(t,\mathbf{x}) = a\nabla^2\psi(t,\mathbf{x}).$$

The Klein-Gordon equation is another relativistic wave equation that is important in traditional quantum mechanics.

The solution of (7.32) in closed form appears to be quite difficult. We shall obtain a solution in the next section only for the case $m = 1$. Although the method employed may possibly be extended to higher dimensions, the work involved is very tedious. However, a related problem which is physically relevant can be easily solved. We call this the momentum problem. Let

$$\phi_n(s) = \sum_j \psi^s(n,j) = \sum_j A_n(q_0,(j,\hat{k}_s)).$$

We call $\phi_n(s)$ the *momentum wave function* and note that it gives the probability amplitude that a particle moves in the $\hat{k}_s$ direction at time n given that it began at q_0 at time 0.

If we sum (7.32) over j, we obtain

$$\begin{bmatrix} \phi_{n+1}(1) \\ \vdots \\ \phi_{n+1}(2m) \end{bmatrix} = M \begin{bmatrix} \phi_n(1) \\ \vdots \\ \phi_n(2m) \end{bmatrix} = M^{n+1} \begin{bmatrix} \phi_0(1) \\ \vdots \\ \phi_0(2m) \end{bmatrix}. \quad (7.37)$$

Since $\phi_0(s)$ is clearly a unit vector in $\mathbb{C}^{2m}$ and M is unitary, it follows that $\phi_n(s)$ is a unit vector in $\mathbb{C}^{2m}$.

Since we know the eigenvalues and eigenvectors for M (see (7.28), (7.29), (7.30)) we can diagonalize M and solve the momentum problem. To facilitate this, define the 2×2 matrices

$$D = \begin{bmatrix} 1 & 1 \\ 1 & 1 \end{bmatrix}, \quad E = \begin{bmatrix} 1 & -1 \\ -1 & 1 \end{bmatrix}.$$

It follows from (7.28), (7.29), and (7.30) that the projections P_1, P_2, P_3 onto the eigenspaces for the eigenvalues

$$\lambda_1 = a - be^{i\theta}$$
$$\lambda_2 = a + be^{i\theta} + 2(m-1)ce^{i\phi}$$
$$\lambda_3 = a + be^{i\theta} - 2ce^{i\phi}$$

respectively, become

$$P_1 = \frac{1}{2} \begin{bmatrix} E & 0 & \cdots & 0 \\ 0 & E & \cdots & 0 \\ \vdots & \vdots & \ddots & \vdots \\ 0 & 0 & \cdots & E \end{bmatrix} \quad P_2 = \frac{1}{2m} \begin{bmatrix} D & D & \cdots & D \\ \vdots & \vdots & \ddots & \vdots \\ D & D & \cdots & D \end{bmatrix}$$

$$P_3 = \frac{1}{2m} \begin{bmatrix} (m-1)D & -D & \cdots & -D \\ -D & (m-1)D & \cdots & -D \\ \vdots & \vdots & \ddots & \vdots \\ -D & -D & \cdots & (m-1)D \end{bmatrix}.$$

Applying the spectral theorem, we have

$$M = \lambda_1 P_1 + \lambda_2 P_2 + \lambda_3 P_3. \tag{7.38}$$

Suppose the initial momentum is in the $\hat{k}_r$ direction. Then $\phi_0(s) = \delta_{sr}$ and by (7.37) we have $\phi_n(s) = (M^n \phi_0)(s)$. Applying (7.38) gives the following explicit solution to the momentum problem

$$\phi_n(s) = (a - be^{i\theta})^n (P_1)_{sr} + \left[a + be^{i\theta} + 2(m-1)ce^{i\phi}\right]^n (P_2)_{sr}$$
$$+ (a + be^{i\theta} - 2ce^{i\phi})^n (P_3)_{sr}.$$

The cubic lattices just considered give very restricted models since the directions of motion are quite limited. For example, in the three-dimensional case, a particle can only move in six directions. However, we can use some of our previous results in the next model. In this model a particle can move in an arbitrary finite number of directions. We shall, however, restrict our attention to the two and three-dimensional cases and use a very simple one-step transition amplitude.

We begin in two dimensions. Let $\hat{k}_1$ be a unit vector in $\mathbb{R}^2$ which we take for concreteness to be in the horizontal direction. Let $n \in \mathbb{N}$ and let $\alpha = \pi/n$ be an angle. Form the unit vectors $\hat{k}_1, \ldots, \hat{k}_{2n}$ where $\hat{k}_j \cdot \hat{k}_1 = \cos(j-1)\alpha$, $j = 2, \ldots, 2n$. Thus, the $\hat{k}_j$ are unit vectors and each forms an angle α with its predecessor. We call a point $x \in \mathbb{R}^2$ *accessible* if it has the form $x = \sum k_j$, $k_j \in \{\hat{k}_1, \ldots, \hat{k}_{2n}\}$. Let V be the set of accessible points and for $x, y \in V$, define $x \perp y$ if $x - y \in \{\hat{k}_1, \ldots, \hat{k}_{2n}\}$. Let $G = (V, E)$ be the graph in which the edges are represented by the straight line between adjacent vertices. The phase space S can be represented as follows:

$$S = \{(x, k) : x \in V, \; k \in \{\hat{k}_1, \ldots, \hat{k}_{2n}\}\}.$$

We now define the simplest nontrivial one-step transition amplitude on S. Let $A : S \times S \to \mathbb{C}$ be a function of the form

$$A((x,k),(x+k,k')) = \begin{cases} a & \text{if } k' = k \\ be^{i\theta} & \text{if } k' \neq k \end{cases}$$

where $a, b > 0$, $\theta \in [0, 2\pi)$ and $A((x,k),(x',k')) = 0$, otherwise. Physically, this may be interpreted as follows. If a particle is moving in direction k, then at the next time step it continues in this direction with amplitude a and it changes to one of the other possible directions with amplitude $be^{i\theta}$.

It is easy to check that A is a one-step transition amplitude if and only if it satisfies

$$a^2 + (2n-1)b^2 = 1 \qquad (7.39)$$

$$a\cos\theta + (n-1)b = 0. \qquad (7.40)$$

Hence, there is only one free parameter which we take to be a. Solving (7.39) and (7.40) in terms of a gives

$$b = \left(\frac{1-a^2}{2n-1}\right)^{\frac{1}{2}} \qquad (7.41)$$

$$\cos\theta = \frac{(1-n)}{a}\left(\frac{1-a^2}{2n-1}\right)^{\frac{1}{2}}. \qquad (7.42)$$

Because of (7.42), a cannot have arbitrary values in the interval $(0,1)$ but must satisfy

$$\frac{(n-1)}{n} \le a < 1. \qquad (7.43)$$

In the limiting case we obtain

$$a = \frac{n-1}{n}, \quad b = \frac{1}{n}, \quad \theta = \pi. \qquad (7.44)$$

In this model, the discrete wave equation is similar to (7.32) except the matrix M is replaced by the $2n \times 2n$ matrix M' with a's on the main diagonal and $be^{i\theta}$ elsewhere. The solution to the momentum problem is similar to our previous work. In fact, all we have to do is replace $ce^{i\phi}$ by $be^{i\theta}$ whenever it appears.

The extention of the above to three dimensions is straightforward. For $n \in \mathbb{N}$ we again let $\alpha = \pi/n$. Form a spherical coordinate system and let $\hat{k}_1$ be a unit vector in the vertical direction. We now have n polar angles, $\alpha, 2\alpha, \ldots, n\alpha$ and for each of the $n-1$ polar angles $\alpha, 2\alpha, \ldots, (n-1)\alpha$ we have $2n$ azimuthal angles $0, \alpha, \ldots, (2n-1)\alpha$. Construct a unit vector $\hat{k}_j$ in each of these $2n(n-1)$ directions. Including the unit vectors $\hat{k}_1, \hat{k}_r = -\hat{k}_1$ we now have $r = 2n(n-1) + 2$ unit vectors $\hat{k}_1, \ldots, \hat{k}_r$. We now define $G = (V, E)$ and $A\colon S \times S \to \mathbb{C}$ in a similar way as we did in two dimensions. The unitarity conditions now become

$$a^2 + [n^2 + (n-1)^2]b^2 = 1 \qquad (7.45)$$

$$a\cos\theta + n(n-1)b = 0. \qquad (7.46)$$

Solving (7.45) and (7.46) in terms of a gives

$$b = \left[\frac{1-a^2}{n^2 + (n-1)^2}\right]^{\frac{1}{2}} \tag{7.47}$$

$$\cos\theta = \frac{n(1-n)}{a}\left[\frac{1-a^2}{n^2 + (n-1)^2}\right]^{\frac{1}{2}}. \tag{7.48}$$

Again, because of (7.48) we must have

$$\left[(n-1)^{-2} + n^{-2} + 1\right]^{-1} \le a < 1.$$

In the limiting case we obtain

$$b = \left[n^2 + (n-1)^2 + n^2(n-1)^2\right]^{-1}, \quad \theta = \pi.$$

7.5 One-Dimensional Lattices and Finite Graphs

In this section we shall find explicit expressions for $A_n(q_0, q)$ in a one-dimensional infinite lattice and in some special finite graphs. But first we shall develop a general technique that can be applied to an m-dimensional cubic lattice. In this technique we employ a Fourier analysis of the problem. Such an analysis simplifies the computation of $A_n(q_0, q)$ by transforming $\ell^2(S)$ to a new representation space.

Let S be the phase space for the graph $G = (V, E)$ where $V = Z^m$ as in Section 7.4. For a function $f \in \ell^2(S)$ we use the notation $f_s(j) = f(j, \hat{k}_s)$, $j \in V$, $s = 1, \ldots, 2m$. Let

$$X = [0, 2\pi]^m = \{x : x = (x_1, \ldots, x_m), x_j \in [0, 2\pi], j = 1, \ldots, m\}$$

and form the Hilbert space $H = L^2(X) \otimes \mathbb{C}^{2m}$. Denote a function $f \in H$ by $f_s(x)$, $x \in X$, $s \in \{1, \ldots, 2m\}$. The inner product in H is taken to be

$$\langle f, g \rangle = \frac{1}{(2\pi)^m} \int_0^{2\pi} f \cdot g \, dx$$

where $f \cdot g = \sum_{s=1}^{2m} f_s \bar{g}_s$, $dx = dx_1, \ldots, dx_m$. For $j \in Z^m$, $x \in X$, define $j \cdot x = j_1 x_1 + \cdots + j_m x_m$. Define the linear transformation $F : \ell^2(S) \to H$ by

$$(Ff)_s(x) = \sum_j f_s(j) e^{ij\cdot x}.$$

It is easy to see that F is unitary and that

$$(F^{-1}g)_s(j) = \frac{1}{(2\pi)^m} \int_0^{2\pi} g_s(x)e^{-ij\cdot x}\, dx.$$

For $s = 1, \ldots, 2m$ define (s) to be the smallest integer that is not smaller than $s/2$. Let $\hat{T}_{sr} = \hat{T}_{sr}(x)$ be the $2m \times 2m$ matrix given by

$$\hat{T}_{sr} = M_{sr} e^{i(-1)^s x_{(s)}}$$

where M is the $2m \times 2m$ matrix corresponding to the one-step transition amplitude A given by (7.27) in Section 7.4. Now $\hat{T}_{sr}$ corresponds to a linear operator $\hat{T}$ on H given by

$$(\hat{T}g)_s(x) = \sum_r \hat{T}_{sr} g_r(x). \tag{7.49}$$

The next theorem shows that $\hat{T}$ is the transform of the propagator T on H.

Theorem 7.22. *If T is a propagator for A, then $\hat{T} = FTF^{-1}$.*

Proof. It is easy to check that the vectors

$$\{\delta_s e^{ij\cdot x} : j \in Z^m, s \in \{1, \ldots, 2m\}\}$$

form an orthonormal basis for H. If we show that $\hat{T}$ and FTF^{-1} agree on these vectors, we are finished. Using the notation $(j,s) = (j, \hat{k}_s) \in S$ we have

$$[T\delta_{(j,s)}](j', s') = A[(j', s'), (j, s)]$$

$$= M_{s's} \delta_{j, j'+\hat{k}_{s'}}$$

$$= M_{s's} \delta_{(j-\hat{k}_{s'}, s')}(j', s')$$

$$= \sum_t M_{ts} \delta_{(j-\hat{k}_t, t)}(j', s').$$

Hence,

$$T\delta_{(j,s)} = \sum_t M_{ts} \delta_{(j-\hat{k}_t, t)}. \tag{7.50}$$

Applying (7.50) and the fact that $F\delta_{(j,s)} = \delta_s e^{ij\cdot x}$ we obtain

$$FTF^{-1} \delta_s e^{ij\cdot x} = FT\delta_{(j,s)} = \sum_t M_{ts} F[\delta_{(j-\hat{k}_t, t)}]$$

$$= \sum_t M_{ts} \delta_t e^{i(j-\hat{k}_t)\cdot x}.$$

Thus,

$$\left(FTF^{-1}\delta_s e^{ij\cdot x}\right)_u(y) = \sum_t M_{ts}\delta_t(u)e^{i(j-\hat{k}_t)\cdot y}$$
$$= M_{us}e^{i(j-\hat{k}_u)\cdot y} = M_{us}e^{i(-1)^u y(u)}e^{ij\cdot y}.$$

Moreover, from (7.49) we have

$$(\hat{T}g)_u(y) = \sum_r e^{i(-1)^u y(u)} g_r(y).$$

Hence,

$$\left(\hat{T}\delta_s e^{ij\cdot x}\right)_u(y) = \sum_r M_{ur} e^{i(-1)^u y(u)} \left(\delta_s e^{ij\cdot x}\right)_r(y)$$
$$= \sum_r M_{ur} e^{i(-1)^u y(u)} \delta_s(r) e^{ij\cdot y}$$
$$= M_{us} e^{i(-1)^u y(u)} e^{ij\cdot y}. \quad \square$$

Corollary 7.23. *If $q = (j,s), q' = (j',s') \in S$, then*

$$A_n(q,q') = \langle \hat{T}^n \delta_{s'} e^{ij'\cdot x}, \delta_s e^{ij\cdot x} \rangle$$

Proof. This follows from Corollary 7.15. $\square$

Instead of finding the n-th power (or spectral resolution) of the infinite-dimensional operator T, Corollary 7.23 reduces the computation of $A_n(q,q')$ to finding the n-th power of the $2m \times 2m$ matrix $\hat{T}_{sr}$. The next corollary reduces this computation even further. The result is similar to (7.25).

Corollary 7.24. *Suppose that the matrix $\hat{T}_{sr}$ has a complete set of unit eigenvectors $\psi_k(x)$ with corresponding eigenvalues $\lambda_k(x)$, $k = 1,\ldots,2m$. If $q = (j,s)$, $q' = (j',s')$, we have*

$$A_n(q,q') = \frac{1}{(2\pi)^m} \sum_k \int \lambda_k^n \psi_{ks} \bar{\psi}_{ks'} e^{i(j'-j)\cdot x} dx.$$

Proof. Let P_k be the one-dimensional projection onto ψ_k, $k =$

$1, \ldots, 2m$. Applying Corollary 7.23, we have

$$A_n(q, q') = \left\langle \sum_k \lambda_k^n P_k \delta_{s'} e^{ij' \cdot x}, \delta_s e^{ij \cdot x} \right\rangle$$

$$= \left\langle \sum_k \lambda_k^n (\delta_{s'} e^{ij' \cdot x} \cdot \psi_k) \psi_k, \delta_s e^{ij \cdot x} \right\rangle$$

$$= \sum_k \left\langle \lambda_k^n e^{ij' \cdot x} \bar{\psi}_{ks'} \psi_k, \delta_s e^{ij \cdot x} \right\rangle$$

$$= \frac{1}{(2\pi)^m} \sum_k \int \lambda_k^n \psi_{ks} \bar{\psi}_{ks'} e^{i(j'-j) \cdot x} dx. \quad \Box$$

We now apply Corollary 7.24 to the one-dimensional infinite lattice $V = Z$. In this case, Equation 7.27 for the one-step transition amplitude A becomes

$$A((j, k), (j + k, k')) = \begin{cases} a & \text{if } k' = k \\ b^{i\theta} & \text{if } k' = -k \end{cases} \quad (7.51)$$

where $a, b > 0$, $\theta \in [0, 2\pi)$ and $A((j, k), (j', k')) = 0$, otherwise. In this case, only (a) and (b) of Theorem 7.18 are applicable. These give $a^2 + b^2 = 1$, $\cos \theta = 0$ and we take $\theta = \pi/2$ for definiteness. The matrix $\hat{T}_{sr}(x)$ is given by

$$\hat{T}_{sr}(x) = \begin{bmatrix} ae^{ix} & ibe^{-ix} \\ ibe^{ix} & ae^{ix} \end{bmatrix}.$$

Notice that $\hat{T}_{sr}(x)$ is a unitary matrix. The eigenvalues of $\hat{T}_{sr}(x)$ are $\lambda_1 = a\cos x + i\alpha(x)$, $\lambda_2 = \bar{\lambda}_1$, where $\alpha(x) = (1 - a^2 \cos^2 x)^{1/2}$. The corresponding unit eigenvectors are

$$\psi_1 = N_1(be^{-ix}, a\sin x + \alpha(x))$$
$$\psi_2 = N_2(be^{-ix}, a\sin x - \alpha(x))$$

where N_1, N_2 are normalization factors given by

$$N_1^{-2} = 2\alpha(x)[\alpha(x) + a\sin x]$$
$$N_2^{-2} = 2\alpha(x)[\alpha(x) - a\sin x].$$

For $q_0 = (0, \hat{k}_1)$, $q = (j, \hat{k}_1)$ we obtain from Corollary 7.24 that

$$A_n(q_0, q) = \frac{1}{2\pi} \int_0^{2\pi} e^{ijx} [\cos n\phi - i a \alpha(x)^{-1} \sin x \sin n\phi] dx$$

where
$$\phi = \tan^{-1} \frac{\alpha(x)}{a\cos x}.$$

For $q_0 = (0, \hat{k}_1)$, $q = (j, \hat{k}_2)$ we obtain

$$A_n(q_0, q) = \frac{1}{2\pi} \int_0^{2\pi} e^{i(j+1)x} \alpha(x) \cos n\phi \, dx$$

where ϕ is given above.

While infinite graphs may be useful for describing the motion of a quantum particle in space, finite graphs may be important in the description of "elementary" particles themselves. For example, in the quark model, mesons are composed of a quark and an antiquark while baryons are composed of three quarks or three antiquarks. Thus, two and three vertex graphs (or multigraphs) may give a useful description of mesons and baryons. In such a model, the vertices represent quarks or antiquarks and the edges represent interaction paths along which gluons propagate between vertices. Gluons are the particles that bind the quarks together and are the mediators of the strong nuclear force. In other particle models, finite graphs with an even number of vertices describe bosons and finite graphs with an odd number of vertices describe fermions. In this section we shall only consider finite graphs of a very simple type called cycle graphs. For more sophisticated multigraphs and models the reader is referred to [Gudder, 1988].

Let $V = \{v_1, \ldots, v_N\}$ be a finite set with $|V| = N \geq 2$. We define $v_j \perp v_{j'}$ if $|j - j'| = 1$ or $N - 1$. Denoting the set of edges by E we obtain an *N-cycle graph* $C_N = (V, E)$. For $N \geq 3$, each vertex has degree 2 and hence precisely two incident edges. We call $\{v_j, v_{j+1}\}$, $j = 1, \ldots, N$ ($j+1$ is taken mod N) the *forward edge from* v_j and $\{v_j, v_{j-1}\}$, $j = 1, \ldots, N$ ($j-1$ is taken mod N) the *backward edge from* v_j. In order to include C_2 we assume that C_2 is a multigraph with two edges denoted v_1v_2 and v_2v_1. Then v_1v_2, v_2v_1 are the forward edges from v_1, v_2, respectively, and v_2v_1, v_1v_2 are the backward edges from v_1, v_2, respectively. The phase space C_N can be represented as

$$S - \{(j,k): j \subset 1, \ldots, N, k = \pm 1\}$$

where $k = 1(-1)$ corresponds to the forward (backward) edge from v_j.

Let $A: S \times S \to \mathbb{C}$ be a one-step transition amplitude of the form (7.51) (with $j + k$ taken mod N). It easily follows that $a^2 + b^2 = 1$ and $\theta = \pi/2$

or $3\pi/2$. For definiteness we take $\theta = \pi/2$. Now $\ell^2(S)$ is a $2N$-dimensional Hilbert space with orthonormal basis $\delta_{(j,k)}$, $j = 1, \ldots, N$, $k = \pm 1$. The propagator T_N satisfies

$$(T_N f)(j,k) = af(j+k,k) + ibf(j+k,-k).$$

It follows that

$$T_N \delta_{(j,k)} = a\delta_{(j-k,k)} + ib\delta_{(j+k,-k)}. \tag{7.52}$$

Relative to the basis $\delta_{(j,k)}$, T_N can be represented by a $2N \times 2N$ matrix whose entries are given by (7.52). Suppose the unitary matrix T_N has eigenvalues $\lambda_1, \ldots, \lambda_{2N}$ with corresponding unit eigenvectors $\psi_1, \ldots, \psi_{2N}$. According to (7.25) we have

$$A_n(q_0, q) = \sum_r \lambda_r^n \psi_r(q_0) \bar{\psi}_r(q). \tag{7.53}$$

We first consider C_2. In this case the matrix T_2 becomes

$$T_2 = \begin{bmatrix} 0 & 0 & a & ib \\ 0 & 0 & ib & a \\ a & ib & 0 & 0 \\ ib & a & 0 & 0 \end{bmatrix}.$$

The eigenvalues of T_2 are $\lambda_1 = a + ib$, $\lambda_2 = a - ib$, $\lambda_3 = -\lambda_1$, $\lambda_4 = -\lambda_2$. The corresponding unit eigenvectors become $\psi_1 = \frac{1}{2}(1,1,1,1)$, $\psi_2 = \frac{1}{2}(1,-1,1,-1)$, $\psi_3 = \frac{1}{2}(1,1,-1,-1)$, $\psi_4 = \frac{1}{2}(1,-1,-1,-1)$. In function form we have $\psi_1(j,k) = \frac{1}{2}$, $\psi_2(j,k) = \frac{k}{2}$, $\psi_3(j,k) = \frac{1}{2}(-1)^{j+1}$, $\psi_4(j,k) = \frac{k}{2}(-1)^{j+1}$, $j = 1,2$; $k = \pm 1$. If $q_0 = (j,k)$ and $q = (j',k')$, then applying (7.53) gives

$$A_n(q_0, q) = \tfrac{1}{4}\left[1 + (-1)^{n+j+j'}\right]\left[(a+ib)^n + (a-ib)^n kk'\right].$$

Letting $\theta = \tan^{-1} b/a$ we obtain

$$A_n(q_0, q) = \begin{cases} \tfrac{1}{2}\left[1 + (-1)^{n+j+j'}\right]\cos n\theta & \text{if } k = k' \\ \tfrac{i}{2}\left[1 + (-1)^{n+j+j'}\right]\sin n\theta & \text{if } k = -k'. \end{cases} \tag{7.54}$$

It follows from (7.54) that

$$A_n(q_0, j') = A_n(q_0, (j',1)) + A_n(q_0, (j',-1)) = \tfrac{1}{2}\left[1 + (-1)^{n+j+j'}\right]e^{in\theta}.$$

Hence,

$$\mu_n(q_0, j') = \tfrac{1}{4}|1 + (-1)^{n+j+j'}|^2 = \begin{cases} 1 & \text{if } n+j+j' \text{ is even} \\ 0 & \text{if } n+j+j' \text{ is odd.} \end{cases}$$

Now suppose that the particle evolves under the influence of a momentum independent potential $v: S \to \mathbf{R}$. Then v has the form $v((1,k)) = \alpha$, $v((2,k)) = \beta$, $k = \pm 1$. According to Corollary 7.17, the propagator T_v for the perturbed one-step transition amplitude A^v satisfies $T_v = e^{-iV} T_2$. We can represent e^{-iV} by the unitary matrix

$$e^{-iV} = \mathrm{diag}(e^{-i\alpha}, e^{-i\alpha}, e^{-i\beta}, e^{-i\beta}).$$

Then the eigenvalues of T_v are

$$\lambda_1 = (a + ib)e^{-i(\alpha+\beta)/2}$$
$$\lambda_2 = (a - ib)e^{-i(\alpha+\beta)/2}$$
$$\lambda_3 = -\lambda_1, \lambda_4 = -\lambda_2.$$

The corresponding unit eigenvectors become

$$\psi_1 = \tfrac{1}{2}(e^{-i\alpha/2}, e^{-i\alpha/2}, e^{-i\beta/2}, e^{-i\beta/2})$$
$$\psi_2 = \tfrac{1}{2}(e^{-i\alpha/2}, -e^{-i\alpha/2}, e^{-i\beta/2}, -e^{-i\beta/2})$$
$$\psi_3 = \tfrac{1}{2}(e^{-i\alpha/2}, e^{-i\alpha/2}, -e^{-i\beta/2}, -e^{-i\beta/2})$$
$$\psi_4 = \tfrac{1}{2}(e^{-i\alpha/2}, -e^{-i\alpha/2}, -e^{-i\beta/2}, e^{-i\beta/2}).$$

In functional form we have

$$\psi_1(j,k) = \frac{d}{2}, \quad \psi_2(j,k) = \frac{kd}{2}$$
$$\psi_3(j,k) = (-1)^{j+1}\frac{d}{2}, \quad \psi_4(j,k) = (-1)^{j+1}\frac{kd}{2}$$

where

$$d = \left[e^{-i\alpha/2}(1 + (-1)^{j+1}) + e^{-i\beta/2}(1 + (-1)^j)\right].$$

The general expression for $A_n^v(q_0, q)$ corresponding to (7.54) becomes quite complicated and it is simpler to divide it into cases.

For the case $q_0 = (1, k)$, $q = (2, k')$, upon applying the analog of (7.53) we obtain

$$A_n^v(q_0, q) = \begin{cases} \frac{1}{2}[1 - (-1)^n] \cos n\phi \exp\left[\frac{-i(n+1)(\alpha+\beta)}{2}\right] & \text{if } k = k' \\ \frac{i}{2}[1 - (-1)^n] \sin n\phi \exp\left[\frac{-i(n+1)(\alpha+\beta)}{2}\right] & \text{if } k = -k' \end{cases}$$

where $\phi = \tan^{-1} b/a$. In particular,

$$A_n^v(q_0, 2) = A_n^v(q_0, (2, 1)) + A_n^v(q_0, (2, -1))$$

$$= \frac{1}{2}[1 - (-1)^n] e^{in\theta} \exp\left[\frac{-i(n+1)(\alpha+\beta)}{2}\right]$$

and

$$\mu_n^v(q_0, 2) = \frac{1}{4}|1 - (-1)^n|^2 = \begin{cases} 0 & \text{if } n \text{ is even} \\ 1 & \text{if } n \text{ is odd.} \end{cases}$$

For the case $q_0 = (1, k)$, $q = (1, k')$ we obtain

$$A_n^v(q_0, q) = \begin{cases} \frac{1}{2}[1 + (-1)^n] \cos n\phi \exp\left[\frac{-i((n+2)\alpha+n\beta)}{2}\right] & \text{if } k = k' \\ \frac{i}{2}[1 + (-1)^n] \sin n\phi \exp\left[\frac{-i((n+2)\alpha+n\beta)}{2}\right] & \text{if } k = -k'. \end{cases}$$

In particular,

$$A_n^v(q_0, 1) = \frac{1}{2}\left[1 + (-1)^n\right] e^{in\theta} \exp\left[\frac{-i((n+2)\alpha + n\beta)}{2}\right]$$

and

$$\mu_n^v(q_0, 1) = \frac{1}{4}|1 + (-1)^n|^2 = \begin{cases} 1 & \text{if } n \text{ is even} \\ 0 & \text{if } n \text{ is odd.} \end{cases}$$

We next consider C_3. The eigenvalues of the unitary 6×6 matrix T_3 are $\lambda_1 = a + ib$, $\lambda_2 = a - ib$

$$\lambda_3 = \lambda_4 = \frac{[-a + i(4 - a^2)^{1/2}]}{2}$$

$$\lambda_5 = \lambda_6 = \frac{[-a - i(4 - a^2)^{1/2}]}{2}.$$

The corresponding unit eigenvectors become

$$\psi_1 = 6^{-1/2}(1, 1, 1, 1, 1, 1)$$

$$\psi_2 = 6^{-1/2}(1, -1, 1, -1, 1, -1)$$

$$\psi_3 = (4 - a^2)^{-1/2}(\lambda_3, -ib, -(\lambda_3 + a), 0, a, ib)$$

$$\psi_4 = (4 - a^2)^{-1/2}(0, -(\lambda_4 + a), -ib, \lambda_4, ib, a)$$

$$\psi_5 = (4 - a^2)^{-1/2}(\lambda_5, -ib, -(\lambda_5 + a), 0, a, ib)$$

$$\psi_6 = (4 - a^2)^{-1/2}(0, -(\lambda_6 + a), -ib, \lambda_6, ib, a).$$

We now compute $A_n(q_0, q)$ using (7.53). For example, let $q_0 = (1,1)$, $q = (j, k)$. Then

$$A_n(q_0, q) = 6^{-1}(\lambda_1^n + k\bar{\lambda}_1^n) + (4 - a^2)^{-1}\left[\lambda_3^{n+1}\bar{\psi}_3(q) + \lambda_3^{n+1}\bar{\psi}_5(q)\right].$$

If $k = 1$, we have

$$A_n(q_0\, q) = 3^{-1}\operatorname{Re}\lambda_1^n + 2(4 - a^2)^{-1}\operatorname{Re}\lambda_3^{n+1}\bar{\psi}_3(q).$$

If $k = -1$, we have

$$A_n(q_0\, q) = i3^{-1}\operatorname{Im}\lambda_1^n + 2i(4 - a^2)^{-1}\bar{\psi}_3(q)\operatorname{Im}\lambda_3^{n+1}.$$

Finally, let us consider C_4. The eigenvalues of the unitary 8×8 matrix T_4 are $\lambda_1 = a + ib$, $\lambda_2 = a - ib$, $\lambda_3 = -\lambda_1$, $\lambda_4 = -\lambda_2$, $\lambda_5 = \lambda_6 = 1$, $\lambda_7 = \lambda_8 = -i$. The corresponding unit eigenvectors become

$$\psi_1 = 8^{-1/2}(1, 1, 1, 1, 1, 1, 1, 1)$$
$$\psi_2 = 8^{-1/2}(1, -1, 1, -1, 1, -1, 1, -1)$$
$$\psi_3 = 8^{-1/2}(1, 1, -1, -1, 1, 1, -1, -1)$$
$$\psi_4 = 8^{-1/2}(1, -1, -1, 1, 1, -1, -1, 1)$$
$$\psi_5 = 8^{-1/2}(1, 1, ia + b, -ia - b, -1, -1, -ia - b, ia + b)$$
$$\psi_6 = 8^{-1/2}(1, 0, ia, -b, -1, 0, -ia, b)$$
$$\psi_7 = 8^{-1/2}(1, 1, -ia - b, ia + b, -1, -1, ia + b, -ia - b)$$
$$\psi_8 = 8^{-1/2}(1, 0, -ia, b, -1, 0, ia, -b).$$

Again, $A_n(q_0, q)$ can be computed using (7.53).

Continuing in this fashion, one can see similarities in the C_n's for n even (corresponding to bosons) and the C_n's for n odd (corresponding to fermions).

7.6 Discrete Feynman Amplitudes

This section presents a discrete dynamical model whose one-step transition amplitudes are a discrete analog of the amplitudes proposed by R. Feynman (Section 2.6). The model is similar to the second model of Section 7.4 except the amplitudes are more complicated (and possibly more realistic). We shall work in a two-dimensional configuration space and the extension to three dimensions will be straightforward.

Let a and n be positive integers that are relatively prime. Let α be an angle with radian measure $2\pi/n$ and let k_0 be a unit vector in $\mathbb{R}^2$ that we take to be in the horizontal direction. Let $k_1, \ldots, k_{n-1}$ be unit vectors in $\mathbb{R}^2$ such that $k_j \cdot k_0 = \cos j\alpha$, $j = 1, \ldots, n-1$. Thus, each k_j forms an angle α with its predecessor, $j = 1, \ldots, n-1$. As in Section 7.4, V denotes the set of accessible points $x = \sum e_j$, $e_j \in \{k_0, \ldots, k_{n-1}\}$ and the phase space S is represented by

$$S = \{(x, k_j) : x \in V, j = 0, \ldots, n-1\}.$$

We define the *Feynman amplitude* $A: S \times S \to \mathbb{C}$ by

$$A((x, k_r), (x + k_r, k_s)) = \frac{1}{\sqrt{n}} e^{ia\pi(s-r)^2/n}$$

and A is zero, otherwise.

Lemma 7.25. *If A is a Feynman amplitude, then A is a one-step transition amplitude.*

Proof. It is clear that

$$\sum_q |A(q_1, q)|^2 = \sum_q |A(q, q_1)|^2 = 1.$$

Suppose that $q_1, q_2 \in S$ with $q_1 \neq q_2$. We can assume that $q_1 = (x_1, k_r)$, $q_2 = (x_2, k_s)$ where $r \neq s$ and $x_1 + k_r = x_2 + k_s$. Since $at \neq 0 \mod (n)$ for any $0 < t < n$ we obtain

$$\sum_q A(q_1,q)\bar{A}(q_2,q)$$

$$= \sum_{j=0}^{n-1} A((x_1,k_r),(x_1+k_r,k_j))\bar{A}((x_2,k_s),(x_1+k_r,k_j))$$

$$= n^{-1}\sum_{j=0}^{n-1} e^{ia\pi(j-r)^2/n} e^{-ia\pi(j-s)^2/n}$$

$$= n^{-1} e^{ia\pi(r^2-s^2)/n} \sum_{j=0}^{n-1}\left[e^{i2a\pi(s-r)/n}\right]^j$$

$$= \left[n^{-1} e^{ia\pi(r^2-s^2)/n}\right]\left[1 - e^{i2a\pi(s-r)/n}\right]^{-1}\left[1 - e^{i2a\pi(s-r)}\right] = 0.$$

In a similar way, it follows that for $q_1 \ne q_2$

$$\sum_q A(q,q_1)\bar{A}(q,q_2) = 0. \quad \square$$

We now show that as n approaches infinity, the above Feynman amplitude approaches the traditional free particle continuum Feynman amplitude in a certain sense. Let

$$p = \{(x_0, k_{j_0}), \ldots, (x_r, k_{j_r})\}$$

be an r-path, where $x_s + k_{j_s} = x_{s+1}$, $s = 0,\ldots,r-1$. Let $\beta_s = 2\pi(j_s - j_{s-1})/n$, $s = 1,\ldots,r$, and suppose that the β_s are small (or close to 2π). This will be the case if n is large and the path does not turn very far. Then the distance between x_{s+1} and x_{s-1} becomes

$$\|x_{s+1} - x_{s-1}\|^2 = \|x_{s-1} + k_{j_{s-1}} + k_{j_s} - x_{s-1}\|^2$$
$$= \|k_{j_{s-1}} + k_{j_s}\|^2 = 2 + 2k_{j_s} \cdot k_{j_{s-1}}$$
$$= 2 + 2\cos\beta_s \approx 2 + 2\left(1 - \frac{\beta_s^2}{2}\right) = 4 - \beta_s^2.$$

Hence, $\beta_s^2 \approx 4 - \|x_{s+1} - x_{s-1}\|^2$. If v_s denotes the particle's "velocity" at time s, we have $v_s^2 \approx \|x_{s+1} - x_{s-1}\|^2/4$. Thus, $\beta_s^2 \approx 4(1-v_s^2)$. Then the amplitude of the path p becomes

$$A(p) = n^{-r/2} \exp\left[ia\pi n^{-1} \sum_{s=1}^{r} (j_s - j_{s-1})^2\right]$$

$$= n^{-r/2} \exp\left[ian(4\pi)^{-1} \sum_{s=1}^{r} \beta_s^2\right]$$

$$\approx n^{-r/2} \exp\left[ian\pi^{-1} \sum_{s=1}^{r} (1 - v_s^2)\right]$$

$$= n^{-r/2} e^{ianr/\pi} \exp\left[-i2n\pi^{-1} \sum_{s=1}^{r} \frac{av_s^2}{2}\right].$$

If we let a correspond to the mass of the particle, then the summation corresponds to the integral of the kinetic energy over the path. In this way $A(p)$ approaches the free particle continuum Feynman amplitude.

We now consider the discrete wave equation

$$\psi_r(q) = \sum_{q'} A(q', q)\psi_{r-1}(q')$$

where $\psi_r(q) = A(q_0, q)$ and A is the Feynman amplitude. This equation has the form

$$\psi_r(x, k_s) = \sum_t A\big((x - k_t, k_t), (x, k_s)\big)\psi_{r-1}(x - k_t, k_t)$$
$$= n^{-\frac{1}{2}} \sum_t e^{ia\pi(s-t)^2/n} \psi_{r-1}(x - k_t, k_t). \tag{7.55}$$

If we define $\psi_r^s(x) = \psi_r(x, k_s)$, $s = 0, 1, \ldots, n-1$, and we define the matrix

$$M_{st} = n^{-\frac{1}{2}}\left[e^{ia\pi(s-t)^2/n}\right], \qquad s, t = 0, 1, \ldots, n-1$$

then (7.55) can be written in matrix notation as

$$\begin{bmatrix} \psi_r^0(x) \\ \vdots \\ \psi_r^{n-1}(x) \end{bmatrix} = M \begin{bmatrix} \psi_{r-1}^0(x - k_0) \\ \vdots \\ \psi_{r-1}^{n-1}(x - k_{n-1}) \end{bmatrix}. \tag{7.56}$$

DISCRETE FEYNMAN AMPLITUDES 295

Defining the momentum wave function $\phi_r^s = \sum_x \psi_r^s(x)$ and summing (7.56) over x gives

$$\begin{bmatrix} \phi_r^0 \\ \vdots \\ \phi_r^{n-1} \end{bmatrix} = M \begin{bmatrix} \phi_{r-1}^0 \\ \vdots \\ \phi_{r-1}^{n-1} \end{bmatrix}$$

or simply $\phi_r = M\phi_{r-1}$. Now it is clear from Lemma 7.25 that M is a unitary $n \times n$ matrix and the momentum problem is solved if we can find the eigenvalues and eigenvectors for M. To find these, the following results will be useful.

Lemma 7.26. *If n is even, then for any $t = 0, \ldots, n-1$ we have*

$$\sum_{s=0}^{n-1} e^{ia\pi(s-t)^2/n} = \sum_{s=0}^{n-1} e^{ia\pi s^2/n}.$$

Proof. For any $t \in \{0, \ldots, n-1\}$ we have

$$\sum_{s=0}^{n-1} e^{ia\pi(s-t)^2/n} = \sum_{s=0}^{t-1} e^{ia\pi(t-s)^2/n} + \sum_{s=t}^{n-1} e^{ia\pi(s-t)^2/n}.$$

On the right side of the above equation, replace $t - s$ in the first sum by $n - j$ and replace $s - t$ in the second sum by j to obtain

$$\sum_{s=0}^{n-1} e^{ia\pi(s-t)^2/n} = \sum_{j=n-t}^{n-1} e^{ia\pi(n-j)^2/n} + \sum_{j=0}^{n-t-1} e^{ia\pi j^2/n}$$

$$= \sum_{j=n-t}^{n-1} e^{ia\pi(n^2-2nj+j^2)/n} + \sum_{j=0}^{n-t-1} e^{ia\pi j^2/n}$$

$$= \sum_{j=0}^{n-1} e^{ia\pi j^2/n}. \quad \square$$

The mathematics in the sequel becomes simpler if we assume that n has the form $n = 2^k$ for some positive integer k. We make this assumption and assume that a is odd. Let

$$I_k = \sum_{j=0}^{2^k-1} e^{ia\pi j^2/2^k}.$$

Lemma 7.27. *Under the above assumptions we have*

$$I_k = \begin{cases} 2^{k/2} e^{ia\pi/4} & k \text{ even} \\ 2^{k/2} e^{ia\pi/4} \operatorname{sgn}\left(\cos a\frac{\pi}{4}\right) & k \text{ odd.} \end{cases}$$

Proof. We first show that I_k satisfies the recurrence relation $I_{k+2} = 2I_k$

$$I_{k+2} = \sum_{j=0}^{2^{k+2}-1} e^{ia\pi j^2/2^{k+2}}$$

$$= \sum_{j=0}^{2^{k+1}-1} e^{ia\pi j^2/2^{k+2}} + \sum_{j=2^{k+1}}^{2^{k+2}-1} e^{ia\pi j^2/2^{k+2}}$$

$$= 2 \sum_{j=0}^{2^{k+1}-1} e^{ia\pi j^2/2^{k+2}}$$

$$= 2 \sum_{j=0}^{2^k-1} e^{ia\pi j^2/2^{k+2}} + 2 \sum_{j=2^k}^{2^{k+1}-1} e^{ia\pi j^2/2^{k+2}}.$$

For $r \in \mathbb{N}$ with $1 \leq r \leq 2^k - 2$ we have

$$\exp\left[\frac{ia\pi(2^{k+1}-r)^2}{2^{k+2}}\right] = \exp\left[\frac{ia\pi(2^{2k+2} - r2^{k+2} + r^2)}{2^{k+2}}\right]$$

$$= e^{-iar\pi} e^{ia\pi r^2/2^{k+2}}$$

$$= (-1)^r e^{ia\pi r^2/2^{k+2}}.$$

where the last equality follows from the fact that a is odd. We then obtain

$$I_{k+2} = 4 \sum_{j=0}^{2^{k-1}-1} e^{ia\pi(2j)^2/2^{k+2}}$$

$$= 2 \sum_{j=0}^{2^{k-1}-1} e^{ia\pi j^2/2^k} + 2 \sum_{j=2^{k-1}}^{2^k-1} e^{ia\pi j^2/2^k}$$

$$= 2 \sum_{j=0}^{2^k-1} e^{ia\pi j^2/2^k} = 2I_k.$$

DISCRETE FEYNMAN AMPLITUDES

Since the initial conditions are $I_2 = 2e^{ia\pi/4}$ and $I_1 = 1 + e^{ia\pi/2}$, we obtain

$$I_k = \begin{cases} 2^{k/2}e^{ia\pi/4} & k \text{ even} \\ 2^{(k-1)/2}(1 + e^{ia\pi/2}) & k \text{ odd}. \end{cases}$$

However,

$$2^{(k-1)/2}(1 + e^{ia\pi/2}) = 2^{(k-1)/2}e^{ia\pi/4}(e^{ia\pi/4} + e^{-ia\pi/4})$$
$$= 2^{(k+1)/2}e^{ia\pi/4}\cos a\pi/4$$
$$= 2^{k/2}e^{ia\pi/4}\text{sgn}(\cos a\pi/4)$$

and the result follows. □

In the sequel we shall assume that a and n satisfy the previous conditions. Since

$$\text{sgn}(\cos a\pi/4) = \begin{cases} 1 & \text{if } a = 1 \text{ or } 7(\text{mod}\,8) \\ -1 & \text{otherwise} \end{cases}$$

we have $I_k = \pm 2^{k/2}e^{ia\pi/4}$ where the minus sign holds if and only if k is odd and $a = 3$ or $5(\bmod 8)$. Applying Lemmas 7.26 and 7.27 we obtain the following corollary.

Corollary 7.28. *For any $t \in \{0, 1, \ldots, 2^k - 1\}$, we have*

$$\sum_{s=0}^{2^k-1} e^{ia\pi(s-t)^2/2^k} = \pm 2^{k/2}e^{ia\pi/4}$$

where the minus sign holds if and only if k is odd and $a = 3$ or $5(\bmod 8)$.

The next result gives the eigenvalues and eigenvectors for M.

Theorem 7.29. *For $j = 0, \ldots, 2^k - 1$, the eigenvalues of M are*

$$\lambda_j = \pm e^{ia\pi/4}e^{-ia\pi j^2/2^k}$$

where the minus sign holds if and only if k is odd and $a = 3$ or $5(\bmod 8)$. The corresponding unit eigenvectors are

$$e_j = 2^{-k/2}(1, e^{-i2\pi aj/2^k}, e^{-i2\cdot 2\pi aj/2^k}, \ldots, e^{-i(2^k-1)2\pi aj/2^k}).$$

Proof. The sth entry of Me_j is

$$\sum_{t=0}^{2^k-1} M_{st}(e_j)_t = 2^{-k} \sum_{t=0}^{2^k-1} e^{ia\pi(s-t)^2/2^k} e^{-iat2\pi j/2^k}$$

$$= 2^{-k} \sum_{t=-s}^{2^k-1-s} e^{ia\pi t^2/2^k} e^{-ia\pi(t+s)\pi j/2^k}$$

$$= 2^{-k} e^{-ia\pi s 2j/2^k} e^{-ia\pi j^2/2^k} \sum_{t=-s}^{2^k-1-s} e^{ia\pi(t-j)^2/2^k}.$$

By Corollary 7.28 the summation on the right side equals

$$\sum_{t=0}^{2^k-1} e^{ia\pi(t-s-j)^2/2^k} = \pm 2^{k/2} e^{ia\pi/4}.$$

It follows that $(Me_j)_s = \lambda_j (e_j)_s$. □

Now that we have given a solution to the momentum problem, let us return to the general discrete wave equation (7.55). This equation can be iterated to find an expression for $\psi_r(x, k_s)$. If $q_0 = (x_0, k_u)$, then initially we have

$$\psi_0(x, k_s) = \delta_{(x_0, k_u)}(x, k_s).$$

Iterating (7.55) and letting $b = a\pi/n$ we have

$$\psi_r(x, k_s) = n^{-\frac{1}{2}} \sum_{t_1} e^{ib(s-t_1)^2} \psi_{r-1}(x - k_{t_1}, k_{t_1})$$

$$= n^{-1} \sum_{t_1, t_2} e^{ib(t_1-t_2)^2} e^{ib(s-t_1)^2} \psi_{r-2}(x - k_{t_1} - k_{t_2}, k_{t_2})$$

$$\vdots$$

$$= n^{-r/2} \sum_{t_1, \ldots, t_r} \exp\left[ib \sum_{j=1}^{r-1}(t_j - t_{j+1})^2\right] e^{ib(s-t_1)^2}$$

$$\times \psi_0(x - k_{t_1} - \cdots - k_{t_r}, k_{t_r}).$$

DISCRETE FEYNMAN AMPLITUDES

This last expression for $\psi_r(x, k_s)$ becomes

$$n^{-r/2} \sum_{t_1,\ldots,t_{r-1}} \left\{ \exp\left[ib \sum_{j=1}^{r-2}(t_j - t_{j+1})^2\right] e^{ib[(s-t_1)^2 + (t_{r-1}-u)^2]} \right.$$
$$\left. : x = x_0 + k_{t_1} + \cdots + k_{t_{r-1}} + k_u \right\}.$$

Defining $\psi_r(x) = \sum_s \psi_r(x, k_s)$, we may interpret $\psi_r(x)$ as the amplitude that the particle is at the position x at time r if it was initially at $q_0 = (x_0, k_u)$. Again, assuming that $n = 2^k$, we have the following expression for $\psi_r(x)$:

$$\pm \frac{e^{ia\pi/4}}{n^{(r-1)/2}} \sum_{t_1,\ldots,t_{r-1}} \left\{ \exp\left[ib \sum_{j=1}^{r-2}(t_j - t_{j+1})^2\right] e^{ib(t_{r-1}-u)^2} \right.$$
$$\left. : x = x_0 + k_{t_1} + \cdots + k_{t_{r-1}} + k_u \right\}.$$

Our last result shows that $\psi_r(x)$ can indeed be interpreted as a probability amplitude.

Theorem 7.30. *For $r = 0, 1, \ldots$, we have*

(a) $\left|\sum_x \psi_r(x)\right| = 1$,

(b) $\sum_x |\psi_r(x)|^2 = 1$.

Proof. (a) Applying our previous expression for $\psi_r(x)$ gives

$$\sum_x \psi_r(x) = \pm \frac{e^{ia\pi/4}}{n^{(r-1)/2}} \sum_{t_1,\ldots,t_{r-1}} e^{ib(t_1-t_2)^2} \cdots e^{ib(t_{r-2}-t_{r-1})^2} e^{ib(t_{r-1}-u)^2}$$

$$= \pm \frac{e^{ia\pi/2}}{n^{(r-2)/2}} \sum_{t_2,\ldots,t_{r-1}} e^{ib(t_2-t_3)^2} \cdots e^{ib(t_{r-2}-t_{r-1})^2} e^{ib(t_{r-1}-u)^2}$$

$$\vdots$$

$$= \pm e^{ia\pi r/4}.$$

Hence, $\left|\sum_x \psi_r(x)\right| = 1$.

(b) We obtain this result by varying the initial condition. Let $\psi_r^{(u)}(x)$ denote the $\psi_r(x)$ with initial condition (x_0, k_u). Then
$$\sum_x |\psi_r^{(u)}(x)|^2 = \sum_x \psi_r^{(u)}(x)\bar{\psi}_r^{(u)}(x)$$
where $\psi_r^{(u)}(x)$ is given by the expression immediately preceding this theorem. By rotational symmetry, we can conclude that
$$\sum_x |\psi_r^{(u)}(x)|^2 = \sum_x |\psi_r^{(v)}(x)|^2.$$
More precisely, there is a bijection between the paths from $x_0 + k_u$ and the paths from $x_0 + k_v$ that preserves the transition amplitudes. We then have
$$n \sum_x |\psi_r(x)|^2 = \sum_{x,u} |\psi_r^{(u)}(x)|^2$$
$$= n^{1-r} \sum_u \sum_{t_1,\ldots,t_{r-1}} \exp\left[ib \sum_{j=1}^{r-2}(t_j - t_{j+1})^2\right] e^{ib(t_{r-1}-u)^2}$$
$$\times \sum_{t'_1,\ldots,t'_{r-1}} \exp\left[-ib \sum_{j=1}^{r-2}(t'_j - t'_{j+1})^2\right] e^{-ib(t'_{r-1}-u)^2}$$
$$= n^{2-r} \sum_{t_1,\ldots,t_{r-1}} \exp\left[ib \sum_{j=1}^{r-2}(t_j - t_{j+1})^2\right]$$
$$\times \sum_{t'_1,\ldots,t'_{r-2}} \exp\left[-ib \sum_{j=1}^{r-3}(t'_j - t'_{j+1})^2\right] e^{-ib(t'_{r-2}-t'_{r-1})^2}$$
$$= n^{3-r} \sum_{t_1,\ldots,t_{r-2}} \exp\left[ib \sum_{j=1}^{r-3}(t_j - t_{j+1})^2\right]$$
$$\times \sum_{t'_1,\ldots,t'_{r-3}} \exp\left[-ib \sum_{j=1}^{r-4}(t'_j - t'_{j+1})^2\right] e^{-ib(t'_{r-3}-t'_{r-2})^2}$$
$$\vdots$$
$$= n^{-1} \sum_{t_1,t_2} e^{ib(t_1-t_2)^2} \sum_{t'_1} e^{-ib(t'_1-t_2)^2}$$
$$= \sum_{t_1} 1 = n.$$
Hence, $\sum_x |\psi_r(x)|^2 = 1$. □

7.7 Notes and References

There is no firm experimental evidence at present that spacetime is discrete. However, this does not logically eliminate such a possiblity. Perhaps we have not yet reached the realm of small enough distances and time intervals and large enough energies for such phenomena to manifest themselves. Perhaps the breakdown of our partial differential equations and other continuum methods in the sense of singularities and infinities is indicating that they have reached their limit of applicability. It is possible that the "true" equations are difference equations and that the partial differential equations are only approximations. If the universe is discrete, it should come as no great surprise. After all, quantum mechanics has already prepared us for such an eventuality. Quantum mechanics has taught us that energy frequently comes in discrete packets or quanta. We have also seen that internal angular momenta (spin) has only discrete values. Observable electric charge comes in integer multiples of the electron charge. "Elementary particles" are described and classified by discrete quantum numbers. Further discussion can be found in the references at the beginning of this chapter.

We have presented various discrete quantum models based on the concepts of amplitude functions and one-step transition amplitudes. The first two sections of this chapter emphasized finite sequences of random variables. These determined corresponding sample paths and their amplitudes. Moreover, certain sequences possessed a quantum Markov property. The subsequent sections showed that beginning with the sample paths and their amplitudes one can derive the important random variables. If the amplitudes are given in terms of a one-step transition amplitude, then the Markov property automatically holds. Simple discrete spacetime models were constructed to illustrate these ideas. One can also construct finite models for "elementary particles". The ground states of these particles are described by two vertex (mesons) and three vertex (baryons) multigraphs. The common particles such as the pions, proton and neutron have no loops while the more exotic particles with strangeness, charm, beauty and truth have various combinations of loops. Moreover, the leptons (electron, muon, tauon, neutrino) are represented by one vertex with various loops. The gluons perform a quantum random walk on these finite multigraphs. Since these models are at a very preliminary stage, we have not presented them here. An important problem is to find the proper one-step transition amplitudes that generate these random walks. One can then check the model against experimental findings such as particle masses, decay modes, decay probabilities, and lifetimes. For more details, we refer the reader to [Gudder, 1988].

The reader may wonder how we can claim that the integer a in Section 7.6 corresponds to a particle mass. This is because we can measure mass relative to an extremely small mass unit (possibly the mass of the electron neutrino).

Bibliography

L. Accardi, "Topics in quantum probability", *Phys. Rep.* **77**, 169–192, 1981.

L. Accardi, A. Fedullo, "On the statistical meaning of complex numbers in quantum mechanics", *Lett. Il Nuovo Cim.* **34**, 161–172, 1982.

D. Atkinson, M. Halpern, "Non-usual topologies on space-time and high-energy scattering", *J. Math. Phys.* **8**, 373–387, 1967.

E. Belifante, *A Survey of Hidden Variable Theories*, Pergamon Press, New York, 1973.

J. Bell, "On the problem of hidden variables in quantum mechanics", *Rev. Mod. Phys.* **38**, 447–452, 1966.

E. Beltrametti, G. Cassinelli, *The Logic of Quantum Mechanics*, Addison-Wesley, Reading, 1981.

P. Billingsley, *Convergence of Probability Measures*, Wiley, New York, 1968.

G. Birkhoff, *Lattice Theory*, Amer. Math. Soc., Providence, 1948.

G. Birkhoff, J. von Neumann, "The logic of quantum mechanics", *Ann. Math.* **37**, 823–843, 1936.

J. Bjorken, D. Drell, *Relativistic Quantum Fields*, McGraw-Hill, New York, 1965.

R. Blumethal, D. Ehn, W. Faissler, P. Joseph, L. Langrotti, F. Pipkin, D. Stairs, "Wide angle election-pair production", *Phys. Rev.* **144**, 1199–1223, 1966.

G. Bodiou, *Théorie Dialectique des Probabilitiés*, Gauthier-Villar, Paris, 1964.

D. Bohm, *Quantum Theory*, Prentice-Hall, Englewood Cliffs, 1951.

D. Bohm, J. Bub, "A proposed solution of the measurement problem in quantum mechanics by hidden variables", *Rev. Mod. Phys.* **38**, 453–469, 1966a.

D. Bohm, J. Bub, "A refutation of the proof of Jauch and Piron that hidden variables can be excluded in quantum mechanics", *Rev. Mod. Phys.* **38**, 470–475, 1966b.

N. Bohr, "Can quantum-mechanical description of reality be considered complete?", *Phys. Rev.* **48**, 696–702, 1935.

M. Born, "Zur Quantenmechanik der Strossvorgange", *Z. Phys.* **37**, 863–867, 1926.

L. Breiman, *Probability*, Addison-Wesley, Reading, 1968.

V. Cantoni, "Generalized 'transition probability'", *Commun. Math. Phys.* **44**, 125–128, 1975.

K. Chung, *Markov Chains with Stationary Transition Probabilities*, Springer-Verlag, New York, 1967.

L. Cohen, "Can quantum mechanics be formulated as a classical probability theory?", *Philos. Sci.* **33**, 317–322, 1966.

T. Cook, "Banach spaces of weights on quasimanuals", *Intern. J. Theor. Phys.* **24**, 1113–1131, 1985.

H. Cycon, K.E. Hellwig, "Conditional expectations in generalized probability theory", *J. Math. Phys.* **18**, 1154–1165, 1977.

E.B. Davies, *Quantum Theory of Open Systems*, Academic Press, New York, 1976.

E.B. Davies, J. Lewis, "An operational approach to quantum probability", *Commun. Math. Phys.* **17**, 239–260, 1970.

P. Dirac, *The Principles of Quantum Mechanics*, Clarendon, Oxford, 1930.

J. Doob, *Stochastic Processes*, John Wiley, New York, 1953.

N. Dunford, J. Schwartz, *Linear Operators, Parts I and II*, Wiley (Interscience), New York, 1958 and 1963.

C.M. Edwards, "The operational approach to algebraic quantum theory I", *Commun. Math. Phys.* **16**, 207–230, 1970.

A. Einstein, B. Podolsky, N. Rosen, "Can quantum-mechanical description of reality be considered complete?", *Phys. Rev.* **47**, 777–780, 1935.

R. Eisberg, R. Resnick, *Quantum Physics*, Wiley, New York, 1985.

G. Emch, *Algebraic Methods in Statistical Mechanics and Quantum Field Theory*, Wiley (Interscience), New York, 1972.

R. Feynman, "Space-time approach to non-relativistic quantum mechanics", *Rev. Mod. Phys.* **20**, 367–398, 1948.

R. Feynman, "Space-time approach to quantum electrodynamics", *Phys. Rev.* **76**, 769–789, 1949.

R. Feynman, A. Hibbs, *Quantum Mechanics and Path Integrals*, McGraw-Hill, New York, 1965.

T. Fine, *Theories of Probability: An Examination of Foundations*, Academic Press, New York, 1973.

H. Fischer, G. Rüttimann, "Limits of manuals and logics", in *Mathematical Foundations of Quantum Theory*, A. Marlow, ed., 127–152, Academic Press, New York, 1978.

D. Foulis, C. Randall, "Operational statistics I: Basic concepts", *J. Math. Phys.* **13**, 1167–1175, 1972.

D. Foulis, C. Randall, "Empirical logic and quantum mechanics", *Synthesis* **29**, 81–111, 1974.

D. Foulis, C. Randall, "Manuals, morphisms and quantum mechanics", in *Mathematical Foundations of Quantum Theories*, A. Marlow, ed., 105–126, Academic Press, New York, 1978.

D. Foulis, C. Randall, "Empirical logic and tensor products", in *Interpretations and Foundations of Quantum Theory*, H. Neumann, ed., 9–20, Wissenschaftsverlag, Mannheim, 1981.

D. Foulis, C. Randall, "Dirac revisited", in *Symposium on the Foundations of Modern Physics*, P. Lahti, P. Mittlestaedt, ed., 97–112, World Scientific Publishing Co., Singapore, 1985.

D. Foulis, C. Piron, C. Randall, "Realism, operationalism, and quantum mechanics", *Found. Phys.* **13**, 813–842, 1983.

G. Gamow, *Thirty Years that Shook Physics*, Doubleday, Garden City, 1966.

R. Geroch, *Mathematical Physics*, University of Chicago Press, Chicago, 1985.

A. Gleason, "Measures on closed subspaces of a Hilbert space", *J. Rat. Mech. Anal.* **6**, 885–893, 1957.

H. Goldstein, *Classical Mechanics*, Addison-Wesley, Reading, 1957.

R. Greechie, *Orthomodular Lattices*, Dissertation, University of Florida, Tallahasse, 1966.

R. Groblicki, *Consistency theorems for phase-space representations of quantum mechanics*, Dissertation, Monash University, Australia, 1984.

S. Gudder, "Dispersion-free states and the existence of hidden variables", *Proc. Amer. Math. Soc.* **19**, 319–324, 1968a.

S. Gudder, "Elementary length topologies in physics", *SIAM J. Appl. Math.* **16**, 1011–1019, 1968b.

S. Gudder, "Quantum probability spaces", *Proc. Amer. Math. Soc.* **21**, 296–302, 1969.

S. Gudder, "On hidden-variable theories", *J. Math. Phys.* **11**, 431–436, 1970.

S. Gudder, *Stochastic Methods in Quantum Mechanics*, North Holland, New York, 1979.

S. Gudder, "Hilbertian interpretations of manuals", *Proc. Amer. Math. Soc.* **85**, 251–255, 1982.

S. Gudder, "An extension of classical measure theory", *SIAM Review* **26**, 71–89, 1984a.

S. Gudder, "Finite quantum processes", *J. Math. Phys.* **25**, 456–465, 1984b.

S. Gudder, "Probability manifolds", *J. Math. Phys.* **25**, 2397–2401, 1984c.

S. Gudder, "Amplitude phase-space model for quantum mechanics", *Intern. J. Theor. Phys.* **24**, 343–353, 1985.

S. Gudder, "Discrete quantum mechanics", *J. Math. Phys.* **27**, 1782–1790, 1986a.

S. Gudder, "Partial Hilbert spaces and amplitude functions", *Ann. Inst. Henri Poincaré* **45**, 311–326, 1986b.

S. Gudder, "Quantum graphic dynamics", *Found. Phys.* (to appear 1988).

S. Gudder, T. Armstrong, "Bayes' rule and hidden variables", *Found. Phys.* **15**, 1009–1017, 1985.

S. Gudder, M. Kläy, G. Rüttimann, "States on hypergraphs", *Dem. Math.* **XIX**, 503-526, 1986.

S. Gudder, J.P. Marchand, "Conditional expectations on von Neumann algebras", *Rep. Math. Phys.* **12**, 317–329, 1977.

S. Gudder, J.P. Marchand, "A coarse-grained measure theory", *Bull. Acad. Pol. Sci.* **XXVII**, 557–564, 1980.

S. Gudder, V. Naroditsky, "Finite-dimensional quantum mechanics", *Intern. J. Theor. Phys.* **20**, 619–643, 1981.

S. Gudder, S. Pulmannová, "Transition amplitude spaces", *J. Math. Phys.* **28**, 376–385, 1987.

S. Gudder, J. Zerbe, "Generalized monotone convergence and Radon-Nikodym theorems", *J. Math. Phys.* **22**, 2553–2561, 1981.

R. Haag, D. Kastler, "An algebraic approach to quantum field theory", *J. Math. Phys.* **5**, 848–861, 1964.

P. Halmos, *Measure Theory*, Van Nostrand Reinhold, Princeton, 1950.

W. Heisenberg, *The Physical Principles of Quantum Theory*, (C. Eckhart, F. Holt, trans.), Dover, New York, 1930.

A. Holevo, *Probabilistic and Statistical Aspects of Quantum Theory*, North-Holland, Amsterdam, 1982.

J. Jauch, *Foundations of Quantum Mechanics*, Addison-Wesley, Reading, 1968.

J. Jauch, C. Piron, "Can hidden variables be excluded from quantum mechanics?", *Helv. Phys. Acta* **36**, 827–837, 1965.

H. Jeffreys, *Theory of Probability*, Clarendon, Oxford, 1948.

P. Jordan, J. von Neumann, E. Wigner, "On an algebraic generalization of the quantum mechanical formalism", *Ann. Math.* **33**, 29–64, 1934.

M. Kläy, C. Randall, D. Foulis, "There exists no Simpson's paradox", (to appear 1988).

J. Keynes, *A Treatise on Probability*, Harper, New York, 1962.

S. Kochen, E. Specker, "The problem of hidden variables in quantum mechanics", *J. Math. Mech.* **17**, 59–87, 1967.

A. Kolmogorov, *Foundations of the Theory of Probability*, Chelsea, New York, 1956.

P. Lahti, P. Mittlestaedt, *Symposium on the Foundations of Modern Physics*, World Scientific, Singapore, 1985.

J. Lamperti, *Probability*, Benjamin, New York, 1966.

A. Landé, *New Foundations of Quantum Mechanics*, Cambridge University Press, Cambridge, 1965.

M. Loève, *Probability Theory*, Van Nostrand Reinhold, Princeton, 1963.

G. Ludwig, *Foundations of Quantum Mechanics, Vols. I and II*, Springer-Verlag, New York, 1983 and 1985.

G. Mackey, *The Mathematical Foundations of Quantum Mechanics*, Benjamin, New York, 1963.

B. Mielnik, "Geometry of quantum states", *Commun. Math. Phys.* **9**, 55–80, 1968.

B. Mielnik, "Theory of filters", *Commun. Math. Phys.* **15**, 1–46, 1969.

P. Mittlestaedt, A. Prieur, R. Schrieder, "Unsharp particle-wave duality in a photon split beam experiment", *Found. Phys.* (to appear 1988).

H. Mullikin, S. Gudder, "Measure theoretic convergence of observables and operators", *J. Math. Phys.* **14**, 234–242, 1973.

C. Pinter, *Set Theory*, Addison Wesley, Reading, 1971.

C. Piron, *Foundations of Quantum Mechanics*, Benjamin, New York, 1976.

I. Pitowsky, "Resolution of the Einstein-Podolsky-Rosen and Bell paradoxes", *Phys Rev. Lett.* **48**, 1299–1302, 1982.

I. Pitowsky, "Deterministic model of spin and statistics", *Phys. Rev. D.* **27**, 2316–2335, 1983a.

I. Pitowsky, *The logic of fundamental processes: nonmeasurable sets and quantum mechanics*, Dissertation, University of Western Ontario, London, Ontario, 1983b.

I. Pitowsky, "A phase-space model for quantum mechanics in which all operators commute", *Fundamental Problems in Quantum Mechanics Conference Proceedings*, CUNY, Albany, 1984.

R. Prather, "Generating the k-subsets of an n-set", *Amer. Math. Monthly* **87**, 740–743, 1980.

E. Prugovecki, *Quantum Mechanics in Hilbert Space*, Academic Press, New York, 1981.

E. Prugovecki, *Stochastic Quantum Mechanics and Quantum Spacetime*, Reidel, Dordrecht, 1984.

S. Pulmannová, "Transition probability spaces", *J. Math. Phys.* **27**, 1781–1793, 1986.

C. Randall, D. Foulis, "Operational statistics II: manuals of operations and their logics", *J. Math. Phys.* **14**, 1472–1480, 1973.

C. Randall, D. Foulis, "Operational statistics and tensor products", in *Interpretations and Foundations of Quantum Theory*, H. Neumann, ed., 21–28, Wissenschaftsverlag, Mannheim, 1981.

C. Randall, D. Foulis, "A mathematical language for quantum physics", in *Transactions of the 25 Cours de Perfectionnement de l'Associatin Vaudoise des Chercheurs en Physique*, C. Gruber, C. Piron, T. Minhtm, R. Weil, ed., 193–226, Montana, Switzerland, 1983.

S. Ross, *A First Course in Probability*, Macmillan, New York, 1984.

B. Russell, *The Analysis of Matter*, Dover, New York, 1954.

G. Rüttimann, "The approximate Jordan-Hahn decomposition", University of Bern, Switzerland, reprint 1987.

E. Schrödinger, *Collected Papers on Wave Mechanics*, (J. Shearer, W. Deans, trans.), Blackie and Sons, London,, 1928.

L. Schulman, *Techniques and Applications of Path Integration*, Wiley (Interscience), New York, 1981.

S. Schweber, *An Introduction to Relativistic Quantum Field Theory*, Row and Peterson, Evanston, Illinois, 1961.

I. Segal, "Postulates for general quantum mechanics", *Ann. Math.* **48**, 930–948, 1947.

D. Shale, "On geometric ideas which lie at the foundation of quantum theory", *Advances Math.* **32**, 175–203, 1979.

D. Shale, "Discrete quantum theory", *Found. Phys.* **12**, 661–687, 1982.

P. Suppes, "The probabilistic argument for a nonclassical logic for quantum mechanics", *Philos. Sci.* **33**, 14–21 , 1966.

A. Uhlmann, "The "transition probability" in the state space of a *-algebra", *Rep. Math. Phys.* **9**, 273–279, 1976.

V. Varadarajan, *Geometry of Quantum Theory, Vols. 1 and 2*, Von Nostrand Reinhold, Princeton, 1968 and 1970.

J. von Neumann, *Mathematical Foundations of Quantum Mechanics*, Princeton University Press, Princeton, 1955.

S. Watanabe, "Pattern Recognition as Information Compression", *Frontiers of Pattern Recognition*, Academic Press, New York, 1972.

J. Wheeler, W. Zurek, *Quantum Theory and Measurement*, Princeton University Press, Princeton, 1983.

E. Wigner, "On the quantum correction for thermodynamic equilibrium", *Phys. Rev.* **40**, 749–759, 1932.

R. Wright, "Spin manuals", in *Mathematical Foundations of Quantum Theory*, A. Marlow, ed., 177–254, Academic Press, New York, 1978.

N. Zanghi, S. Gudder, "Probability models", *Il Nuovo Cim.* **79**, 291–301, 1984.

J. Zerbe, S. Gudder, "Additivity of integrals on generalized measure spaces", *J. Comb. Theor. A* **39**, 42–51, 1985.

Index

accessible 281
action 41, 63
additive class 90
additive integral 189
additive partition manual 91
additive product 185
adjacent 264
admits hidden variables 92, 199
almost stationary 252
amplitude 160, 265
   density 229, 232
   function 113, 122, 144, 166
   matrix 246
   n-chain 246
   space 122
A-set 141
associative ortho-algebra (AOA) 75
atomic 51
   state 51
A-transition amplitude
   conditioned by B 153
automorphism 147
axiom of choice 211
axis 73

backward edge 287
backward operational product 94
Bayes' rule 9, 98, 99
Bayesian states 200
Bell's inequalities 179
Bell, J.S. 177
binomial random variable 12
Borel function 4, 5

Borel manual 88
Borel subset 5
Borel-Cantelli lemma 15, 25
Born, M. 42
bounded 150

C^*-state 153
calculus of variations 41
cartesian product 94
center 189
central limit theorem 33,34
chain 207
Chapman-Kolmogorov equation
   20, 138, 262, 266
characteristic function 28, 29
Chebyshev's inequality 22
choice function 211
classical trajectory 41
closed quantum n-chain 250
closed subspace 119
coarsening 108
coherent 74
   coarsening 110
combined system 48
comparable 207
compatible 53, 82, 186, 187
component state 97
composite system 48, 49
compound operation 94
conditional amplitude 166
   matrix 246
conditional expectation 10, 11
conditional gps 183

311

conditional probability 9, 10, 11, 54, 183
conditional state 54, 199
conditioning set 166
configuration space 41
confirmed 73
conjugate map 85
constrained n-chain 247
continuum hypothesis 212
convergence almost surely 22
convergence in mean 22
convergence in measure 22
convergence in probability 22
converges weakly 25
convex combination 86
convolution 17
coordinate functions 37
coordinates 37
correlation 100
correspondence principle 44
countable Λ-partition 91
countable Σ-partition 88
cycle graph 287
cylinder set 11

definite n-chain 247
degree 264
degrees of freedom 37
density 6, 18
  operator 51
deterministic 89
dimension of a tas 146
Dirac superposition 125
Dirac's equation 279
Dirac, P. 42
direct sum 48, 49
discrete wave equation 273, 278, 294
disjoint events 51
disjoint union 50, 51
disjointness 50
dispersion 46
dispersion-free state 86
distribution 5, 82
  function 5

domain 75
dual 121
dynamic transition amplitude 138, 140
dynamical group 38
dynamical variable 40

edges 264
embedded 172
energy 40
  observable 43, 45
  operator 53
E-outcome 161
event 2, 5, 50, 72, 166
expectation 3, 82, 217
experimental proposition 81
extendable 182
extended gps 182
extremal state 86

fair 11
Feynman amplitude 292
Feynman, R. 57, 62
filter 139
final vector 246
finite measurable partition 9, 10
first element 76
force 39
form 149
forward edge 287
forward operational product 94
Fourier transform 44, 234
frame manual 89, 133
frame subquasimanual 135
Fubini's theorem 17, 18, 224
function of one variable 234
fuzzy 219

generalized measure space (gms) 170
generalized probability space 169
generated 75
geometric random variable 13
g-harmonic 221

INDEX

Gleason's theorem   52, 133, 190, 201
Gleason, A.   51
global expectation   219
globally measurable   216

Hahn extension theorem   5, 12
Hamilton's equation   241
Hamiltonian   40
   form   40
   formulation   41
Heisenberg uncertainty
   principle   172
   relation   47, 53
hidden variables   177, 207
Hilbert space probability   50
Hilbert space quantum mechanics   140
hook   87

identity function   5
immediate predecessor   208
immediate successor   208
incident   264
inclusion-exclusion   2
   formula   14
incoherent coarsening   109
incomparable   207
increasing   208
independent   11
   random variables   14, 16
indicator function   3
inductive   211
infinitesimal generator   39
influence   96
initial segment   207
initial state   250
initial vector   246
integral   40
interference   99
   pattern   59, 61
interfering alternatives   59, 61
inverse Fourier transform   234
irredundant   74
isomorphic   80, 147, 208

isomorphic partial Hilbert space   119
isomorphism   85, 147, 208

joint distribution   7, 8, 16
   function   7, 16
jointly observable   162

kinetic energy   40
   observable   45
Klein-Gordon equation   279
Kolmogorov's theorem   18, 23
Kolmogorov, A.   8, 34
Krein-Milman theorem   204
kth vector   246

Λ-partition   91
Lagrangian   41, 63
   equations of motion   42
   formulation   41, 63
   state   41
last element   75
least action principle   41
Lebesgue space   223
limit ordinal   213
local expectation   219
local structure   215
locally identically distributed   221
locally independent   221
locally measurable   216
loops   264

magic sequence   196
manual   72, 74
marginal distribution   7
marginal state   97
Markov amplitude n-chain   247
Markov chain   19
Markov process   138
Markov property   265
mass   39
maximum likelihood estimator   104
mean   3
measurable   187
   space   1

minimal 86
  upper bound 76
mixed state 43, 51
moment 22
momentum 39
  problem 280
  wave function 280
morphism 85
multigraph 264

negation 73
negative binomial 13
no influence 96
nonlimit ordinal 213
nonoccur 73
normal 122
  distribution 29
  random variable 6
  state 82
n-path 253, 265
n-step amplitude expectation 270
n-step expectation 270
n-step transition amplitude 265
n-step transition probability 21, 265
null sequence 149

observable 43, 50, 52, 82, 162, 166, 232
occur 73
Ω-dense 86
one-parameter group 38
  of automorphisms 148
one-parameter semigroup 38
one-step transition amplitude 265
operation 72, 161, 166
operational complement 72
operational logic 79
operational proposition 74, 79
operational statistics 71
operationally perspective 72
orbit 38
order determining 92
order-n Bayesian 200
order-preserving 208

ordinal numbers 212
orthocoherent 78, 80
orthocomplementation 76
orthocomplemented poset 76
orthogonal 72, 76, 141, 166, 185
orthomodular 76
outcome 72, 161, 165
  preserving 85
  set 72

P-amplitude density 230
parallel classical n-chain 256
Parseval's equality 48, 140
partial binary operation 75
partial Hilbert space 118
partition 90
  quasimanual 90
path 41
  n-chain 253
  integral 65
  space 253
path-integral formalism 62, 63
Pauli matrices 55
p-coin 11, 12
P-expectation 233
phase space 219, 264
physical operation 72
Pitowski, I. 180, 221
Planck's constant 43
point closed 160
Poisson random variable 6
poset 75
positive 85
  definite 29, 146
  measure 122
postconditioned 96, 98
potential energy 39
  observable 45
P-probability 230, 232
Prather, R. 173
preconditioned 96, 98
premanual 74
probability amplitude 48, 61, 160
probability manifold 215
probability measure 2

INDEX

probability of A
    along $X(x,y)$ 216
    at x 216
probability space 2
product measure 16
product rule 139, 260, 265
product state 95
propagators 268
proper 161
proposition 50
pure amplitude function 123
pure state 43, 52, 86

Q-amplitude density 230
Q-expectation 233
Q-probability 231, 232
quantum n-chain 249
quantum events 51
quantum interference term 61
quantum logic 71, 81
    approach 75, 76, 81
quantum probability space 160
quantum process 260
quantum system 43
quasimanual 72
quasistochastic matrix 247

Radon-Nikodym theorem 11, 149
random phase transformation 255
random variable 3
reduced state 97
refuted 73
regular amplitude density 234
regular state 83
relative frequency 34, 35
relatively compact 27
representation 143
resolution of identity 43
reversible 38
Riemann-Stieltjes integral 6
Riesz lemma 121

same cardinality 212
sample paths 246
sample points 160, 165

sample space 2
Schrödinger's eigenvalue equation 46
Schrödinger's equation 66, 67, 240
section 208
selection process 140
separating 86, 127, 254
σ-additive class 90, 169
σ-additive partition manual 91
σ-algebra 1
σ-homomorphism 82
σ-orthocomplete 81
simple functions 190
Simpson's paradox 102, 103
smaller cardinality 212
special orthogonal group 107
spectral measure 43, 52
spectral theorem 46
spectrum 82
spin 54, 106
    down state 56
    operator 55
    up state 56
spin-j system 54
spin-$1/2$ function 225
spin-$1/2$ system 55
s-set 118
standard deviation 46
state 38, 43, 50, 51, 83, 84, 268
static transition amplitude 138
stationary 19, 247
    Markov chain 20
Stern-Gerlach experiment 106
stochastic 219
    amplitude matrix 247
    matrix 19
    quantum mechanics 219
Stone's theorem 53, 149
strong 141
    dual 121, 129
    law of large numbers 21, 23, 24
strongly separating 135
subquasimanual 135
subtas 155

sum of tas's 154
summable 160, 189
superposition principle 69, 118
superselection rule 48, 49

tail field 23
tail σ-algebra 15
tensor product 48, 95
  of tas's 157
tensorial 95
test operation 73
tight 27
time transition amplitude 48
time transition probability 48
time-dependent
  Schrödinger equation 45
topological dual 121
total 141
  expectation 219
  probability 98, 216
  spin operator 55
totally measurable 216
totally ordered subset 207
trajectory 41
transfinite induction principle 209
transfinite recursion theorem 213
transition amplitude 47, 52, 62, 113, 141, 268
  space (tas) 141
transition matrix 19
transition probability 19, 47, 52, 140, 268

transmission amplitude 139
transmission probability 139
trivial outcome 161
twice differentiable 39
two-hole experiment 106
two-stage operation 94
Tychonoff's theorem 203

uncertainty 46
unital 86, 126
unitarily equivalent 148

Varadarajan, V. 187
variance 7, 22, 46
velocities 38
velocity functions 38
vertices 264
von Neumann formula 54, 100, 202
von Neumann, J. 42

wave function 66, 273
weak law of large numbers 21, 23, 24
weakly perspective 73
well ordered 208
well ordering theorem 212
Wright triangle 87

0-1 law 15, 23
Zorn's axiom 211

PROBABILITY AND MATHEMATICAL STATISTICS

Thomas Ferguson, *Mathematical Statistics: A Decision Theoretic Approach*
Howard Tucker, *A Graduate Course in Probability*
K. R. Parthasarathy, *Probability Measures on Metric Spaces*
P. Révész, *The Laws of Large Numbers*
H. P. McKean, Jr., *Stochastic Integrals*
B. V. Gnedenko, Yu. K. Belyayev, and A. D. Solovyev, *Mathematical Methods of Reliability Theory*
Demetrios A. Kappos, *Probability Algebras and Stochastic Spaces*
Ivan N. Pesin, *Classical and Modern Integration Theories*
S. Vajda, *Probabilistic Programming*
Robert B. Ash, *Real Analysis and Probability*
V. V. Fedorov, *Theory of Optimal Experiments*
K. V. Mardia, *Statistics of Directional Data*
H. Dym and H. P. McKean, *Fourier Series and Integrals*
Tatsuo Kawata, *Fourier Analysis in Probability Theory*
Fritz Oberhettinger, *Fourier Transforms of Distributions and Their Inverses: A Collection of Tables*
Paul Erdös and Joel Spencer, *Probabilistic Methods of Statistical Quality Control*
K. Sarkadi and I. Vincze, *Mathematical Methods of Statistical Quality Control*
Michael R. Anderberg, *Cluster Analysis for Applications*
W. Hengartner and R. Theodorescu, *Concentration Functions*
Kai Lai Chung, *A Course in Probability Theory, Second Edition*
L. H. Koopmans, *The Spectral Analysis of Time Series*
L. E. Maistrov, *Probability Theory: A Historical Sketch*
William F. Stout, *Almost Sure Convergence*
E. J. McShane, *Stochastic Calculus and Stochastic Models*
Robert B. Ash and Melvin F. Gardner, *Topics in Stochastic Processes*
Avner Friedman, *Stochastic Differential Equations and Applications, Volume 1, Volume 2*
Roger Cuppens, *Decomposition of Multivariate Probabilities*
Eugene Lukacs, *Stochastic Convergence, Second Edition*
H. Dym and H. P. McKean, *Gaussian Processes, Function Theory, and the Inverse Spectral Problem*
N. C. Giri, *Multivariate Statistical Inference*
Lloyd Fisher and John McDonald, *Fixed Effects Analysis of Variance*
Sidney C. Port, and Charles J. Stone, *Brownian Motion and Classical Potential Theory*

Konrad Jacobs, *Measure and Integral*
K. V. Mardia, J. T. Kent, and J. M. Biddy, *Multivariate Analysis*
Sri Gopal Mohanty, *Lattice Path Counting and Applications*
Y. L. Tong, *Probability Inequalities in Multivariate Distributions*
Michel Metivier and J. Pellaumail, *Stochastic Integration*
M. B. Priestly, *Spectral Analysis and Time Series*
Ishwar V. Basawa and B. L. S. Prakasa Rao, *Statistical Inference for Stochastic Processes*
M. Csörgö and P. Révész, *Strong Approximations in Probability and Statistics*
Sheldon Ross, *Introduction to Probability Models, Second Edition*
P. Hall and C. C. Heyde, *Martingale Limit Theory and Its Application*
Imre Csiszár and János Körner, *Information Theory: Coding Theorems for Discrete Memoryless Systems*
A. Hald, *Statistical Theory of Sampling Inspection by Attributes*
H. Bauer, *Probability Theory and Elements of Measure Theory*
M. M. Rao, *Foundations of Stochastic Analysis*
Jean-Rene Barra, *Mathematical Basis of Statistics*
Harald Bergström, *Weak Convergence of Measures*
Sheldon Ross, *Introduction to Stochastic Dynamic Programming*
B. L. S. Prakasa Rao, *Nonparametric Functional Estimation*
M. M. Rao, *Probability Theory with Applications*
A. T. Bharucha-Reid and M. Sambandham, *Random Polynomials*
Sudhakar Dharmadhikari and Kumar Joag-dev, *Unimodality, Convexity, and Applications*
Stanley P. Gudder, *Quantum Probability*

...YSICS LIBRARY
351 LeConte   642-3122